Proceedings of the Conference on Promoting Undergraduate Research in Mathematics

Proceedings of the Conference on Promoting Undergraduate Research in Mathematics

Joseph A. Gallian
Editor

American Mathematical Society
Providence, Rhode Island

Cover photography courtesy of Cindy Wyels, CSU Channel Islands, Camarillo, CA, and Darren Narayan, Rochester Institute of Technology, Rochester, NY.

This material is based upon work supported by National Security Agency under grant H98230-06-1-0095.

ISBN-13: 978-0-8218-4321-5
ISBN-10: 0-8218-4321-4

Table of Contents

Part II. Descriptions of Summer Enrichment Programs

Part III. Articles

Part IV. Summaries of Conference Sessions

Part V. Surveys of Undergraduate Research Programs

Part VI. Conference Program

Preface

In 1999, the AMS received a grant from the National Security Agency to sponsor a three-day conference on Summer Undergraduate Mathematics Research Programs for the purpose of bringing together mathematicians from across the country who were involved in summer mathematics research programs for undergraduates to exchange ideas, discuss issues of common concern, establish contacts, and gather information that would be of value to others interested in establishing similar programs. By 2006, the numbers of NSF REU programs had doubled, the NSA had greatly increased its support for summer research programs, the MAA had initiated a national REU for underrepresented students and began a minigrant program to support conferences for undergraduates to showcase their research, more students wanted to participate in the poster sessions at the joint meetings than could be accommodated, PhD granting institutions with VIGRE grants were integrating undergraduates into research groups with graduate students and faculty, and many of the most talented undergraduate mathematics students felt that they needed to have an REU experience to be admitted to the top graduate programs. With this great increase in interest in promoting research by undergraduates, it seemed to me to be time for another conference for the same purpose as the 1999 conference. A conversation with Jim Schatz, a phone call to Barbara Deuink, and an email to John Ewing eventually led to a grant proposal to the NSA and a three-day conference in late September 2006 at the Westin O'Hare Hotel in Rosemont, Illinois.

The 2006 conference featured plenary talks, panel discussions, and small group sessions. The topics included summer research programs, academic year research opportunities, diversity issues, assessment methods, and perspectives from alumni of research programs. In this volume we compile articles written by many of the participants as well as articles I solicited from people who did not attend but had something valuable to contribute. Also included are the surveys about the activities of program participants and a summary of a survey of students who had participated in an REU between 1997 and 2001.

The people responsible for planning and conducting the conference were Frank Connolly, Barbara Deuink, John Ewing, Joseph Gallian, Aparna Higgins, Ellen Maycock and Ivelisse Rubio. We are grateful to the National Security Agency for providing the funding.

I wish to thank Vickie Ancona, Sam White and Peter Sykes of the AMS staff for their excellent work in producing the volume and to Phil Matchett Wood for taking notes throughout the conference. Michael Dorff has generously assisted me by arranging to have several articles prepared in LaTeX. The proceedings of the

1999 conference are available at
http://www.ams.org/employment/REUproceedings.pdf.

DECEMBER 30, 2006

JOSEPH A. GALLIAN, UNIVERSITY OF MINNESOTA DULUTH, DULUTH, MN 55812
E-mail address: jgallian@d.umn.edu

Part I

Descriptions of Summer Research Programs

The AIM REU: individual projects with a common theme

David W. Farmer

The principle behind the American Institute of Mathematics (AIM) REU is that mathematicians with an active research program have ideas for more projects than they have time to work on. By enlisting the help of enthusiastic undergraduates, it is possible to broaden the range of problems in which they are actively involved.

The AIM REU is run by David Farmer and Brian Conrey, along with one or two of their current postdocs. Conrey and Farmer, and their postdocs, have research interests in analytic number theory: the Riemann zeta-function, modular forms, random matrix theory, elliptic curves, etc. The REU projects all come directly from their ongoing research programs, so there are natural relationships between the projects. The fact that there is an underlying theme between the various student projects is one of the strengths of the program. There is a wide variety of interaction among the students and the advisers, with all the students developing an interest in each other's projects, and all the advisers being able to work with all the students. The students become part of an active research community.

1. Key features of the AIM REU

Individual projects. Every student chooses their own individual research problem. This allows the student to take full responsibility for their own project and more closely models traditional mathematics research.

The students all work on related projects, although it often takes them many weeks to understand the connections! Through weekly progress reports and a variety of daily interactions with the advisers and other students, the participants spend a lot of time talking to each other, explaining their work, and learning about each other both on a mathematical and a personal level.

By having all the projects in related areas we are able to foster a sense of community where everyone feels a part of a larger endeavor. This allows us to have all the students work on separate projects without the likelihood of feeling isolated.

Immediate work. Students choose their problem on the first day and begin working immediately, without any preliminary background reading. This feature is somewhat unusual, considering that the problems are in an area that is considered

Received by the editor November 30, 2006.

The REUs described in this paper were supported by AIM and by grants from the NSF, including the Focused Research Group grant DMS 0244660.

to require a reasonable amount of background. We elaborate on this in the next two sections of this paper.

Ongoing research. The student projects arise naturally from the research programs of the mathematicians running the REU. One consequence is that the specific projects cannot be finalized until the last minute. While it is possible to give a general idea of the program (students usually want to know the possible projects and how their particular project will be decided), we always have to consider recent developments before the beginning of the REU.

Postdoctoral training. The postdocs are involved in all aspects of the REU: everything from selecting appropriate projects to helping the students prepare their talks in the final week. This is valuable training because there is an increasing expectation for junior faculty to advise undergraduate projects (especially at liberal arts colleges).

This "training" begins shortly before the program when the list of possible projects is developed. It is not required that the postdocs contribute a problem (although they usually do). The senior mathematicians and postdocs have 2 or 3 meetings which start with a discussion of ideas for projects. The senior mathematicians examine all aspects of the proposed problems, being as explicit as possible about which properties make a problem suitable, or not, for the REU. By the final meeting, the discussions turns to the critical issue of how to introduce the problems to the student so that they can make a reasonable choice for their project and begin working immediately. This process is not easy to capture on paper, so it is a big help for the postdocs to see it in action.

Regular meetings between the directors and the postdocs occur during the program. These are used to monitor student progress and also to discuss issues such as: what is the right amount of help to give a student who is "stuck," how to prepare a student to give a talk, etc. These meetings are in addition to the daily interaction of the mentors and the students, which gives the postdocs an opportunity to observe the more experienced mentors and to call on help as needed.

Mathematica. Many of our REU projects begin with experimentation in *Mathematica*, and in many cases the work is done primarily with the help of *Mathematica* and other computer-algebra systems.

Standard topics. The AIM REU has many facets which feature in most REUs, but we mention them for completeness. Throughout the program there are lectures on material related to the student projects. All the students learn LaTeX and write a final report in the style of a research paper. Also, the students give a formal talk on their work and are encouraged to give a talk when they return to their home institution. The mentors and the students all see each other several times every day and often eat lunch together. Each student has their own desk in a shared office.

2. Identifying appropriate projects

Prior to the REU we must choose projects which allow our students to select a problem on the first day and begin work immediately, after only a few verbal preliminaries and without any background reading. We have identified three key criteria for deciding if a problem is appropriate for this approach, and one "non-criterion" which is best ignored.

1. How would I start to work on the problem?   The problem is part of my research, so obviously I find it interesting. Suppose I had time to work on the problem: *what would be my first step?*   If I don't have an answer to that question, then clearly the problem is not appropriate for a student. On the other hand, if I have a good idea of what I would do, then it is possible that the problem may be good for the REU.

2. Could an undergraduate execute my first steps on the problem? I know what I would try if I were to work on the problem. Would an undergraduate be able to do that work? For example, if I would start with some numerical experiments in *Mathematica*, or I would start by considering the problem for some low-degree polynomials, or if I would enumerate the first few cases by brute-force, then perhaps a student would be able to do it, too.

Note: this question only asks if the student could do the work, at this point we don't consider whether or not the student could understand *why* that work relates to the problem.

3.   Is there something related to the problem which the student could begin doing immediately?   An effective way to engage a student on a problem is to have them begin work on the first day: work which will lead them to see something interesting. This only needs to be in the same general area as the intended problem, because its purpose is to quickly put the student into a research-like environment and to start them towards an understanding of the problem. Once they have seen something new (to them), they want to understand their observations (which motivates them to learn background material), and they will want to keep going (which motivates them to get to the real problem).

For example, I have had several students do research on modular forms. A good first task for them is: use *Mathematica* to expand the following infinite product into a sum:

$$(1) \qquad\qquad q \prod_{j=1}^{\infty} (1 - q^j)^{24} = \sum_{n=1}^{\infty} \tau(n) q^n.$$

It doesn't take much knowledge of *Mathematica* to truncate the product on the left and expand it into a sum, which gives the first several values of $\tau(n)$. The student will "discover" that $\tau(2)\tau(3) = \tau(6)$ and $\tau(2)\tau(5) = \tau(10)$, and in general $\tau(n)\tau(m) = \tau(nm)$ if n and m are relatively prime. The proof of this result is accessible to an undergraduate (although it is not found in the undergraduate curriculum), and the student is motivated to see the proof.

4.   You don't need to fill in the gap.   After the student has started, you still have to chart a course to their chosen problem. While it is helpful to have an idea of this before the REU begins, I have not found this to be critical. You will work together with the student to find a path from the initial task to the problem. You have already determined that both ends of that path are accessible to the student, so you have to trust that you can connect them in an understandable way. Often you can identify a huge chasm which seems hopeless for the student to breach. In these cases a few carefully chosen "black boxes" (which the student will have to accept on faith) can be helpful. Your goal is not to give an entire graduate course in the subject, but to put them in a position to work on their problem.

It takes a lot of work to bring a student up to speed on a problem, and this is best done in partnership with them because you need to take into account their particular strengths and weaknesses.

3. The beginning of the project

Ten weeks (or less) is a very short amount of time to accomplish meaningful research, so it is important for students to begin right away. In the AIM program a variety of possible projects are described on the first morning. Shortly after lunch each student chooses a problem and begins work. Students are never asked to read or learn anything before beginning their project, and the students are not formally introduced to any of the background of their project until their work is well underway.

Real work begins the first day. This is the aspect of the program which many people find surprising, including the students. The first morning all the mentors describe possible projects. Technical terms are suppressed as much as possible, and the emphasis is on the *methods* that the project will involve. For example, some projects involve a lot of computer work, others have a more geometric flavor, etc. By lunchtime all the projects have been described, and the students are told that after lunch they can ask as many questions as they want, after which they must choose a project and begin working immediately. They are not forced to stay with that project if they don't like it, but they have to give it a serious try. Should two people want to do the same project, the mentors can usually find two independent sub-projects with the same theme. The mentors and students eat lunch together and then return to AIM.

The after-lunch meeting usually takes an hour, and after the student questions are answered we leave them alone for some time to talk it over. Some students are uncomfortable with this process, and many explain that they would like to learn more about the problems before choosing. We tell them: "You have spent years learning mathematics, and you will have plenty of time to learn more mathematics during the semester. If you spend too much time learning the background, then you won't have time to *do* the mathematics."

Starting a project. How are they are supposed to start working without any background? The answer is to give them specific tasks which they can do using only what they already know, and which bear some relation to their project. The tasks should lead them to see some interesting phenomenon. This could involve a pen-and-paper calculation, or a computer calculation (we use *Mathematica* a lot), or drawing a picture, etc. There is no need to explain the connection between the task and the project – that will come later.

The students complete their first task, which hopefully leads them to an interesting observation. Now they are hooked. They want to move on to what is next, they want to know the relationship to their project, they want to understand the background, etc. Of course the student will have to learn some background at some point, but that can be done in parallel with the "real" work, and the research provides the motivation. You have asked them to trust you and begin without knowing where they are going, but once they get started you have to provide them the means to fill in the gaps.

I intend this approach to be an explicit rejection of the following method, which has no place in an REU: "Go read these books and papers and when you understand them come back and I'll tell you about your project."

I do not claim that every research project can be started in this way. But a very large number of those *which are suitable for REU projects* can indeed fit our model. It takes some time and creativity, and this is a key part of the process of identifying suitable projects.

Students have individual projects. Each student has his or her own separate project. This project is under the primary supervision of one of the mathematicians, although at some point during the summer every mathematician works with every student. This helps expose the student to different perspectives and also makes it possible for the mathematicians to go to conferences without compromising the attention given to the students.

4. Miscellaneous

Some features of the AIM REU which we recommend for other programs:

Conference call prior to the summer. I arrange a conference call with all the students who have an offer to join our REU. This is efficient because there are many questions which nearly all students have, so I can answer them all at once. It also helps the students to find out a little about each other and to begin building a sense of community.

Weekly reports. Every Monday morning all the students give progress reports. We specify that the reports should be self-contained and should not assume the audience recalls the details of their previous report. What is their project and why is it interesting? What did they do last week? What is the overall plan and what specifically will they do this week?

We meet with the students before the report to discuss what they will say. Usually they need encouragement to be expansive about the "big picture." We meet with them again after the report to offer constructive criticism, usually phrased in the form: "Next week you might try..."

Initially the students do not like giving these reports, but in our exit surveys they tell us that the reports were valuable and they are glad they did them. In addition to making them more comfortable about giving a formal presentation of mathematics, they specifically indicate that the reports help them to keep their project in perspective and to plan their week.

LaTeX at the first opportunity. Have the students learn LaTeX the first time they do any work that could be part of their final report. This could be original work, or it could be some standard calculation that they worked through which could be part of the introduction in their paper. Give them a simple template file, and just have them type something small. Don't ask for a formal writeup or introduction: just have them start learning how to LaTeX. Encourage them to keep writing up anything that could be useful.

Preparing final talks. We host a mini-symposium during the final two days of the program where all the students give 40-50 minute talks. We treat this just like a session at a conference.

Sally Koutsoliotas and I have developed an effective approach to preparing students to give talks. We prepared a guide, available at `aimath.org/mathcommunity/`. Here is a summary: it is based on a series of four meetings.

The first meeting is just a conversation with the students. We identify the main point of their talk and work backward to determine the absolute minimum amount of information needed to *understand* and *appreciate* their main point. The student goes away and prepares a rough outline of their talk, usually in the form of handwritten drafts of slides on paper. (It is critical that the student prepare something very rough, because it will probably have to be completely taken apart and you don't want them to resist your suggested changes).

The second meeting begins with the student summarizing their main point, *and then* presenting their rough draft by laying it all out on a table. The mentor and student work together to identify the different parts of the talk, stressing the importance of the *transition* between each part, and the *flow* of ideas. The student goes away to make a complete version of their slides.

The third meeting involves laying out the slides on a big table (use printouts if the talk is on the computer) and grouping them according to the different parts of the talk. We evaluate the total number of slides, the content and layout of each slide, and the all-important transition between the different parts of the talk. The student goes away and makes final versions of the slides.

The fourth meeting is the first time the student actually gives a practice talk. Tell them to just talk and don't worry about how long it will take (you don't want them rushing at the end: if material has to be cut, it probably isn't at the end of the talk). Then sit down together at a table with all the slides laid out, and try to discuss only the 4 or 5 most important things they need to change. If appropriate, that is, if they have to make extensive revisions, schedule another practice talk. If there are minor changes, such as the talk was slightly too long, you may want to suggest that they practice in front of a friend.

Sample papers. To help the students in their final writeup we prepare a folder of sample papers. These are research papers from the general subject area of the REU and which have good overall organization, a well-written abstract, and an accessible introduction. These papers are models to show the student what their final paper should look like.

5. The benefits, mathematical and otherwise

Running an REU is a rewarding experience, particularly when you can quickly bring students up to speed and make them part of your research group. However, it is a huge amount of work, and to do it right you have to be willing to dedicate the majority of your time to it.

Is the total mathematical output larger than than it would have been if, instead of the REU, I just spent all day working on my own research? I think the answer is 'yes,' but in some sense the question is meaningless because I would not have the stamina to focus all-day-every-day on my research. The REU definitely increases the breadth of projects I am involved in, and that makes me a better mathematician.

AMERICAN INSTITUTE OF MATHEMATICS, PALO ALTO, CA 94306-2244
E-mail address: `farmer@aimath.org`

The Applied Mathematical Sciences Summer Institute

Erika T. Camacho and Stephen A. Wirkus

1. Introduction

AMSSI is an intensive seven-week summer research program in applied mathematics that focuses on diversifying the research experience through the creation of a strong community and the education of the whole person. The topics of AMSSI include the theory and applications of abstract algebra, finite fields, probability, statistics, stochastic processes, and differential equations. In addition, students learn the mathematical software packages of Matlab, Maple, and Mathematica as well as the document processing system LaTeX. These academic tools equip the students for their mathematical research. For their participation, students receive travel to and from Los Angeles, room and board for the duration of the seven-week program, and a stipend. AMSSI, while modeled after successful REUs, has some distinguishing features that help it achieve its goals.

In setting up the structure of AMSSI, we highlight a few of the main objectives of the NSF REU program: expanding student participation in all kinds of research; developing a *diverse*, internationally competitive, and globally-engaged science and engineering workforce; drawing on the integration of research and education *to attract a diversified pool of talented students* into careers in science and engineering; reaching broadly into the student talent pool of our nation to help increase the numbers of women, underrepresented minorities, and persons with disabilities in research; and involving students in research who might not otherwise have the opportunity, particularly those from academic institutions where research programs are limited.

We feel that diversity in every aspect of our society is essential and we use the REU as a way to focus on three underrepresented groups: women, underrepresented minorities, and those that do not have such opportunities in their home institutions. Diversifying the mathematical sciences through the incorporation of these groups has been a top priority for us.

Even though AMSSI 2005 was the first year of the program, its structure has drawn on years of personal and life-changing experiences with similar programs.[1]

Received by the editor December 1, 2006.

[1]Some aspects of AMSSI were modeled after the REUs MTBI and SIMU. MTBI was founded by Herbert Medina (Loyola Marymount University) and Carlos Castillo-Chavez (Arizona State University) and SIMU was founded by Ivelisse Rubio (University of Puerto Rico, Humacao) and

EC has participated in four distinct parts of the REU—as a student in 1996, as a TA from 1998-2002, as a faculty in 2003, and as a co-director and faculty from 2005-present; SW participated as a TA from 1996-1998, as co-director/summer director from 1999-2003, and as co-director and faculty from 2005-present. The pre-AMSSI experience of both was in the Mathematical and Theoretical Biology Institute (MTBI), now operating out of Arizona State University. Together with Ed Mosteig (2005-2006), Randy Swift (2005-2006), and Mercedes Franco (2006), we guide the students in research and mentor them throughout the program.[2] The collective experiences of the faculty provide a unique perspective on the structure of AMSSI and its goal of diversification.

2. The Academic Structure of AMSSI

From the first day of the program, we stress that AMSSI is a team effort and that helping others learn fosters the building of a learning community from which everyone can benefit. The first week of AMSSI is particularly intense with "math boot camp" lasting from 8 a.m.–12 midnight. We have four hours of interactive lecture and problem sessions in the morning in which the students learn some of the basic material that will form the basis for each of the projects. (This prepares the students for work on their research project, but also gives them common ground from which they can talk with their fellow students about each other's project.) In the afternoon, we have a four-hour interactive computer lab in which students work in different pairs each lab and learn the basics of Matlab, Maple, and Mathematica. During the evenings, the students have a fairly difficult homework assignment. Only by working with each other and with the help of the graduate research associates and faculty can the students finish the assignment in a reasonable amount of time. All the students who have finished voluntarily stay around to help the others. This schedule lasts for the first four days of AMSSI.

At the end of the first week, the students go through a LaTeX tutorial with the graduate research assistants as the faculty meet to break the students into groups of four to begin more focused background lectures for the student research projects. During the second and third weeks, the four research groups will make the transition from classroom theory to independent mathematical research. While the daily schedules of the second through sixth week is not as structured as the first week, the intensity is maintained. A complete written draft of the group's work is due at the end of the sixth week. The students are then given one day of unstructured free time. This "down time" is necessary for the students to re-charge their batteries for the equally intensive seventh week. Each draft is carefully read by two AMSSI faculty and constructive written comments are made on each project. Over the next few days, the students make the relevant changes to their written work and work on their oral and poster presentations. At the end of the seventh week, AMSSI holds two colloquia, one at California State Polytechnic University, Pomona (Cal Poly Pomona) and the other at Loyola Marymount University (LMU), during which time the students give oral presentations of their summer's work. The students submit final copies of all of their work on Friday, thus concluding the academic portion of AMSSI.

Herbert Medina (Loyola Marymount University). Some aspects were modeled from the late Janet Anderson's summer research program at Hope College.

[2]See www.amssi.org for more information on each of the AMSSI faculty.

3. Diversifying the REU Academic Environment

It is clear that strong efforts need to be made to draw from the underrepresented communities (women, underrepresented minorities, and those that might not otherwise have the opportunity) in order to maintain excellence and diversity of ideas. But immersing students in an REU or similarly unfamiliar environment is not enough to guarantee a lasting effect as illustrated by the current data of Ph.D.s awarded in math; see Medina (2004), the February *Notices*, and Peterson (2005). Simply placing such a student in an REU that provides academic/research opportunities but does not attempt to reach out to the individuals' experiences is not an effective way of diversifying. Confrey states that

> "allowing mathematics to continue to require students to disengage from their personal sources of experience and to learn a system of rituals that makes little sense to them but which will admit them to the ranks of the elite is one of the most effective ways of maintaining this oppression" (Confrey, 1995)

in discussing how the oppressive view of abstraction only preserves the status quo.

We address Confrey's concern by exposing our students to activities that cultivate their academic and personal/emotional growth. We "stretch" students by taking them out of their normal element while still providing certain components of familiarity to promote their growth and empowerment. In addressing their personal and emotional growth, we create a strong component of familiarity and community involving fellow students, faculty, and visiting mathematical researchers (both industrial and academic). AMSSI's goal is to develop Ph.D. mathematicians from these underrepresented groups that will impact the culture and diversity of the U.S. while strengthening our communities.

• **Split Campuses:** AMSSI is held for three weeks at Cal Poly Pomona and four weeks at LMU. The schools are approximately 50 miles apart and their locations within the greater Los Angeles area also provides the students with two different academic climates and experiences.[3] We thus build a strong network of support for the students by having the students interact with as many supportive faculty and potential employers as possible.

• **Partnerships with NSF IGERT[4] Programs in the Mathematical Sciences:** As neither Cal Poly Pomona nor LMU have Ph.D. programs, it is essential that our students have sufficient interaction with supportive faculty at Ph.D. granting institutions in order to help facilitate their access to these programs. To help with this, we have arranged partnerships with certain IGERT programs in the mathematical sciences that we feel may be of potential benefit to our students. Various IGERT directors come to AMSSI to give a research colloquia and to distribute information about their respective programs.

• **Recruitment of Students from Non-research/Non-selective Schools:** In order to achieve a diversified group of participants, it is essential to recruit from a very wide pool of talented students. In seeking applicants, we target schools with a large percentage of women and/or underrepresented minorities including all-women colleges, historically black colleges, Native American Tribal Colleges, public universities near large cities (especially in California, Texas, and Florida),

---

[3]Cal Poly Pomona is a large public school while LMU is a samller private school.

[4]Integrative Graduate Education and Research Traineeship

and community colleges. We use flyers, e-mails, the local newspaper, and especially personal contacts to inform faculty and students about this opportunity. The SAC-NAS[5] Conference, the joint meetings of the MAA/AMS[6], the Infinite Possibilities Conference, the Andrew Mays Undergraduate Summer Conference, and the Ford Foundation Conference of Fellows serve as excellent forums to recruit a diverse set of students and disseminate information about our program. The students (and faculty) who attend these conferences are not just diverse in their socio-economic and ethnic background but also in their educational interests and field of study. The active participation of the AMSSI faculty in these conferences creates opportunities to talk one-on-one with the students and faculty who can help contribute to AMSSI's sustainability and growth.

The selected dates of AMSSI also play favorably in our recruitment effort. With numerous schools throughout the U.S. on the quarter system, it is essential that REU programs be aware of the potential students that may not be able to apply simply because of the program dates. The AMSSI applications are also reviewed by all the AMSSI faculty, maximizing the potential for choosing a diverse student group and giving the faculty ownership in the program.

• **Exposure to Industrial, Academic, and Non-traditional Applications of Mathematics:** We make the wide-range of career possibilities known to the students. Presentations by the RAND Corporation and Aerospace Corporation, as well as tours of the DreamWorks facilities, the Jet Propulsion Laboratory (JPL), and UCLA Department of Human Genetics gives the students non-traditional applications of mathematics. Besides the non-academic opportunities, we expose the students to more traditional academic career paths and host colloquia by many mathematicians who are known to be great mentors and role models. They each have the opportunity to interact with the students and talk with them frankly about graduate school and Ph.D. programs at their respective institutions. They also talk about the factors, including their personal experiences, that made them pursue a Ph.D. in their respective fields of study.

• **Open Weekly Research Meetings:** Even though the student groups meet and work with their advisors on a daily basis, they also present their work at weekly research meetings to the full AMSSI faculty, the visitors (including colloquium speakers), and campus faculty. These meetings give the students the opportunity to practice their oral presentation skills. Because there are usually first-time visitors present, students give a brief background of their problem and then give an update of current results. Each of these helps the students reinforce what they've learned and better understand their research problem. The meetings also give them an opportunity to think on their feet. Both AMSSI faculty and visitors ask questions (often throughout the presentations) and give feedback to the student groups.

• **Fostering a Collaborative Learning Community:** The positive role of collaborative learning cannot be stressed enough because it teaches the students how to work with others and also builds a sense of community that is essential for success in the summer program and beyond. It is important for the students to understand the type of environment in which they perform best and are happiest.

---

[5]Society for the Advancement of Chicanos and Native Americans in Science (www.sacnas.org).

[6]American Mathematical Society (www.ams.org) and Mathematical Association of America (www.maa.org).

To help provide students with information about graduate school we have designed a comprehensive and thorough four-hour interactive session led by an expert on the graduate school application process. Supportive fellow students, faculty, and visitors also provide encouragement to pursue higher degrees.

Because we focus on the recruitment of students who might not otherwise have the opportunity to conduct research, we believe that there must also be a concerted effort to use all of these academic tools in the greater context of overall student growth. Confrey's 1995 research suggests that many of our students would typically have no choice but to "disengage from their personal experiences" in their efforts to pursue their goals of attaining a higher degree. This in turn will have detrimental effects in their learning capacity. Recognizing this possibility, from the beginning we try to ensure that the AMSSI experience for our students is one in which their emotional growth parallels their academic growth. We agree that

> "the introduction of emotional intelligence into discussions of mathematics education allows one to assert that both facilitating and debilitating emotions play a significant role in learning, and that emotional qualities of classroom interactions will exert a significant influence on what is learned" (Confrey, 1995).

A key step in the direction of incorporating the emotional growth in the AMSSI experience is the weekly sharing meetings, which are attended by student participants and the AMSSI faculty. In these meetings, each member shares some of their personal side, including significant life experiences, dreams, fears, obstacles they have overcome, upbringing, family background, role models, hobbies, etc. Only AMSSI students and faculty attend the meetings to maintain a culture of confidentiality, trust, and respect that we carefully build throughout the program. These meetings establish a non-academic connection between the students that carries over into their work, as they now have a better understanding of each other from a non-academic perspective. The meetings also foster a very strong bond between students and faculty that further establishes a sense of community and creates a solid mentoring relationship. By the end of the program, everyone has had the opportunity to ask questions, have questions asked of them, and share a personal side of them that otherwise would not have been possible to share outside of this setting. The emotional intensity of the weekly sharing meetings complements the intensity we require for the academic aspects of AMSSI. Both students and faculty also participate in a weekly extracurricular activity.[7]

4. Preliminary Results

Even though AMSSI 2005 was the first year of our program under this REU model, we have had many successes. The eight manuscripts from 2005-2006 were compiled into a bound Technical Report volume; see www.amssi.org. The students also gave poster presentations at the SACNAS conference and at the joint meetings of the AMS/MAA. At the latter conference, two of the groups won poster awards for their work. Multiple students also presented their research at their home institutions and local undergraduate conferences. One of the posters also won an

---

[7]A large percentage of AMSSI's targeted students come from low-socioeconomic backgrounds and thus we do not require them to pay for mandatory activities; funds to cover the students in any mandatory extracurricular activities come from the generosity Cal Poly Pomona's Dean of the College of Science, Donald Straney, who recognizes the importance of a complete REU experience.

award at the SIAM[8] Southeast Atlantic Section Annual Meeting and again at the MGE@MSA/WAESO[9] Conference. One student presented his group's work at the 2006 Joint Meetings AMS Session on Probability and Stochastic Processes (#262, "Birth-Death Processes with Polynomial Transition Rates"). The manuscript entitled "Alcohol's Effect on Neuron Firing" was accepted for publication to the refereed journal *The Mathematical Scientist* and is scheduled to appear in vol. 32, no. 1, June 2007. The paper entitled "On the Representation of Birth-Death Processes with Polynomial Transition Rates" has been accepted for publication to the refereed *Journal of Statistical Theory and Practice*. We anticipate similar dissemination and success of the work of AMSSI 2006.

A key to AMSSI's success is the dedication of its faculty. As mentioned earlier, a sense of ownership in the program is fostered from the beginning, which is essential for success in this time-intensive environment of learning and mentoring. Educating and mentoring the whole person includes sharing life experiences, listening, strengthening self-worth, connecting emotions and familiarity to new knowledge, fostering individual growth, building trust, creating a strong community, and making work a team effort and experience where everyone contributes and benefits.

5. Acknowledgements

AMSSI is supported from grants given by the National Science Foundation (DMS-0453602), the Department of Defense (through its ASSURE program), and the National Security Agency (MSPF-04IC-227, MSPF-06IC-022). Additional financial and moral support was also provided by Donald Straney, Dean of the College of Science at Cal Poly Pomona, the Department of Mathematics at LMU, and the Department of Mathematics & Statistics at Cal Poly Pomona.

References

1. Confrey, Jere (1995). A Theory of Intellectual Development, Part III, *For the Learning of Mathematics*, 15(2): 36-45.
2. Gallian, Joseph A. (Ed.) (2000). Summer Program Survey, in *Proceedings of the Conference on Summer Undergraduate Mathematics Research Programs*, American Mathematical Society, Providence, RI, 2000.
3. Medina, Herbert A. (2004), Doctorate Degrees in Mathematics Earned by Blacks, Hispanic/Latinos, and Native Americans: A Look at the Numbers, *Notices of the AMS*, 51(7): 772-775.
4. *Notices of the American Mathematical Society*, Feb. issues, 1999-2005.
5. Petersen, Mark R., Kraus, Barbara E., & Windham Thomas L. (2005, March). Striving Toward Equity: Underrepresented Minorities and Mathematics, *SIAM News*, 38(2).

DEPARTMENT OF MATHEMATICS, LOYOLA MARYMOUNT UNIVERSITY, 1 LMU DRIVE, SUITE 2700, LOS ANGELES, CA 90045
E-mail address: ecamacho@lmu.edu

CALIFORNIA STATE POLYTECHNIC UNIVERSITY, POMONA
E-mail address: swirkus@csupomona.edu

---

[8]Society for Industrial and Applied Mathematics
[9]More Graduate Education at Mountain States Alliance (MGE@MSA)/Western Alliance to Expand Student Opportunities

Proceedings of the Conference on
Promoting Undergraduate Research in Mathematics

Promoting Research and Minority Participation via Undergraduate Research in the Mathematical Sciences. MTBI/SUMS-Arizona State University

Carlos Castillo-Chavez, Carlos W. Castillo-Garsow, Gerardo Chowell, David Murillo and Melanie Pshaenich

Mathematical Theoretical Biology Institute (MTBI) and the Institute for Strengthening Understanding of Mathematics and Science (SUMS)

The *primary goal* of the Mathematical Theoretical Biology Institute[1] and the Institute for Strengthening Understanding of Mathematics and Science[2] (MTBI/SUMS) is to increase the number of US Residents or Citizens, particularly members of US underrepresented minority groups, who complete a Ph.D. in the mathematical sciences. The cornerstone of this effort is MTBI/SUMS' summer REU program. MTBI and SUMS have instituted a series of carefully refined mentorship and apprenticeship programs, from the high school[3] to the postdoctoral level, that promote mathematical learning and, in the process, help diversify participation in the mathematical sciences.

Traditionally, MTBI/SUMS has provided research mentorship training for students who want to work at the interface of applied mathematics and theoretical and computational biology[4]. The REU summer program, which has been held for the past eleven years, in conjunction with Cornell University (1996 − 2003), Los Alamos National Laboratory (2003 − 2005), and Arizona State University (2004-Present)[5] focuses on the applications of mathematics to problems at the interface of the natural and social sciences. MTBI/SUMS programs are aimed at increasing

---

Received by the editor December 1, 2006.

[1]http://mtbi.asu.edu

[2]http://www.asu.edu/mshp/index.htm and winner of a 2003 *Presidential Award for Excellence in Science, Mathematics and Engineering Mentoring.*

[3]SUMS through its Mathematical Sciences Honors Science Program has mentored 2098 high school students who have begun to participate in MTBI/SUMS' summer REU undergraduate program early in their college studies.

[4]Research mentorship is provided during the summers or throughout the year (at the host institution) with the support of MTBI/SUMS Sloan Pipeline Program.

[5]MTBI/SUMS efforts have been supported through grants by the National Security Agency, the National Science Foundation, the Sloan Foundation, Los Alamos National Laboratory, and the offices of the provosts of Cornell University and Arizona State University.

the participation of students from diverse educational, cultural, racial and socioeconomic backgrounds in areas where the mathematical sciences play a fundamental role. A large percentage of summer selected participants come, by design, from colleges and universities with limited research opportunities. Not surprisingly, these institutions often tend to serve large numbers of individuals from US underrepresented minority groups.

MTBI/SUMS summer research programs are run like NSF-sponsored workshops[6]. New students take three and half weeks of intense training in dynamical systems (broadly understood to include stochastic processes) by modeling in the biological and social sciences while becoming familiar with tools like MATLAB[7], MATHEMATICA[8], XPP-AUTO[9], other computational packages, and LATEX[10]. At the end of the initial training period participants form groups of $3 - 4$ students around a project of their own choice. In other words, students set each the research agenda each summer. This philosophy accounts for the diversity of research manuscripts that have been produced[11] over the past decade. Each group gets assigned a faculty advisor and is provided with appropriate graduate student support. Between 20% and 33% of the undergraduates participate in two summers and graduate student participants (14) have participated at least two times with some participating as many as seven times. MTBI/SUMS summer workshop produces an average of 10 technical reports per year. The research productivity of each summer group of young investigators has instigated or re-invigorated the research efforts of a class of participants that includes students, faculty and postdoctoral students.

MTBI/SUMS Sloan Pipeline Program (MTBI/SUMS-SLPP) is at the heart of MTBI/SUMS mentorship activities at the graduate level. MTBI/SUMS alumni[12] who are enrolled at MTBI/SUMS home institution[13] receive partial support towards the completion of their graduate studies (leading towards the Ph.D.'s) in the forms of fellowships, research assistantships and MTBI/SUMS research assistantships. The area of Ph.D. study has been dictated exclusively by the students' interests and objectives. Recent alumni (Cornell) have received their Ph.D.'s in numerical

---

[6]Our efforts have been carried out with the active collaboration of the Santa Fe Institute (NM) and the Center for Statistical Genetics and Genome Science Training Program at the University of Michigan.

[7]MATLAB is a product of MathWorks (http://www.mathworks.com/).

[8]It is a product of WOLFRAM RESEARCH (http://www.wolfram.com/).

[9]http://www.math.pitt.edu/ bard/xpp/xpp.html

[10]http://www.latex-project.org/

[11]MTBI/SUMS alumni are prolific writers. *Am I too fat? Bulimia as an epidemic* has appeared in the Journal of Mathematical Psychology ($47(2003)515 - 526$); and the article *Effects of education, vaccination and treatment on HIV transmission in homosexuals with genetic heterogeneity* has appeared in the Journal Mathematical Biosciences ($187(2004)111 - 133$). These are but two examples of what our students produce.

[12]The term MTBI/SUMS alumni has been used throughout to identify primarily those who participated in one or more summer research experiences. However, MTBI/SUMS does not focus *exclusively* on summer research experiences. In fact, it has had a large number of support programs that involve sponsored research at the interface of the mathematical and natural and social sciences as well as the administration of various human resource programs including RTGs, Sloan Fellowships, Mentorship Programs and Additional Educational and Training Programs K-20 and beyond.

[13]Cornell University (1996−) and Arizona State University (2004+)

analysis, mathematics of finance, statistics, computational biology, mathematical physiology, mathematical epidemiology, biometry and dynamical systems. At Cornell University, sixteen MTBI/SUMS alumni were awarded Sloan Fellowships while six MTBI/SUMS alumni were supported via a RTG[14] grant (pre-IGERT) in computational biology. MTBI/SUMS has worked hard to establish similar forms of support at ASU. *Nine* Sloan fellowships have been awarded to underrepresented US minorities who are enrolled in the mathematics department at ASU[15]. Professor Carlos Castillo-Chavez[16], MTBI/SUMS Director (MTBI-founder) has led MTBI since 1996 and SUMS since 2004.

Success Stories

In the years 2001 and 2002, prior to MTBI/SUMS producing Ph.D. graduates, the U.S. awarded an average of 10 Ph.D.s to Latinos[17]. MTBI/SUMS efforts have significantly increased the national rate of production of U.S. Ph.D.'s. In 2005, MTBI/SUMS alumni received 10 Ph.D.s in the mathematical sciences, 7 of which were awarded to members of underrepresented[18] US minority groups. This is almost *a fourth* of the national total output for that year. Of those, 6 were Latino, *one third* for that year (6 out of 18). Of the 10 total MTBI/SUMS alumni Ph.D. graduates in 2005, 7 took on prestigious postdoctoral positions and one became an Assistant Professor at the University of Puerto Rico, Mayaguez campus. Looking at female graduates, MTBI help produce *one third* (5 out of 15) of the total female underrepresented minority groups for 2005. Four of those five were Latinas, over half of the national production (4 out of 7). In 2006 10 MTBI/SUMS alumni received their Ph.D. including 7 from US underrepresented minority groups and their record of success is similar to those who completed their degrees in 2005.

MTBI/SUMS has mentored and supported 277[19] undergraduate students, and 31 graduate students, of which 14 had participated previously in MTBI/SUMS as undergraduate students throughout eleven summers[20].

Since 1996, MTBI/SUMS alumni[21] have co-authored 111 technical reports during the summers[22]. Revised reports are continuously being published in refereed

---

[14]I was the either the PI or co-PI of an NSF RTG grant at the interface of mathematics and biology for over a decade.

[15]The rate of growth in its minority graduate student population would have not been possible without the support programs instigated by the Hispanic Research Center led by Gary Keller and the openness of the mathematics faculty that has embraced MTBI efforts to increase and maintain diversity.

[16]Winner of the 1997 Presidential Award for Excellence in Science, Mathematics, and Engineering Mentoring.

[17]The data for national Ph.D. graduates was obtained from the AMS notices http://www.ams.org/notices/200602/05firstreport.pdf.44444444

[18]US Residents who are Latino (the overwhelming majority are Mexican Americans or Chicanos and Puerto Ricans but there are some whose heritage is from Peru or El Salvador) or African-American or Native Americans.

[19]This figure includes students from the summer of 2006 who completed their work on July 30, 2006.

[20]It has also mentored dozens of students throughout the regular academic year at either Cornell University and Arizona State University.

[21]MTBI/SUMS alumni refers to those who have participated in its summer research programs or who are mentored throughout the year by MTBI/SUMS

[22]This number already includes the reports of the summer of 2006.

journals. The references at the end of this manuscript list *ten recent representative refereed publications*, the result of the collaborative work between MTBI alumni, staff and visitors over the past 10 years.

MTBI/SUMS alumni[23] have, or currently are attending universities across the United States, Colombia, Argentina, Britain and Mexico. MTBI/SUMS sequential summer programs have helped establish large communities of underrepresented U.S. minorities at Cornell University[24] (totaling twenty-four[25], eighteen[26] of which are members of underrepresented US minority groups), at the University of Iowa[27] (totaling seventeen[28], fourteen of which are members of underrepresented US minority groups) and Arizona State University (totaling thirty-four[29], twenty-four of which are members of underrepresented US minority groups[30]).

MTBI/SUMS has sent 130 students from *underrepresented minority groups* to graduate school over its *first ten years*[31] and a total of 169 students overall. Furthermore, 52% have been females, including 65 from minority groups.

MTBI/SUMS Alumni have established a community of minority scholars at ASU and in the process, their presence has facilitated the recruitment of increasing number of minorities to its graduate programs. We currently have *at least* 29 US

---

[23]That is, individuals who participated in MTBI/SUMS summer programs as graduate or undergraduate students or who received mentorship support from the MTBI/SUMS Sloan Pipeline Program.

[24]Most of the minority students in the mathematical sciences since 1996 have come from the pool of MTBI/SUMS alumni. Four of them completed their Ph.D. in 2005 (a fifth transferred to Florida and also completed her Ph.D. in 2005); two more in the summer of 2006; and two more are expected to graduate by December of 2006. The success rate (obtained their Ph.D.) at Cornell University when it comes down to MTBI/SUMS alumni (the only data that we have) is over 80%.

[25]That is, 24 MTBI/SUMS alumni have enrolled in a mathematical sciences program at Cornell University. Students have enrolled in the departments of biological statistics and computational biology, statistics, the center for applied mathematics and the department of theoretical and applied mechanics.

[26]Three US minorities received a MS degree prior to or on 2000 and left. Seven US Minorities have completed their Ph.D.s with two more expected by December of 2006. Five more will receive their degrees over the next two years and one will transfer to the Ph.D. program in Arizona State University.

[27]The Mathematics Department at the University of Iowa has created a model in which most if not all the members of the mathematics faculty participate. It recruits heavily in Puerto Rico, Historically Black Colleges and MTBI/SUMS. The Mathematics Department at the University of Iowa received a 2004 Presidential Award for Excellence in Science, Mathematics, and Engineering Mentoring.

[28]That is, 17 MTBI/SUMS alumni (who participated in its summer program) have enrolled in a mathematical sciences program at the University of Iowa, most in the mathematics department.

[29]34 MTBI/SUMS alumni have enrolled at ASU. Thirty-two of them have enrolled in the mathematical sciences (mathematics and statistics) and two in the sciences. This group includes 28 permanent residents or US citizens and six international students. Currently, there are 32 enrolled as two received a masters degree and left.

[30]July 1st 2006-data.

[31]This number does not include the admission to graduate school of members of the summer of 2006 MTBI class. However, we are to a good start. eight MTBI/SUMS alumni from the 2006 class will be attending graduate school in the fall of 2006 or the Spring of 2007.

minorities[32] in the mathematics department at ASU including 24 US Latinos and 5 African-Americans. ASU[33] will add three MTBI/SUMS alumni[34] (US underrepresented minorities) to its mathematics graduate program in the spring of 2007[35]. ASU graduate mathematics program will continue to host the largest US minority student population (32) in the nation[36]. Finally, it is worth noticing that ASU campuses have hired *three MTBI alumni to its mathematics and statistics faculty, all from underrepresented minority groups and two in tenure-track appointments.*

Philosophy and Goals

MTBI/SUMS believes that this community of minority scholars is *vital* to increasing the number of underrepresented minorities in the sciences. It is not enough that MTBI/SUMS creates an increase in the population of minority scholars. This change must become self-generated. In order for MTBI to have a lasting impact outside the scope of its own individual focus programs, MTBI/SUMS alumni must begin to make their own changes, their own waves, and their own recruitment initiatives. In terminology that our own program alumni would be instantly familiar with, we want to create an *epidemic* of minorities in the sciences. Now that MTBI/SUMS alumni are beginning to take faculty positions and in fact, there is some evidence of secondary recruitment[37].

MTBI/SUMS philosophy adheres to the principles of the *New American University*[38] that is, MTBI/SUMS is an institute that, like its home institution, ASU[39], wants to be judged by the quality of the research and academic accomplishments of its students and alumni rather than by the academic pedigree or prior access to *selective* educational settings of its participants. MTBI/SUMS wants to be an institute whose alumni, while pursuing their scholarly and scientific interests, "also consider the public good."[40] MTBI/SUMS wants to be an institute whose students,

---

[32]Twenty-seven minority students in the mathematics department, most of them have been recruited by MTBI/SUMS. All are members of its Sloan Pipeline Program and hence are consider MTBI/SUMS members.

[33]The premier graduate program in the mathematical sciences is Richard Tapia's at Rice University. Richard who received a 1996 Presidential Award for Excellence in Science, Mathematics, and Engineering Mentoring has been the primary mentor of US minority Ph.D.'s over the past two decades. Additional strong efforts led in part by Raymond Johnson, have been carried out at the University of Maryland. In December of 2000 alone, three African-American women received their Ph.D. in applied mathematics from the University of Maryland.

[34]This group includes four US-underrepresented minorities, a US female student and a female international student.

[35]We expect that these three students will accept ASU support offers. Their admission will bring the number of US underrepresented minorities within the mathematics department at ASU to 30 and the number of MTBI/SUMS alumni who have enrolled at ASU will have reached 37.

[36]The mathematics department at the university of Iowa has had an average number of 23 US minority students over the past few years (David Manderscheid, chair mathematics department, personal communication), has graduated $2 - 3$ US minorities per year over at least the past two years and may currently have 26 minority graduate students.

[37]The establishment of the *Applied Mathematical Sciences Institute, http://www.amssi.org/* by MTBI alumni Erika Camacho and MTBI graduate mentor and former summer Director Steve Wirkus are indicative of things to come.

[38]http://www.asu.edu/president/newamericanuniversity/arizona/

[39]Here, we are paraphrasing ASU's mission but in the context of the work that is being carried out at MTBI/SUMS.

[40]http://www.asu.edu/president/newamericanuniversity/arizona/

alumni, faculty, and staff "transcend the concept of community service to accept responsibility for the economic, social, cultural, and environmental vitality of the communities they serve."[41]

Acknowledgments

MTBI/SUMS efforts have not been carried alone. MTBI received extraordinary support by the Cornell University's administration[42], the Center for Applied Mathematics and the Biological Statistics and Computational Biology Department. MTBI/SUMS has had no less support at ASU[43]. We have established a highly effective partnership with the Hispanic Research Center[44]. ASU's Mathematics and Statistics Department has not only embraced our efforts but have actively joined them. MTBI/SUMS successes have been possibly because of the leadership and hard work of all our partners, supporters, its staff and its summer faculty. However, at the end of the day it is the continuous funding by NSA, NSF and the Sloan Foundation[45] that have kept this effort alive long enough time to make a difference.

References

[1] Baojun, S., Garsow-Castillo, M, Marcin M., Henso, L., and C. Castillo-Chavez. Raves Clubs, and Ecstassy: The Impact of Peer Pressure. *Journal of Mathematical Biosciences and Engineering* Volume 3, Number 1, January 2006 1-18.

[2] Gjorgjieva, J., Smith K., Chowell, G., Sanchez, F., Snyder J., and C. Castillo-Chavez. The Role of Vaccination in the Control of SARS. *Journal of Mathematical Biosciences and Engineering* Volume 2, Number 4, October 2005 753-769

[3] Kribs-Zaleta, C., Lee, M., Román, C., Wiley, S., Hernández-Suárez, C.M. The effect of the HIV/Aids epidemic on Africa's truck drivers. *Journal of Mathematical Biosciences and Engineering* 2, (4), 771-788. 2005.

[4] Chowell, G., Cintron-Arias, A., Del Valle, S., Sanchez, F., Song B., Hyman, J. M. and C. Castillo-Chavez. Homeland Security and the Deliberate Release of Biological Agents. *In: Modeling The Dynamics of Human Diseases: Emerging Paradigms and Challenges. Gumel A. (Chief Editor), Castillo-Chavez, C., Clemence, D.P. and R.E. Mickens* American Mathematical Society (in press).

[5] Yakubu, A-A, Saenz R., Stein, J., and L. E. Jones. Monarch butterfly spatially discrete advection model. *Journal of Mathematical Biosciences and Engineering* 190, 183-202, 2004.

[6] Rios-Soto, K.R., Castillo-Chavez, C., Neubert, M., Titi, E.S., and A-A Yakubu. Epidemic Spread in Populations at Demographic Equilibrium. *In: Modeling The Dynamics of Human Diseases: Emerging Paradigms and Challenges. Gumel A. (Chief Editor), Castillo-Chavez, C., Clemence, D.P. and R.E. Mickens,* American Mathematical Society (in press).

[7] Sanchez, F., Engman, M., Harrington, L. and C. Castillo-Chavez. Models for Dengue Transmission and Control. In: Modeling The Dynamics of Human Diseases: Emerging Paradigms and Challenges *In: Modeling The Dynamics of Human Diseases: Emerging Paradigms and Challenges. Gumel A. (Chief Editor), Castillo-Chavez, C., Clemence, D.P. and R.E. Mickens,* American Mathematical Society (in press).

---

[41]http://www.asu.edu/president/newamericanuniversity/arizona/

[42]Malden Nesheim, Don Randel, Biddy Martin, Frank Rhodes, David Call, Hunter R Rawlings III and W. Kent Fuchs.

[43]Michael Crow, Milton Glick, David Young, Maria Allison, Marjorie Zatz, Jon Fink, Andrew Webber, Peter Crouch, Elizabeth Capaldi and Marjorie Zatz who have done everything possible to help the goals and the vision of MTBI/SUMS.

[44]Albert McHenry, Gary Keller, Antonio García and Michael Sullivan are the kind of university citizens that every university dreams to have.

[45]The encouragement and confidence given to MTBI by Barbara Deuink, Lloyd Douglas, Ted Greenwood, Jim Schatz and Michelle Wagner have played a critical.

[8] Del Valle, S., Morales Evangelista, A., Velasco, M.C., Kribs-Zaleta, C.M., Hsu Schmitz, S.F. Effects of education, vaccination and treatment on HIV transmission in homosexuals with genetic heterogeneity. *Journal of Mathematical Biosciences and Engineering* 187, 111-133. 2004.

[9] González, B., Huerta-Sánchez, E., Ortiz-Nieves, A., Vázquez-Alvarez, T., Kribs-Zaleta, C. Am I too fat? Bulimia as an epidemic. *Journal of Mathematical Psychology* 1.47, 515-526, 2003.

[10] Chowell, G., P. W. Fenimore, M. A. Castillo-Garsow and C. Castillo-Chavez. SARS Outbreaks in Ontario, Hong Kong and Singapore: the role of diagnosis and isolation as a control mechanism. *J. of Theoretical Biology* 224, 1-8, 2003.

CASTILLO-CHAVEZ, CASTILLO-GARSOW, MURILLO, PSHAENICH, DEPARTMENT OF MATHEMATICS AND STATISTICS, ARIZONA STATE UNIVERSITY, TEMPE, AZ 85287
  E-mail address: `chavez@math.asu.edu, cwcg@asu.edu, dlm35@mathpost.la.asu.edu,`
`MelanieMSP@asu.edu`

CHOWELL, THEORETICAL DIVISION (MS B284) LOS ALAMOS NATIONAL LABORATORY, LOS ALAMOS, NM 87545
  E-mail address: `chowell@lanl.gov`

Summer Mathematics Research Experience for Undergraduates (REU) at Brigham Young University

Michael Dorff

1. Introduction

Brigham Young University (BYU) hosts an NSF-funded 8-week summer Research Experience for Undergraduates (REU) in geometric analysis and mathematical physics. The objectives of our program are: (1) to provide undergraduate students with the opportunity to experience doing mathematical research; (2) to encourage undergraduate students to attend graduate school in mathematics; and (3) to prepare participants to be successful in graduate school. We are specifically interested in recruiting female students and students who are from institutions without a graduate program in mathematics. Our experience with undergraduate research in mathematics has shown us that students are effective in learning about mathematics and in doing mathematical research if they are working on challenging problems in a supportive but structured environment. Hence, we offer many components in our program. To help students experience the process and excitement of doing research, students are first given a series of introductory lectures. Then they work individually and as a group on challenging research problems with faculty who have been successful in mentoring undergraduates. At the end of the REU, students present a research talk and write a final research report. Also, we want to provide the students with skills that will help them in their research and in graduate school. To help achieve this, we offer training sessions in computers, seminars on needed research skills, and discussions on graduate school. Further, we provide social and recreational activities that foster interaction and collaboration in a relaxing environment. All of this is done to ensure that the students have a meaningful research experience and develop the skills to help them be successful in graduate school.

2. The BYU summer REU

2.1. The research groups. During the BYU summer REU, there are two research groups each consisting of 4-5 undergraduate students, 1 faculty mentor, and 1 graduate student assistant. Each research group is further divided into two subgroups of 2-3 undergraduates; the subgroups work on a specific problem related

Received by the editor November 3, 2006.

to their main research area. There are two research topics selected from geometric optimization, minimal surfaces, and mathematical physics. The type of research problems the REU participants work on are graduate level mathematics.

In each research group, there will be a common overall problem or method. The problem will be divided into several components with each student working on a specific component. This setup has been effective in our undergraduate research groups – it fosters both collaboration and individual work. Because there is one common problem or method in each research group, students will naturally discuss their work with each other. However, having each student working on a specific component of the problem will provide each student with specific assignments to do.

2.2. Sample research problems.

Geometric Optimization: The basic problem in geometric optimization is to minimize length, area, or some other quantity, among curves or surfaces satisfying a given constraint. A well-known example is that a circle has least perimeter among curves enclosing a given area. Some problems that students have considered are: (1) the octahedron problem which asks what is the least area surface spanning the edges of the octahedron. The conjectured solution is a beautiful, piecewise-planar soap film consisting of twelve triangles and six kites; (2) large minimizing networks in the hyperbolic plane; that is finding the shortest path connecting a set of points in the hyperbolic plane; and (3) Melzak's conjecture which asks what polyhedron with unit volume in Euclidean 3-space has the shortest edgelength (for example, a cube with volume 1 has total edge length 12; there is a polyhedron that has a shorter edgelength–can you figure it out?).

Minimal Surfaces: At each point p on a regular surface $M \in \mathbb{R}^3$, we can compute a normal vector n. Any plane that contains n will intersect the surface in a curve c. For each curve c, we can compute its curvature. As we rotate the plane through the normal n, we will get a set of curves on the surface each of which has a value for its curvature. Let k_1 and k_2 be the maximum and minimum curvature values at p, respectively. The mean curvature of M at p is $H = (k_1 + k_2)/2$. Then M is a minimal surface if the mean curvature equals zero at every point. We can use ideas from complex analysis to investigate minimal surfaces. In particular: (1) we can use the Schwarz-Christoffel formula from complex analysis to derive analytic functions that map onto convex polygonal regions. We can then shear these functions and derive the corresponding minimal surfaces and see how they are related to Jenkins-Serrin minimal surfaces which project to convex polygons; (2) we can shear a specific elliptic integrals of the first kind to get a family of minimal surfaces that range from Scherk's doubly-periodic to the helicoid. What families of surfaces do we get, when we shear other elliptic integrals?; and (3) we can represent minimal surfaces with specific properties as solutions to a system of DE's. This system of DE's can be described as a manifold in $\mathbb{R}^n$. By computing its Lie symmetries with the help of Maple or Mathematica we get a continuous family of minimal surfaces that have the origianl specific property.

2.3. The schedule. The mathematical component of the REU can be somewhat divided into two parts. The first part occurs during the initial weeks and consists of activities to help the students be prepared to do research, such as introductory lectures on the research topic and problem sets related to the research lectures. Also, there are training sessions in computers (e.g., LaTeX, Maple, and Mathematica) and seminars on needed research skills (e.g., "How to read a math paper," "How to write a math paper," "How to give a math talk," "Available research tools on the web and in the library"). In addition, we start having discussions on graduate school (e.g., "Common misconceptions about graduate school in mathematics," "How to choose a graduate program," "What you should do now to prepare for graduate school.").

The second part begins after the students have learned enough to begin to do actual research (near the end of the second week). At this point, the formal introductory lectures transition into discussions of research problems and assignments of components that the students can work on. During this time, there are also afternoon colloquia on other areas of mathematics so to broaden the students' exposure to mathematics and presentations on graduate school. Some of these presentations are given by BYU mathematicians. Others are given by non-BYU female mathematicians whom we invite to the REU to not only give research presentations, but to also interact with the REU students and to answer their questions about graduate school. These colloquia have included such areas as geometric group theory, algebra, graph theory, mathematical biology, number theory, and algebraic geometry.

Also, we believe that social and outdoor activities are beneficial to the students participating in the REU. These activities are not only enjoyable, but also help the students to interact and promote collaboration. We arrange barbeques about every other week, several local hikes to waterfalls and springs in the nearby mountains, and one major weekend trip. In 2006, we organized an excursion to Arches National Park followed by a day-long white-water rafting trip.

2.4. Research presentation and research paper. Near the end of the REU, the students give a formal presentations about their research results. Giving research presentations is a skill that can be improved with practice. We give a list of positive aspects of their talks plus suggestions on how to improve specific aspects of it. Our experience is that this helps them to be better prepared to give a presentation at a regional or national conference, an activity that we feel is very beneficial for them as future mathematicians. We have found that undergraduates are excited about attending and participating in conferences. This is an important experience for them and we provide funds for the REU students to attend and present their research at such a conference. Also, we require the students to submit a final written report. Ideally, this will form the basis of a research publication for the students.

3. Recruitment and logistics

Students are recruited nationally through various ways. First, we mail a flyer announcing our program to mathematics departments at institutions in the local region. Also, we maintain a webpage (http://math.byu.edu/reu) that includes a flyer announcing the program, suggestions for background information about research topics, information about the previous programs, photos, and students' comments and information about applying for the program. Further, the proposed REU program is listed in the AMS and NSF listings of Summer REU programs. Recruitment has been successful. For the 2006 REU we received about 100 applications. Originally, NSF provided funds for 8 undergraduates, at least 6 of whom are non-BYU students, to participate in the BYU summer REU. Then in 2006 BYU agreed to separately fund a 9th student who is not from BYU.

Participants are given a stipend of $2750, up to $450 for travel to BYU, free housing in off-campus apartments during the 8-week program, and $400 travel allowance to attend a conference during the following academic year to present their research.

4. Project Evaluation and Reporting

Our evaluation plan has several components: (1) papers authored or co-authored by undergraduate participants; (2) presentations by participants at conferences; (3) percentage of participants who attend graduate school in mathematics; and (4) multi-stage surveys.

The multi-stage surveys are based upon the ideas of the statisticians Adhikari and Nolan[1] concerning the assessing of student learning in undergraduate research projects. The questionnaire contains both open-ended questions and questions that ask for a numerical rating. It is important to have a baseline survey that gives us an accurate reading of where the students are starting from and a way to measure long-term impact. Hence, we administer the questionnaire during the REU and one year later. Also, we send a brief survey to the faculty members who wrote letters of recommendation for the participants asking these faculty members to assess the effect the REU program had on the student's work in mathematics and their plans for attending graduate school.

In 2005, the first year of the BYU summer REU, there were 8 undergraduates (6 female) who participated. Before attending the REU, 36% of the participants were planning on attending graduate school in mathematics. After attending the 2005 REU, 100% of the participants had either applied to at least one graduate program in mathematics for next fall or are juniors who are planning on applying during their senior year. Also, 100% of the participants have presented their research at a mathematics conference. Last year, one group published their research result in the Illinois Journal of Math, in which mathematicians from research institutions publish.

DEPARTMENT OF MATHEMATICS, BRIGHAM YOUNG UNIVERSITY, PROVO, UTAH 84602
E-mail address: mdorff@math.byu.edu

---

[1]A. Adhikari and D. Nolan, "But what good came of it at last?"–How to assess the value of undergraduate research, *Notices of the AMS*, **49** no. 10 (2002), 1252-1257.

Introducing Undergraduates from Underrepresented Minorities to Mathematical Research: the CSU Channel Islands/California Lutheran University REU, 2004 - 2006

Cindy Wyels

History and Funding

In 2003 the Mathematical Association of America (MAA), through its Strengthening Underrepresented Minority Mathematics Achievement (SUMMA) program, obtained funding from the NSF and the NSA to initiate summer REUs for mathematics majors from underrepresented minorities. Three sites were funded in 2003; in 2004 California Lutheran University (CLU) joined the original sites and two others in hosting an REU. CLU again hosted an REU in 2005 (and the Moody's Foundation also began providing funding), and in 2006 the REU migrated to CSU Channel Islands. In 2004 and 2005 four undergraduates worked on individual (but related) research projects under the direction of a faculty member, assisted by a volunteer graduate student. Six students participated in 2006: they worked in pairs on research projects, receiving guidance from two faculty members. This report will focus on the 2006 REU, as lessons learned during 2004 and 2005 were implemented in 2006, making 2006, in the opinion of the faculty involved, the most successful REU (in terms of student learning).

Program Goals

The following program goals were developed by the faculty and shared with the students on the first day of the REU. After group discussion and modification, the goals were posted prominently and referred to periodically throughout the REU.

(1) Raise the mathematical maturity level of the program's participants.
(2) Get participants excited about doing mathematical research.
(3) Create a learning community.
(4) Help participants develop the confidence to succeed in ongoing mathematical studies.
(5) Increase participants' skills in communicating mathematics.
(6) Extend the participants' abilities to read, understand, construct, and write proofs.

Received by the editor March 19, 2007.

(7) Acquaint participants with the culture and activities of research mathematics.

(8) Develop participants' skills in reading professional-level mathematics.

(9) Give the participants technical tools for future mathematical learning and research (e.g. MathSciNet, LaTeX, presentation software).

Goals #1 - 7 and #9 were met substantively; some progress was made towards #8[1].

Participant Recruitment and Selection

The MAA (SUMMA) REUs follow an atypical procedure for proposal submission. A call for proposals goes out in the fall; proposals are due in January or February. At the time the proposal is submitted, the proposer must have identified the students who will participate. Qualified students[2] are sought through e-mails to faculty and undergraduate advisors at colleges and universities in the region, and through announcements via the web and at sectional MAA meetings. Funding for travel is not included in the MAA (SUMMA) grants, so only regional students are recruited. Given that the MAA (SUMMA) wishes to reach minority students at a critical point in their career path - midway through their undergraduate programs - the ideal student is a sophomore who is at a transition point between lower and upper division mathematics.

Students apply by providing basic information about their mathematical background and motivation. Once the proposal has been approved, "finalists" are asked to provide letters of recommendation from faculty. For various reasons, not least of which is the scarcity of students from underrepresented minorities who have both sufficient mathematical background to undertake intensive research as well as the desire to do so, recruiting a minimum of four qualified students annually has been very challenging. Each year, at least one student named at the time of proposal submission has dropped out, and replacements were found only through strenuous efforts. Selectivity has thus been low: each year every student who actually completed the application process was accepted to the program - in some cases, despite faculty reservations about the student's readiness. In 2006 the most advanced courses taken by participants were Calculus II (one student), one proof-based course (two students), and two or three proof-based courses (three students).

Description of Research Area

Students addressed open questions in graph pebbling during the 2004 and 2005 REUs; in 2006 the questions were in the area of graph labeling, specifically radio labeling. Radio labeling is motivated by the FM channel assignment problem, which - in greatly simplified form - states that two channels that are geographically close must receive channel assignments with a large frequency difference. We model this situation with (simple) graphs: a radio labeling of a graph G is an assignment c of positive integers ("labels") to the vertices of the graph satisfying the condition $d(u,v) + |c(u) - c(v)| \geq \text{diam}(G) + 1$ for every pair of vertices u, v of G. The

---

[1]Overall, participants' skills in reading professional-level mathematics was only somewhat enhanced: students struggled with journal articles early in the REU, and for most this aspect was put aside as they began tackling their own research problems.

[2]The call for proposals specifies that students must be members of the following minority groups: African Americans, Latino Americans, American Indians, and Native Pacific Islanders.

radio number of a graph is then the smallest integer used as a label, taken over all radio labelings of the graph. Graph labeling provides an excellent source of research questions for beginning researchers: students can almost immediately begin creating examples, seeking patterns, and making and testing conjectures. After determining the radio number of some "easy" families of graphs, and exploring what it took to prove these numbers were correct, the group as a whole brainstormed over a dozen research questions. Students then formed pairs to work on three different research questions; after exploring a few, each group decided to work on determining the radio number of one or more families of graphs. Two pairs of students completely resolved their problems; the third pair had resolved all but a few small cases by the end of the REU. Some approaches and proof techniques developed by one pair of students or by the faculty mentors were found to have applicability in the work of other pairs. It should be noted that between start-up time (about one week of reading literature, getting a feel for the research area, brainstorming and exploring potential research questions) and wind-down time (about a week of writing up results and polishing presentations), students actually had only a little over four weeks to concentrate on their problems.

Nature of Activities

An observer arriving on a typical day in the middle of the program would see students working on their research questions, whether by generating examples on the boards or on paper, discussing ideas with their partners, constructing or revising proofs, or reading literature looking for ideas they could adapt to their own problems. The observer would see the two faculty strategically spending time with each pair of students, providing feedback, guidance, and encouragement[3]. The observer might notice that one student seemed to be in charge, reminding the other students to take breaks, or to get back to work as necessary. In fact, the students' time was highly structured: the faculty recognized that it's unrealistic to ask anyone to concentrate deeply for eight hours daily, and so created daily and weekly schedules featuring blocks during which students were to have different "focus" levels[4]. A different student served as team captain each week, responsible for some logistics and also for keeping the entire REU team on task.

On days 13, 19, and 24 (of 29), faculty from other institutions visited the REU. These faculty gave informal presentations of their own research, so that the REU students gained knowledge of other active areas of research. The students then made formal presentations of their work-in-progress to the faculty visitors. Following the presentations and some discussion, the entire group went to lunch together. The faculty visitor aspect provided students practice presenting formally as well as talking informally and answering questions about their research. Visitors gave feedback on presentations, suggested ideas for next steps, and shared their own research topics. Over lunch and informally the visitors also provided different

---

[3]The two faculty typically spent 10 hours (total) daily working directly with the students, and other hours reading students' work and working privately on ideas that would help the students make progress on their research questions.

[4]From the document **Expectations of Students**, discussed and posted: "Focus Level 1: maximum intensity concentration and work. No social conversations, cell phones calls, e-mail, etc. Focus Level 2: OK to do work that requires less intensity here (e.g. solicit teammates' opinions, literature search, write up results, etc.)."

perspectives on becoming and working as mathematicians and gave advice on applying to and choosing graduate schools, doing mathematical research, succeeding in upper division courses, etc.

The REU schedule also included breaks in the form of social events and workshops on various topics. July 4 occurred during the middle of the REU: mathematics was put aside as most of the participants, the faculty, and associated friends enjoyed a barbeque and relaxation time at a nearby beach. Students frequently ate lunch together and some planned after-hours get-togethers. The faculty hosted workshops on several topics, e.g. applying different proof techniques to some well-known results in graph theory (three workshops), why we bother with proofs, deadlines and application procedures for graduate study, learning LaTeX, using PowerPoint effectively, searching with MathSciNet, creating posters, preparing and delivering talks, the nature of the mathematical community, and a viewing of *N is a Number*.

Evaluation

Evaluation of all the MAA (SUMMA)-sponsored REUs is being conducted at a program-wide level. An independent researcher at the University of Oregon has developed and implemented an evaluation plan consisting of student pre-program and post-program surveys, project manager surveys, e-mail contact with students, interviews with students, and selected site visits. Preliminary results suggest that the MAA (SUMMA)-sponsored REUs are having a result in terms of increasing the likelihood that participants pursue graduate study in the mathematical sciences. Personal interviews and tracking of the participants in the CLU/ CSUCI Mathematics REU correlate with these preliminary results and are summarized below.

2004 Participants

- Post-bachelor's plans upon applying to the REU: unclear (3), grad school (1)
- Post-bachelor's plan upon finishing the REU: grad school (3), work with possible grad school later (1)
- Current status: in grad school (3), working (1)

2005 Participants

- Post-bachelor's plans upon applying to the REU: unclear (1), grad school (2), teaching credential (1)
- Post-bachelor's plan upon finishing the REU: unclear (1), grad school (2), teaching credential (1) (Interestingly, two students interchanged "unclear" and "grad school" plans.)
- Current status: in grad school (1), working on teaching credential (1), finishing bachelors (2)

2006 Participants

- Post-bachelor's plans upon applying to the REU: unclear (3), grad school (1), teaching credential (2)
- Post-bachelor's plan upon finishing the REU: grad school (5), credential and possible grad school later (1)
- Current status: finishing bachelors (6)

A second measure of the effectiveness of an REU is the participants' progress on their research questions. Each year students made sufficient progress to present their results at sectional MAA meetings[5] and at the Southern California Conference on Undergraduate Research. In 2006, all three groups, working jointly with the faculty mentors, solved problems significant enough to submit for publication. Specifically, one team resolved the radio number of gear graphs, another determined the radio number of three types of generalized prism graphs, and the last team found the radio number of cross products of cycles. Additionally, the techniques developed in this work show promise in terms of adaptability for finding radio numbers of other families of graphs.

Lessons Learned

The faculty director's practices have evolved over the three years of this REU program, guided by reflection on what worked and what didn't, by conversations with REU participants of this and other programs, and by advice - through consultations, articles, and conference presentations - of other REU directors. The following lessons learned apply specifically to REUs whose student participants are midway through their undergraduate mathematical studies, and whose faculty recommenders rate them in the top half, but not the top quarter, of their peer groups. Some of the lessons apply as well to students from underrepresented minorities, particularly those who may be the first in their family and/or neighborhood to attend college.

- Be explicit about both goals and expectations. Discuss each with students and modify when appropriate. Return to goals and expectations midway through the REU and discuss progress.
- Structure the students' working time carefully. Build in intense working time, more moderate working time, interaction, and breaks, and be clear about what activities are encouraged during each.
- Spend time getting to know the participants as people.
- Expect a lot of the participants, and encourage them to set high goals for themselves.
- Take advantage of friendly faculty at the host institution and at nearby institutions: ask them to listen to talk with students about their mathematics, to ask students questions about their projects, and to provide the students advice about everything from upcoming courses to graduate study to careers especially about developing the habits of mind of a mathematician.
- Have students work in teams of 2 - 3 per project. Students seem to work more consistently and enthusiastically when responsible to a teammate for carrying out their share of a project. This also allows faculty to rotate between fewer projects, with the benefit of being able to spend more time with each team of students than would be possible to spend with individuals.
- Provide the students a working space of their own. Ideally it would have computers, bountiful white boards and tables, and a couch or two.
- If size of the REU permits, have multiple faculty serve as co-directors. Directors can then share the load, cover for each other should one be called

---

[5]A 2004 participant won the top prize for his presentation.

away, bounce ideas off each other, and more easily maintain enthusiasm for
the work required to keep the students making progress on their projects.
- Should the students' results be of sufficiently quality to submit for publication, recognize that helping the students write to the level necessary
  will take countless hours, even after the REU ends.

DEPARTMENT OF MATHEMATICS, CSU CHANNEL ISLANDS
E-mail address: cindy.wyels@csuci.edu

The REUT and NREUP Programs at California State University, Chico

Colin M. Gallagher and Thomas W. Mattman

Since 2003, we have been running a summer REU program at California State University, Chico. We place particular emphasis on participation by women and minority students as well as high school teachers. Our six week program is divided into phases modelled on Polya's four stages of problem-solving. Our evaluation plan includes a case-control observational study of student career choices and a Leikert-scale survey that attempts to measure effects on perception of mathematics. Following a brief overview of the history and goals of our program, we describe below the nature of participant activities, recruitment, and our evaluation plan and outcomes.

History and Goals

In the summer of 2003, CSU, Chico served as a pilot site for the MAA's National REU Program (NREUP). In addition to successfully renewing NREUP funding every year since then, 2004 was the first year of a three year REU award from the NSF. We are currently seeking to renew this funding.

With NREUP support we recruit students from minority groups (i.e., African Americans, Latino Americans, American Indians, and Native Pacific Islanders). As this grant does not include travel funds for participants, we target students from the North California region. A total of fifteen undergraduates have joined us in the four summers of NREUP support.

The focus of the NSF REU award is the integration of undergraduates and high school teachers. Hence, we refer to this as an REUT (i.e., a combined REU and RET, or Research Experiences for Teachers). Each summer we engage six undergraduates and two teachers to work in teams on research projects. We feel that the skills and experiences of these two populations complement each other. In particular, the teachers are adept at presenting the results of the research and can tap into a certain mathematical maturity. The undergraduates, on the other hand, have had more recent exposure to higher mathematics and tend to bring a lot of enthusiasm and drive to the enterprise.

We have the following specific objectives for our REU programs:

Received by the editor December 1, 2006.

- Encourage undergraduate students, especially those from underrepresented groups, to pursue careers in sciences and engineering, including teaching.
- Help to better prepare students to pursue advanced degrees and careers in the sciences.
- Provide in-service teachers with a research experience that will foster excitement about mathematics, increase content understanding, and inspire pedagogical innovation in their classrooms.
- Promote and enhance mathematical research involving undergraduates at CSU, Chico.

We will next discuss how participant activities are designed to support these objectives.

Nature of participant activities

Our program lasts for six weeks each summer. This is mandated by NREUP funding, but also works well for the teachers who typically must balance many demands on their schedule during the summer months. The relatively short time frame requires well thought out research problems that allow participants to "hit the ground running." For the first two years of the program, the research focus was knot theory. In 2005, participants were split into two groups, one investigating knots and the other statistics. We again had two research groups in 2006, knot theory and mathematical modelling.

We typically have twelve participants each summer who are split into four groups of three. Each team is led by a faculty member who helps participants progress from dependent learners to independent investigators by modelling and explicitly discussing Polya's four stages of problem-solving: understanding the problem, devising a plan of attack, carrying out the plan, and reflecting on the work. We recognize that students progress at different rates, and there will be overlap between successive stages; however, our activities are structured to facilitate the smooth progression from one stage to the next at a pace appropriate to the competence of each participant.

Following that model, in the initial stage, the team leader offers a mini-course related to the team's mathematical focus area. This provides an opportunity for the team leader to give participants additional relevant background material, introduce computer software, evaluate the competence of each participant, foster a supportive team environment, and in general ensure the group has the necessary tools to carry out the research project. The team leader concurrently begins to introduce specific or general open research problems in the appropriate content area; students are encouraged to begin their own exploration right away. In their first step towards independence, the participants are asked to select a research problem. At this point the participants may be given some research or expository articles to read in order to gain understanding of their particular problem. These research articles can be discussed with the faculty leader in a group setting; participants begin to learn how to read journal articles, a skill that usually requires much practice. These informal discussions help the faculty member ensure that each member of the team "understands the problem" and is thus ready for the next stage.

In the second stage, the faculty member takes a step back by gradually transitioning from group director to group member, allowing the team more flexibility in deciding its own path. The team is responsible for "devising a plan of attack," which includes developing research directions and allocating responsibilities. The faculty member contributes to discussions during informal meetings and helps guide the group in fruitful directions without imposing his own ideas. The faculty leader also ensures that each individual is contributing to the development of the team's plan and has a reasonable share of the responsibilities in carrying out the attack. This stage culminates in a presentation of the team's research problem and plan of attack.

In the third stage, the faculty member steps back even further; having helped guide the group in the development of a plan, he now allows the group to carry out that plan with minimal assistance and acts primarily as an advisor as the team becomes self-sufficient and takes ownership of all aspects of their particular problem. Here, both teachers and undergraduates are confronted with their lack of experience in doing mathematical research. They can help each other to overcome this hurdle and learn to be independent mathematical explorers.

In the fourth and final stage the group "reflects" by jointly authoring a written report and preparing a presentation on their research including: a clear statement of their problem, their plan of attack, any obstacles which were encountered, results they obtained, and perhaps directions for further investigation. This stage begins at the end of the fifth week with two short workshops: one on writing technical reports and a second on presenting research to a mathematically literate audience. In this stage the faculty member's role is primarily to give advice and answer questions that may arise during this process. The reflective stage will likely continue after the term of the REUT, as students work with their team leader to prepare a manuscript for publication. As appropriate, we submit their papers to research journals or journals for undergraduate research.

These team efforts are complemented by a weekly series of invited talks which all students attend. The guests are usually professors from research universities in the region. These invited talks are paired with student presentations and followed by dinner. This informal setting often leads to lively table conversation as students share their progress with our guest. In addition to making two oral reports during the course of our program, many of our students also present their research at national and regional conferences.

Participants are housed in an eight bedroom apartment-style dormitory that provides a large kitchen, dining room, and living area. In addition to on-going opportunities for interaction due to the shared living quarters and lab space, we organize social activities such as softball games and visits to local theatres and cinemas.

Recruitment

Our focus on minority students from our region and high school teachers present particular challenges for recruitment. We have been successful in attracting teachers from around the country by placing advertisements in the NCTM Bulletin. We also advertise in the AWM Bulletin and on the SACNAS (Society for the Advancement of Chicanos and Native Americans in Science) web-site in an effort to recruit women and minority students. However, we have found that the most effective way of recruiting local minority students is to send e-mails directly to the math undergraduates of institutions in our area.

Program Evaluation and Outcomes

In addition to the desirable and measurable outcomes of participant publications and talks or posters given at professional meetings, we also evaluate the success of our project by measuring its impact on the participants. For the purposes of assessing how well we are meeting our project goals we separately consider:

- effect on participants' perception of mathematics and research
- effect on undergraduates' career choices
- effect on teachers' instructional strategies.

Our primary assessment method is self-reporting surveys.

To measure the effect on participants' perception of mathematics and research we have all participants complete a Leikert-scale questionnaire at the start of their experience, at the end of their experience, and during the following academic year. In addition, we ask teachers to report on new in-class activities or modifications to classroom practices in the year following their summer experience

As a way to measure the effect on undergraduates' career choices, we have designed a case-control observational study. To select participants we group desirable applicants into pools of three from which we select one student to participate. In this way we have, for each participating student, two non-participants with similar (as similar as possible) backgrounds. After the REU experience we mail surveys to both the participants and their non-participating counterparts. Participants are also given a survey at the start of the six-week program and another at the end. To give the reader some idea of these assessment techniques and the survey results we discuss results from the summer 2005 program. The teachers who participated in the RET also filled out surveys during the program and during the next academic year.

Surveys were mailed to all seven of the undergraduate participants as well as twelve non-participants. To encourage response a gift certificate for purchasing books from an on-line vendor was included with the survey. This was highly successful as a motivator, despite the fact that we did not require filling out the survey in order to receive the certificate. Of the seven participants all responded and eight of the twelve non-participants returned a survey.

Effect on undergraduates' career choices: Not too surprisingly, most of the students who applied to the program ended up applying and being accepted to a graduate program in the mathematical sciences. Of the seven participants from summer 2005, all but one had applied to a graduate program in a mathematical science (one became a senior in fall 2006). Of the six participants who applied to a program, all but one has been accepted to at least one program. The non-participants have similar numbers: six of eight applied and were accepted to a graduate program in a mathematical science; one has become a high school teacher and the other has taken a job in industry.

Effect on perceptions of mathematics research: In an attempt to measure participants' perception of mathematics research, a survey was designed which consisted of quotes about mathematics and mathematics research – some from famous mathematicians. Students were asked to indicate the degree of their agreement with the statement (strongly disagree, disagree, neutral, agree, or strongly agree). This survey was filled out three times: once at the start of the program (before we said anything), once at the end of the six-weeks and once near the end of the following academic year. We won't report a full analysis of the data, but give an example. One item stated, "Mathematicians rarely make guesses." Comparing responses from the first day to those on the last day of the program, five of the seven students had moved one step in the strongly disagree direction. Four of the seven were neutral on this statement on the first day of the program, on the last day these four disagreed with the statement and by the time of the mailed survey two of these had moved to strongly disagree. Statistical analysis indicates a significant change in the response to this item in the pre and post survey. This seems to indicate a change in perception of mathematics research due to the experience. The responses to this item on the first day survey are also statistically different from those on the mailed survey nearly one year later. This indicates that the changed perception lasted or was reinforced elsewhere.

Effect on teachers: Three teachers participated in the 2005 RET. What is probably most informative is the responses from the survey given near the end of the following academic year. All three teachers had incorporated specific activities in their classroom that were based on their summer research. Two of the three teachers said the experience changed how they teach in that they incorporated more discovery based activities in class. All three teachers indicated they enjoyed working with undergraduates.

Conclusion

In closing, we remark that in addition to being a positive experience for all participants and faculty involved, the REU programs are having an important effect on the research life of the CSU, Chico math department. In 2003, we were virtually alone as proponents of undergraduate research. Today, the idea of collaborating with undergrads has been taken up by many in the department and a contingent of five of us have been involved in requesting renewed funding. We also make a point of holding a few positions open each summer for our own students. These students return to their classes in the fall eager to share their experiences with peers and with a huge boost in self esteem and confidence. We would like to take this opportunity to thank the funding agencies (the National Science Foundation, the National Security Agency, and the Moody's Foundation) for all that these programs have done for participants, faculty, and our department.

Mathematical Sciences, Clemson University, O-104 Martin Hall, Clemson, SC 29634-0975, U.S.A
E-mail address: `cgallag@clemson.edu`

Department of Mathematics and Statistics, California State University, Chico, Chico CA95929-0525, U.S.A.
E-mail address: `TMattman@CSUChico.edu`

Proceedings of the Conference on
Promoting Undergraduate Research in Mathematics

Undergraduate Research at Canisius
Geometry and Physics on Graphs
Summer 2006

Stratos Prassidis

1. Overview

This is the second year we have hosted the R.E.U. program at Canisius College. Two faculty members of Canisius College, Terrence Bisson and myself, were responsible for the program. It lasted for eight weeks and we had eight students participating.

The main focus of the program was research on graphs using methods from category theory, algebraic topology and physics. The students were divided in two groups of four students. The first group, directed by Terrence Bisson, worked on spectral properties of directed graphs and their covers. In particular, the group worked on the following problems:

- to what extent does the spectrum of the Cayley graph of a finite abelian group determine the graph up to graph isomorphism? Abelian groups were chosen because the eigenvalues of their Cayley graphs can be expressed as sums of roots of unity. The problem was completely solved when the generating set has a small cardinality.

- In the previous summer, divisibility properties of Chebychev polynomials (of first and second type) where derived by looking at covers of directed path and circle graphs. Variations of this method produced more divisibility properties for Chebyschev - like polynomials. This summer one student derived closed formulas for the coefficients of these characteristic polynomials by reduction to a collection of solvable combinatorial problems.

- What is the effect on the spectrum of directed graphs under graph operations? Different graph operations (endo-functors on the category of directed graphs) were described (addition, multiplication, exponentiation, doubling) and the effect on the spectrum of the result was studied.

- What type of information can be derived from the spectra of the family of coset graphs of a finite group, and the coverings between them? The case for the symmetric group on 4 letters was studied in detail. The coset graphs of a group G were also applied to Burnside-Polya methods for computing the number of different colorings of figures with a G action.

Received by the editor September 5, 2006.

The second group, directed by me, worked on understanding the connections between the spectrum of the Laplacian of a graph and its combinatorial properties. More specifically the problems considered were the following:

- Give the definition and the properties of a Kazhdan graph. Kazhdan groups are defined as groups whose trivial representation is isolated in the space of representations. That puts certain restrictions on the spectrum of the Cayley graphs of the group. In particular, the spectrum of any finite cover of a graph covered by the Cayley graph of the group has a 'spectral gap' (the difference between its first and second eigenvalue) that only depends on the group and not on the cover. That property was used in the construction of infinite families of expanders. In this project the students defined a Kazhdan graph to be a graph that satisfies a similar spectral property, they constructed non-Cayley graphs that are Kazhdan graphs, and they prove the expansion properties of finite graphs covered by them.

- Generalize the definition of the zig-zag product of graphs. Zig-zag products were used to construct infinite families of expanders combinatorially, without using algebraic methods. The students gave a categorical description of the construction that generalizes the classical construction. Then, they computed the spectral properties of the construction. Also, they described the covering properties of the generalized zig-zag products and used it to construct infinite families of graphs, each covering the previous one, that have the same spectrum.

- Define and compute Ihara's zeta functions for certain infinite graphs.In this project, the students computed the Ihara zeta function for certain infinite covers of graphs that can be expressed as limits of finite covers. In particular, they generalized Bass' formula of the zeta function to infinite graphs, by replacing the determinant by the trace of an operator to certain operator algebra.

- Describe the combinatorial Laplacian of "homesick" random walks on graphs. In this project, the students studied the combinatorial Laplacian of the homesick random walk on the infinite path and they showed that it is equal to the Laplacian of the regular random walk on a regular tree, when the parameter is an integer. Using this result, they computed the spectrum of the Laplacian and the heat kernel of the homesick random walk on lattices.

- Describe vanishing results for repeated exotic Nil-groups in algebraic K-theory. One of the students recruited had a strong background in algebra and category theory. The project that he worked on was on certain computations in algebraic K-theory. More specifically, when the K-theory of push-out diagrams of rings is computed, there is an "error term" that determines how far is K-theory from being a homology theory. In this case, this groups are called Waldhausen's Nil-groups. The computations done this summer are on Waldhausen's Nil-groups when the base ring is also a push-out in the category of rings. Repeated Nil-groups are derived this way and certain vanishing results are proved.

The students maintained a wiki site where they recorded most of the calculations, experiments and partial results they derived. There is a plethora of data and speculations in the site: http://wiki.canisiusmath.net.

2. The Program

The recruitment of the students was done by distributing a flyer to the nearby colleges and universities. To reach more students, flyers were distributed in conferences attended by Canisius faculty, like the Nebraska Conference for Women in

Mathematics. Four of the students were from the western part of the country, one from the south and three from the wider area around Buffalo. Four students were female and four male. The stipend was $2000. Accommodations were included in a dormitory at Canisius College (Eastwood Hall). Also, meals were included and they were provided by the food services of the college. The College provided for us classrooms for our meetings and computer access for all our needs. The weekends were spent in social gatherings, organized by the Canisius faculty or the organizers, or by visiting local attractions like Niagara Falls.

Even though this is the second year we are hosting the program, we manage to recruit very strong students. Three out of the six seniors are applying to graduate programs in mathematics and they have very good chances to be accepted and succeed. The other three seniors will follow a career as math educators. The two who were not seniors have the interest and the ability to become researchers in mathematics.

The students would meet with us every morning and work together until late afternoon, with a lunch break. Usually, they discussed the work they did last evening and they asked question on how to continue their research. The two groups worked separately but they interacted and exchanged ideas all the time. There were invited speakers that gave talks to the students on different subjects. The speakers were from Canisius, SUNY Buffalo and the University of Rochester.

3. Conclusion

Overall, we are very satisfied with our recruitment process and the quality of the work produced this summer. Two papers have been submitted for publication. One of them and a third paper are posted in the archive preprint series. The students presented their result in the Young Mathematicians Conference at Ohio State University in early August. They received very positive comments on their presentations.

Judging from the participants' comments, the organizers believe that the program had a strong impact in their decision to pursue a career as research mathematicians or educators.

DEPARTMENT OF MATHEMATICS & STATISTICS, CANISIUS COLLEGE, 2001 MAIN STREET, BUFFALO, NY 14208

E-mail address: prasside@canisius.edu

The NSF REU at Central Michigan University

Sivaram Narayan and Ken Smith

1. Introduction

The summer NSF REU at Central Michigan University is directed by Sivaram Narayan and Ken Smith, focusing on research topics in matrix analysis, graph theory and algebraic combinatorics.

Narayan and Smith have been invloved in undergraduate research in mathematics since the early 1990s, originally working with strong honors students from Central Michigan University (CMU). During the 1990s, Narayan worked individually with students in an independent study format and began seeking internal and external funding for his students. Initially Narayan received support from the Dean of the College of Science and Technology at CMU to support the students during summer projects. Beginning in 1998, some of his students received funding through the CMU Summer Research Scholarship program. Meanwhile, in the summer of 1992, Smith worked with undergraduate students at the Director's Summer Program at the National Security Agency in Maryland. This was an exciting experience for him and he returned to Central Michigan eager to work with undergraduate math majors in research projects in algebra and combinatorics. He also guided honors projects and undergraduate independent studies.

In 2000, Narayan wrote a grant application to the NSF for an REU site and received funding for the summer of 2002. During that eight week summer program, Narayan, Smith and a colleague, Yury Ionin, worked with six students supported by the NSF and five students supported by Central Michigan – three faculty working on five projects with 11 students!

Since 2002, CMU has been funded each summer to run an NSF-REU site. (Our current grant runs through 2008.) Each summer, eight students are selected to participate in the program with three faculty mentors. The NSF-supported students are joined by two to four students who are supported by CMU. The students work on research problems in teams of two or three. Each faculty mentor supervises one or two research teams. The program begins on the first or second Monday in June and runs for eight weeks.

Recently CMU has partnered with Coppin State University, the University of Richmond and Olin College to create the LURE (Longterm Undergraduate Research Experience) program, funded by the NSF. In that program, each institution

Received by the editor November 28, 2006.

will work with their own students in two-year projects. (The LURE program is described more fully in a separate article.)

2. NSF-REU site program

The goal of the CMU REU program is to stimulate talented undergraduate students to pursue graduate work in science and mathematics by providing accessible, challenging, and unsolved research problems in mathematics. Emphasis is placed on learning new mathematics, the excitement of discovery, and interactions with other students and faculty mentors. Communication of ideas and results are emphasized through oral presentations and written reports.

In summer 2002 Narayan and Smith and Ionin worked with eleven students on five projects. Lisa DeMeyer, Narayan and Smith mentored thirteen students on five research projects in summer 2003. Ionin, Narayan, and Smith mentored twelve students on four research projects during summer 2004. During the summer of 2005, fifteen students were mentored by DeMeyer, Narayan and Smith on six research projects. This past summer, Narayan, Smith and Boris Bekker mentored ten students on five projects. Over five summers, our program has worked with 61 students (28 men, 33 women). Thirty-eight students (18 men, 20 women) have been supported by the NSF, 21 have been supported by CMU and two have been supported by another institution. The NSF students come from a variety of institutions, some large (Berkeley, Rutgers, Yale, Princeton, Michigan, Arizona State, Kansas, Georgia, Nebraska) and others small and private (Susquehanna University, Ashland University, Bates College, Grinnell College, Providence College, Oberlin College.) We have deliberately mixed students from small colleges with those from larger schools. We have also worked to take relatively new and inexperienced students (freshmen, sophomores) and mix them with one or two students who have previously been to an REU.

Central Michigan University's REU program has been very successful. The following is a quote from the external evaluation report of Professor Carl Cowen (IUPUI). He states, *"Overall, the students and the mentors seem happy with the program. Moreover, the results in terms of progress on the project problems, increasing the interest of the students in research mathematics, and providing support for further development of the students as future scientists all seem to be meeting the project goals. I think the Department of Mathematics, Central Michigan University, and the National Science Foundation should all be pleased with this project and interested in continuing to support it in the future."*

3. Mathematical activities of the NSF-REU site

The REU problems involve basic mathematical research in the areas of algebra, combinatorics, graph theory, matrix theory, and number theory. The students work on unsolved problems that focus on improving our understanding of the very structure of mathematics, problems whose solutions will give insight into the nature of commutative and noncommutative mathematics, the structure of graphs, properties of numbers, and the theory of linear algebra.

Every effort is made to involve students with problems that are considered significant by both the students and the faculty. Problems are contemporary, important, and usually of interest to a wide mathematical audience. Only a modest mathematical background at the sophomore level is necessary to understand most of the open questions that are presented to the participants. Some problems require a knowledge of abstract algebra. This allows the participants to be at different stages of undergraduate mathematical preparation. All the problems require that the students rapidly learn a fair amount of mathematics.

During the past five summers, students have found the minimum semi-definite rank of a variety of graphs (under the supervision of Narayan) used group representations and algebraic number theory to construct difference sets (Smith), solved conjectures in graph pebbling (Narayan) and computed the "summability number" of families of graphs (Narayan). Other teams investigated zero divisor graphs of semigroups (DeMeyer), tight subdesigns of symmetric designs (Ionin), addition graphs and nonabelian Cayley graphs (Ionin), matrix stability (Narayan), distance regular graphs (Smith), cellular automata (Smith), and Polya's fenced garden problem (Bekker).

On the first two days of the REU program, there are introductory lectures by faculty mentors followed by descriptions of four or five targeted problems. On the third day there are usually one or two lectures to provide needed background material followed by discussions about how to conduct mathematical research. The students are encouraged to ask questions and participate in these discussions. At the end of the third day, the students are asked to describe their two favorite problems and the reasons for their choices. Based on this information, the students are divided into teams. A typical team consists of one faculty mentor and two students. Students then meet with their mentor for morning lectures which examine topics in matrix analyis, group representation theory, finite geometries, and graph theory.

Typically, the students work on problems from 9:00 a.m. to noon and 1:30 to 4:30 p.m. on weekdays. Faculty mentors are actively involved with the students during these working hours. The day typically begins and ends with the team reviewing the students' progress.

We make every effort to involve students in publishing their research. Although there is no guarantee that any particular research leads to publication, students are required to write their results and submit a final report at the end of the program. The written report will use some version of TEX and follow the article format of the most appropriate journal. Moreover, students are asked to present their work as talks and/or posters at the Mathematical Association of America meeting (Math Fest) in August or at the American Mathematical Society-Mathematical Association of America joint meetings in January or at the annual undergraduate mathematics research conferences held at different locations.

4. Additional student activities at the NSF-REU site

Participants live on campus in university apartments. These apartments are across the street from Pearce Hall, which houses the Mathematics Department and the computer labs. The Park Library and the Bovee University Center are a five-minute walk from Pearce Hall. The Multicultural Center located in the University Center sponsors various activities throughout the year. The apartments are also within walking distance of the Student Activity Center, tennis and basketball courts, and downtown Mount Pleasant.

Participants live in two-bedroom apartments with a shared living room, kitchen, and bathroom. Two male or two female students share an apartment. Each apartment is furnished with a bed, mattress, dining table, chairs, dresser, cable TV access, and telephone. Each student is provided with a notebook computer during the eight weeks of the program. They have access to the Internet and are therefore able to conduct library searches from their apartments.

Apart from working on their research projects, the students participate in seminars given by faculty mentors, seminars by guest speakers on topics of interest (such as "millennium problems" or applications of math in biology) and seminars given by other students. REU students participate in a one-day mid-summer conference with undergraduates similarly engaged in mathematics research at other Michigan institutions (Grand Valley State University and Hope College). This conference takes place at a location convenient for all schools. This event provides a good opportunity for the REU students to interact with students doing research in other areas. The Michigan REU conference was hosted by CMU in 2003 and 2006, by Hope College in 2004, and by Grand Valley State University in 2002 and 2005.

REU students also participated in many social activities including a day trip to the Mackinac Island. Anant Godbole (East Tennessee State), Carl Cowen (IUPUI) and Harriet Polletsek (Mt. Holyoke College) were external evaluators of the CMU REU site in 2002, 2003 and 2004 respectively. Charles Johnson (College of William and Mary) and Hugh Montgomery (Michigan) were the keynote speakers at the 2003 and 2006 Michigan REU conferences held at CMU.

Every participant in our REU program has given an oral or poster presentation at one or more regional meetings (such as the Michigan Undergraduate Mathematics Conference, Young Mathematicians Conference at Ohio State, Hudson River Valley Math Conference or the Nebraska Conference for Women in Mathematics) or at national meetings (such as the joint meetings of the American Mathematical Society and the Mathematical Association of America, the MAA Summer Math Fest or the annual meeting of the American Association for the Advancement of Science). Published or submitted journal articles based on the REU projects from 2002-05 are listed below.

5. Publications and Awards

Awards

As of fall 2006, six articles from our program have been accepted for publication.

- J. Muntz, S. K. Narayan, N. Streib, K. VanOchten, *Optimal Pebbling of Graphs*, to appear in Discrete Mathematics.
- Y. Jiang, L. Mitchell, S.K. Narayan, *Unitary Matrix Digraphs and Minimum Semidefinite Rank*, to appear in Linear Algebra and its Applications.
- P. Hackney, B. Harris, C. R. Johnson, M. Lay, L. Mitchell, S. K. Narayan, A. Pascoe, K. Steinmetz, B. Sutton, W. Wang, *On the minimum rank among positive semidefinite matrices with a given graph*, to appear in SIAM J. of Matrix Analysis and its Applications.
- S. Klee, L. Yates, *Tight subdesigns of the Higman-Sims design*, RHIT Undergraduate Math Journal, Volume 5(2), Fall 2004.
- C. Berkesch, J. Ginn, E. Haller, E. Militzer, *A Survey of Relative Difference Sets*, RHIT Undergraduate Math Journal, Volume 4(2), Fall 2003.

- B. Cheyne, V. Gupta, C. Wheeler, *Hamiltonian Cycles in Addition Graphs*, RHIT Undergraduate Math Journal, Volume 4(1), Spring 2003.

At least six manuscripts based on the results of REU projects are in preparation for submission in the academic year 2006-07. A number of students have won awards for their poster presentation.

- Christine Berkesch, *Relative Difference Sets*, MAA Undergraduate Poster Session, Baltimore, MD, January 2003.
- Josh Whitney, *Subgraph Summability Number of Graphs*, MAA Undergraduate Poster Session, Phoenix, AZ, January 2004.
- Margaret Lay, Amanda Pascoe, *The Minimum Positive Semi-definite Rank of a Graph*, MAA Undergraduate Poster Session, Atlanta, GA, January 2005.
- Steve Klee and Leah Yates, *Tight Subdesigns of the Higman-Sims Design*, MAA Undergraduate Poster Session, Atlanta, GA, January 2005.
- Jessica Muntz and Kelly VanOchten, *The Optimal Pebbling Number of a Graph*, Student Poster Session, American Association for Advancement of Science, Washington D.C., February, 2005.

6. Summary

Five summers of energetic math research with eager undergraduates have been exciting. It is an art to find good undergraduate research problems! The problems need to be genuine, arising out of one's own explorations, and they need to be pitched at the right level. The students need to be given the tools: math concepts, broad mathematical principles and a survey of recent results, yet this must be provided rapidly in a format which does not overwhelm the students. It is important that faculty monitor student progress and provide the appropriate amount of support (in individual conversations or in group seminars). At times, this daily mentoring can be exhausting and time-comsuming.

There is a strong social component to undergraduate research. Bright eager students may need to be directed into productive areas of study; timid insecure students need encouragement and instruction. At times, the faculty member may need to work a little ahead of the students, scouting the research terrain, while still leaving the problem unsolved. Occasionally the faculty mentor must admit that the research team is at a dead end, and must suggest alternative problems – while encouraging the team members that "dead ends like this are typical in research!"

An effective undergraduate research program requires faculty mentors who love mathematics and enjoy working with undergraduates. It also requires significant institutional support. We are grateful to the support of both the NSF and CMU for our program.

DEPARTMENT OF MATHEMATICS, CENTRAL MICHIGAN UNIVERSITY, MT. PLEASANT, MI 48859
E-mail address: Sivaram.Narayan@cmich.edu, Ken.W.Smith@cmich.edu

Claremont Colleges REU, 2005–07

Jim Hoste

In 2005 the Claremont Colleges, with Pitzer College acting as lead institution, began a summer Research Experiences for Undergraduates program in mathematics. The Claremont Colleges include five undergraduate liberal arts colleges: Claremont McKenna, Harvey Mudd, Pitzer, Pomona, and Scripps (a women's college). The five campuses are contiguous and the physical size of the consortium is comparable to a typical university campus. Roughly 85% of the funding for the REU comes from the National Science Foundation with the balance divided among the five Claremont schools.

The primary goal of the REU is to expose students to the real-world nature of mathematical research and the practicalities of an academic career in mathematics. In doing so we hope to encourage students to go into careers in the mathematical sciences. A secondary goal of the program is to provide a uniting focus for the already existing summer mathematical activities in Claremont, which heretofore had been occurring at each college separately and without the benefit of shared activities.

Previous to 2005 the Claremont colleges have not operated a shared REU, or equivalent, program. However, the Claremont colleges place a high value on undergraduate instruction and have a long history of fostering undergraduate research. Students majoring in mathematics are required to write a senior theses (or "senior exercise") at three of the five colleges and must choose between writing a thesis or participating in a research clinic at a fourth college. A majority of the faculty have supervised student research and many have co-authored mathematical papers with undergraduates. Research which does not lead to publications will often still result in a poster presented at a conference.

Ten students participate in our REU each summer for a period of eight weeks (June and July). Students work in teams of two or three with one faculty advisor. (Teams of four students with two faculty advisors may also occur.) The advisors propose the research topics in advance of the summer and each topic that will run that summer is advertised on our website. When students apply they are asked to indicate which research group they would like to participate in.

Research topics for 2005 and 2006 were: Lissajous knots and billiard knots, dynamics of three-population interactions, robustifying clustering methods for microarray data, analysis on metric trees, generalized continued fraction representations

Received by the editor November 30, 2006.

of certain algebraic numbers, and Fourier analysis on groups. Topics for 2007 will include knot theory, lattice theory, geometric combinatorics, and mathematical biology.

Students are primarily recruited through the website and by mailing out flyers to around 400 mathematics departments. Flyers and letters are also sent to a small collection of colleagues who we think may have access to interesting students, especially ones from typically underrepresented groups. Faculty who will be leading research projects also promote the program by speaking at various venues, mostly neighboring institutions in Southern California. In our first year, we received about 95 applications and accepted 6 men and 4 women. In our second year applications increased to nearly 200 and we accepted 5 men and 5 women. Most are between their Junior and Senior year with some one year younger. In accepting students, we are primarily guided by a desire to have

- A diverse group of students, yet one that can work well together;
- Students with the background necessary to successfully engage in the research topic;
- Students who are considering a career in the mathematical sciences for whom this could be a pivotal experience.

During the summer, students not only work on their specific research project, but also attend a variety of workshops as well as a weekly colloquia. The workshops are designed to teach basic skills needed by every research mathematician:

- How to use library resources including MathSciNet, arXiv, etc.
- How to use TEX, LATEX, etc.
- How to give a mathematics talk.
- How to prepare a mathematics poster.

We also hold a workshop on how to apply to graduate school. In addition to the workshops our weekly colloquium series draws on talented expositors throughout Southern California who can introduce the students to a wide variety of current research topics.

Each student receives a stipend of $2800, a free single dorm room, a weekly food allowance of $100, a travel allowance (to and from Claremont) of $250, and an allowance of $350 for attending a conference following the summer. We encourage all of our students to attend the Joint Meetings in January and to present their findings at the Undergraduate Poster Session. Each faculty advisor receives a stipend of $3500.

To evaluate the program we survey the students both before and after the program, asking them a variety of questions to assess their attitudes and skills related to mathematics and mathematical research. The surveys were designed by our evaluation consultant who is a developmental psychologist at a local university. The consultant also leads a group focus session twice during the summer. We also survey the faculty advisors regarding the attitudes and skills of the students. It is too early to report on the data, except to say that, in general, the students attitudes do not change much (they already had a high regard and interest in mathematics before coming) and their skills improved (although many of them already had rather advanced skills before coming). At least two students (one from each summer) have decided, as a result of learning what it is like to do research in mathematics, that it is not for them. One has decided to apply to law school the other to medical school. The others all seem to have been drawn closer to a career in mathematics by the experience.

In addition to all the mathematical activities that take place during the REU, we also plan weekly social events that allow a high level of contact between REU students and faculty and with other students and faculty who are spending the summer in Claremont doing research. Some of these activities have been in collaboration with a Computer Science REU taking place at the same time at Harvey Mudd College. The students also have adjacent rooms in the residence halls, usually eat lunch and dinner together, and the faculty eat lunch with the students at least once a week.

Our website, `www.pitzer.edu/mathREU`, contains more details of the program including summaries of the 2005 and 2006 programs.

PITZER COLLEGE CLAREMONT, CA 91711

E-mail address: `jhoste@pitzer.edu`

The First Summer Undergraduate Research Program at Clayton State University

Aprillya Lanz

The Bachelor of Science in Mathematics degree is a new program at Clayton State University (CSU) that started in Fall 2004. We recently produced our first two graduates in the Summer of 2006. Currently there are 20 mathematics majors in the department. As a part of the curriculum, each student is required to do a research project and present his/her result in the Math Department Colloquium Series. The mathematics faculty maintains strong commitment in providing research training as an essential part of undergraduate experience for students. I have been actively engaged in undergraduate research since my arrival at CSU, and this inspired me to seek funding from the Mathematical Association of America through their National Research Experience for Undergraduates Program (NREUP). This program is funded by the National Science Foundation, the National Security Agency, and the Moody's Foundation. The grant application requires the Principal Investigator to list students who may participate provided that there will be funds. I received the grant in April 2006 to support four minority students during the Summer of 2006. The program ran for six weeks, starting from May 30. This is the first summer undergraduate research program housed by the Department of Mathematics at CSU. In addition to gaining the experience of organizing such program, I enjoyed my time working with students.

The MAA NREUP. Quoting from the MAA NREUP website, this project "supports the participation of mathematics undergraduates from underrepresented groups in focused and challenging research experiences to increase their interest in advanced degrees and careers in mathematics." The project has grown significantly, from only four programs in 2003 12 programs in 2006. The program is directed for underrepresented minorities in the mathematical sciences. It provides financial support both for the principal investigator and the students with a maximum total of $25,000.

Recruitment. For the NREUP, the recruitment process has to be done before the grant is awarded. I spent about three weeks prior to the due date of the proposal to recruit students on campus. Although approximately 63% of CSU student population is composed of minorities, they make up only about 10% of CSU's math majors. Because of this fact, I did have some difficulties in recruiting students.

Received by the editor January 2, 2007.

However, I also found it helpful to talk to my colleagues to seek recommendations from them of students they are familiar with. In addition, my involvement in our Math Club helped me greatly, since I was able to get in touch directly with students. I was able to recruit three minority students from Clayton State to work along with me for six weeks in the Summer of 2006.

Program Overview. The topic of this summer research experience for undergraduates program was the investigation of solutions of boundary value problems in differential equations. In participating in this summer project, students learned about different types of boundary value problems and some of the classical problems in differential equations, primarily the uniqueness and existence of solutions. Using different methods, such as solution matching and shooting methods, students investigated the third order boundary value problems satisfying different boundary conditions. Based on this research topic, each participant was required to have taken Calculus II.

Because of the nature of the grant application procedure, the program started with a pre-program meeting. Since there was no guarantee of program funding at the time students signed up, this meeting was scheduled to prevent any future conflicts of students' summer plans. I felt that it was important to have students' full commitment to the program and for them not to take any courses during the duration of the program. In this meeting, I explained the scheduling and the expectations of the program.

My intention in working with undergraduate students is to provide them with the experience of doing research. In addition to gaining mathematical knowledge, they would also learn each aspect of research, from looking for ideas, searching for references, finding and summarizing the results, to finally presenting them both through formal presentations and articles. With this in mind, for this summer research program, each student was expected to accomplish the program goals which include

- to gain a working understanding of boundary value problems in differential equations,
- to propose their own boundary value problems,
- to document their work in a journal-formatted paper,
- to present their project in a seminar or conference environment both on campus and off campus.

During the first two weeks of the program, students were given intensive introductory lectures on differential equations and boundary value problems. The group had a Q&A session and discussion in each meeting. At the end of each meeting, an assignment was given for discussion on the following meeting. Students worked together on their assignments in the process of gaining familiarity with mathematical proofs and some theories of differential equations.

Following the introductory weeks, students worked on individual research problems. Each week, students brainstormed together to develop individual research problems. Although each student had his/her own individual research problem, students helped each other throughout the program. They discussed their research project with the group, in addition to discussing the progress with me in the daily meeting.

While working on the individual research problem, each student was expected to turn in a draft article at the end of each week; these papers were then returned, with feedback, at the beginning of the following week. Often, students would also discuss the feedback with the me. By having the papers typed in as the program progressed, it helped avoid delays at the end of the program and students actually became more familiar and more comfortable with their work.

I also think that peer review is an important part of the research process, hence, prior to the final presentation of their research, students distributed their papers among the group and were asked to give each other feedback. To prepare for the final presentation, they also gave practice presentations where their peers listened and gave input.

The program was concluded with a formal presentation on campus. The event was announced in the campus paper, and we invited the Provost, Associate Provost and Dean of Retention from the Office of Academic Affairs, faculty members and students of College of Information Technology and Mathematical Sciences (CIMS). We also invited family and friends of the students to witness their proud moment of accomplishment.

In their presentations, each student discussed their research project and answered questions from the audience. The participants expressed their appreciation of the opportunity and the experience during the intensive six weeks, where they learned what mathematics research means, starting from learning some basic concepts and concluding with proving mathematical problems.

Guest Speakers. I think it is important for students to see how the topic that they are learning is being applied or to meet other mathematicians who do similar research. I invited two local guest speakers; the first speaker, Jason Lanz, was a thermal analyst from Lockheed Martin Marietta who uses the finite difference method in his simulation for the F-22 project, the second speaker was Jeff Ehme of Spelman College whose research expertise is differential equations. Students had the opportunities to meet both guest speakers and spent lunch together with them.

Social Activities. As a part of the program, each week ended with an informal gathering. We planned different activities, including going to the movies, visiting the World of Coca Cola, and going bowling. During the duration of the program, I also liked to sit down and join the students for lunch. We went to different restaurants on several different occasions. This turned out to be a nice way to get to know the students and for the students to get to know me. As a celebration, at the end of the program, students invited their families and friends to go white water rafting at the Ocoee river in the Southeastern Tennessee.

Project Evaluation. The program ran with a total of three students. One student decided that research is not for her, while the other two said that they enjoyed this summer experience despite some frustrations that they experienced along the way. This experience also inspired them to pursue a graduate degree once they finished at Clayton State. Both students will highly recommend this type of program to other undergraduate mathematics majors. The program gave them experiences that went beyond what they learned in the classroom, as well as the opportunity to work alongside their professors and peers. Perhaps, more importantly, they gained research experience that broadened their mathematical knowledge.

I asked students to give feedback regarding the program, to include the research content and the structure of the program. One of the student commented *"The program was a rewarding experience, not just financially, but mostly academically... The program taught me how to work together with others. During the six weeks, I experienced frustration when I couldn't figure the proofs out or I got stuck, however I also experienced excitement and learned how to be patient when doing research, which helped tremendously. The best part of completing this program was that I felt extremely proud of myself for such an achievement because there were times when I doubted myself... Overall, I would most definitely do another program like this with or without the financial reward."* Another said that it was very rewarding to learn how to apply a method in a different approach from the journal articles he read. He was able to understand how the method works and applied it on his own research problems.

The students will present their results at conferences. One student will be presenting a poster at the Joint Mathematics Meeting in New Orleans and plans to attend the Mercer University Undergraduate Research in Mathematics Conference in February 2007. Students also plan to attend the Harriett J. Walton Symposium on Undergraduate Mathematics Research hosted by Morehouse College in April 2007.

Conclusion. I cannot place a value on my experience working with these students this past summer. I hope to continue working with students in a similar program environment in the future. Because this is our first summer research program, there are many areas that we can improve. To have a more successful program, we will need to recruit more students and have more participation from our faculty members. Collaborative work among the faculty members, not only as faculty mentors but also as the program administrators, will help the program to run more smoothly. It is also important to have a better screening process of the participants. Improvements under consideration include requiring recommendation notes or letters from faculty members, an unofficial transcript, and for students to list their summer plans. I also found it crucial for students not to take any courses while participating in an intensive research program like this.

We hope to have more opportunities like this for the students at Clayton State University.

CLAYTON STATE UNIVERSITY
E-mail address: `alanz@clayton.edu`

Proceedings of the Conference on
Promoting Undergraduate Research in Mathematics

Clemson REU in Computational Number Theory and Combinatorics

Neil Calkin and Kevin James

1. Program History

We began supervising a summer REU program in 2002. We initially received funding for one year through an NSF REU grant. We supervised eight undergraduates for eight weeks. These eight neatly divided themselves into three teams of three and worked on projects related to elliptic curves, partitions and chess. We also invited a couple of our graduate students to attend our daily meetings and interact with the participants. The involvement of graduate students in the research projects worked very well. It was exciting for the graduate students and helpful to the undergraduates and to us.

We enjoyed this so much that we immediately wrote another proposal requesting three additional years of funding and increasing the number of participants to nine. Since the involvement of graduate students worked so well during the initial year, we requested funding to cover the summer salaries of two graduate student mentors. The use of graduate student research team leaders has been very beneficial to the supervisors, the undergrads and graduate students themselves. The undergraduates benefit from having a team leader who is more accessible than the supervisors while the graduate students gain valuable experience in research and in research advising. In fact, several of our summer research projects have continued through to publication and have become a part of graduate students' thesis research.

During our last year of funding under the second grant, we decided that we would like to experiment with a larger group of students. We requested supplemental funding from the NSF to cover the cost of adding more undergraduates and an additional faculty supervisor. We asked David Penniston from nearby Furman University to help us supervise this larger group of students. Unfortunately, we were not notified that our request was granted until the after the 2005 program had begun which made the larger program impossible that year. We were successful in obtaining a new REU grant which will fund our program through summer 2010. The current funding is for two faculty, three graduate students and nine undergraduate participants. During summer 2006, we were able to use the supplemental funding to our previous grant along with the new grant to support twelve

Received by the editor January 20, 2007.

undergraduates, three graduate students and Dave Penniston (and ourselves) as a faculty supervisor. Finally our REU had grown beyond the point which a particular faculty member could closely follow all of the research projects. This made the 2006 program a bit more hectic. However, this program was very productive and we will pursue additional funding to maintain the larger size.

2. Program Goals.

In NSF terms, the main goal of our REU is to help students attain a higher level of independence in mathematical research by taking part in a significant and interesting research project. Put more simply, we want our participants to know firsthand what it is like to be a professional mathematician, so that they can make informed decisions about future careers in mathematics.

We achieve this by including participants into our own research programs and giving them open problems which are important to us and of interest to other mathematicians. We emphasize the importance of the problems which we introduce to participants and state that one of our goals is to publish a paper in a mathematical journal. Although we don't always achieve this lofty goal, the statement of such a goal is extremely important because it causes participants to change their perception of their roles in mathematics. They begin to see themselves as active researchers as well as students. They begin to consider the possibility of engaging in meaningful research and publishing the results, and they begin to believe for the first time that this is an attainable goal. This is a new and exciting possibility for these young mathematicians, and they embrace it eagerly, which creates an exciting atmosphere in which they work harder and learn more than ever before.

While choosing research topics, we look for problems which have some computational aspect. Thus students can start work immediately and they always have work to do even while at a standstill on developing new theorems. We also prefer problems for which we have a plan of attack. We want to give participants every opportunity to be successful in achieving new results.

As a complement to working on their research, participants attend a weekly colloquium series. These lectures are given by mathematicians who have been recognized as leaders in the various fields related to the students' research problems and who have demonstrated the ability to clearly communicate mathematical ideas to audiences of varying levels of mathematical maturity. This not only provides our participants a chance to significantly broaden their mathematical horizons but also provides a networking tool for our students as well as for us. It brings the leaders of our field together with bright and promising undergraduate students. In addition to giving really interesting talks, our speakers typically spend at least a couple of days interacting with our participants. This interaction is a source of great encouragement and inspiration to our participants and is one of the most popular aspects of our REU program (based on anonymous evaluation forms).

We also place great emphasis on the development of communication skills, both oral and written, and a familiarity with the various technologies available for typesetting and presentations. Each participant gives at least two lectures during the program and receives feedback and encouragement on these lectures. We have a workshop on 'how to give a good talk'. During the last week of the program, each research group presents their findings in a talk which is open to all faculty and graduate students. We see vast improvement during the course of the summer in all participants' abilities relate nontrivial mathematics to an audience of their peers.

We also conduct workshops on writing and each research group submits a report on their summer research in latex format. We collect rough drafts near the end of the program and give the participants some guidance and feedback on their writing. We also encourage participants to continue their research beyond the summer and when appropriate we help them to write up their results and submit them for publication.

We are committed to recruiting a diverse group of participants. Of our 47 past REU participants, 21 are female, 3 are African American, 3 are Hispanic and 1 is physically challenged. In addition, two other African American students were successfully recruited but then were unable to attend the program for personal reasons. We also strive to identify students who would benefit tremendously from our program but might be missed by other REU recruiters. For instance, we have recruited students from smaller colleges and Universities such as UNC-Asheville, Wake Forest, University of Puerto Rico-Humacao, University of Western Alabama, Bellarmine University, Appalachian State University, Bucknell University, Skidmore College, and University of Nebraska-Omaha. We also try to identify non-traditional students who could benefit from our program.

3. Activities

During the first week, our daily schedule is typically as follows. Each day, we give two morning lectures. The purpose of these lectures is to introduce the problems. By the end of the first week, the participants divide themselves into groups of 2 or 3 and choose a problem on which to focus for the remainder of the REU. During the second week, the morning lectures provide a deeper understanding of the selected problems and the significance of these problems while the participants are already making progress in their research. We schedule daily meetings with each research team to discuss their progress and any hurdles they have encountered. We use this time to give advice on how to proceed and to offer encouragement. The students spend the remainder of the day working with their team on their chosen problem.

After the second week, we decrease the frequency of the meetings with supervisors to allow participants more time for independent work. We continue throughout the summer to meet for morning lectures on various topics. Some of these are given by the participants on topics not necessarily directly related to their research problems. These morning meetings lend structure to our days and help ensure that everyone is making the most of our short time together. We then meet with one or two of the research groups and discuss their progress, meeting with each team two to three times per week.

A notable exception to the typical day described above is any day on which we have a colloquium visitor in town. On these days we go to lunch and dinner with our guest in addition to our normal activities. This gives our participants a chance to interact with some really interesting professional mathematicians.

In addition, we host cook on Memorial Day and Independence Day at our homes. Memorial Day is typically very early in the REU and the cook provides an informal setting in which participants can get to know each other and interact with faculty and graduate students.

In previous REUs, participants have organized various outings including hiking, Six Flags over Georgia, Atlanta Braves games, college baseball playoffs, ultimate frisbee, movies, etc. While the dynamics of each years' groups varies, subsequent participants will be encouraged to get to know each other beyond working on research.

4. Judging our Success

As mentioned earlier, publication of journal articles is stated as one of our goals, and although such publications are indicators of success in research, this is not our primary measure of success. We are primarily interested in whether our participants can make an informed decision about continuing their education in graduate school and whether they develop a realistic view of mathematical research. Since our first program, 28 of our past participants have graduated and at least 25 of these have gone on to graduate school. Most of these 25 have gone to grad school in math at Universities like North Carolina State, MIT, Stanford, Dartmouth and University of South Carolina to name a few. However, some chose other areas such as economics, computer science, mathematics education and law school. One of the 3 who chose not to go on to grad school is currently participating in the Teach For America program. She is teaching in an inner city school in Chicago and will likely enter graduate school in math or math ed. after completing this program.

Our participants should learn a good bit of new mathematics while in our program which they do. They should interact with many professional mathematicians which is achieved through our weekly colloquium series. Another indicator of a successful program is the impact of the program on the research of the undergraduate and graduate students and supervisors. Since 2002, our past participants have had 6 journal articles accepted for publication in Acta Arithmetica, Mathematics of Computation, Integers and Congressus Numerantium, and there are another 4 papers in preparation. Two of our graduate mentors collaborated on some of these and another graduate student had an article accepted to the Ramanujan Journal on research which began as a result of our 2004 REU. Also research begun in our REU program has been the topic of at least 15 conference and seminar talks during the past 5 years.

DEPARTMENT OF MATHEMATICAL SCIENCES, CLEMSON UNIVERSITY, CLEMSON, SOUTH CAROLINA 29634

E-mail address: `calkin@math.clemson.edu, kevja@clemson.edu`

Research with Pre-Mathematicians

Charles R. Johnson

1. Introduction

The purpose of these remarks is to record the author's experience with and re-sulting thoughts about research with "pre-mathematicians." By this I mean young (generally pre-graduate school) people who have the ability to enter a career in high-level mathematics. Since such a career will require substantial, independent, creative mathematical thought, an honest and serious acquaintance with mathemat-ical research is a valuable complement to conventional course work. Fortunately, a taste of research can help to seduce bright people into mathematics. The appeal of mystery and empowerment given by success in research can be most addictive.

Much of my experience in research with pre-mathematicians has been via NSF support, both individual grants and site grants for summer REU programs, since 1989. A brief history and account of that activity may be found in [1].

The remainder of these remarks is organized into topical sections, beginning with a short update on the REU program at W&M. Then, we discuss where the most essential ingredient, the problems, come from, the value of undergraduate research, and some examples of recent research. In a future piece, I hope to record the talk I give "Doing Research in Mathematics" to REU students. It is designed to prepare them for the inevitable frustration inherent in serious mathematical research.

2. Update on REU at W&M

Our concept of summer REU at W&M has always been collaborative research in small groups (at least one pre-mathematician and at least one mentor, no more than 4 total) on serious research problems whose outcome will be of interest to others. The unifying theme has been matrix analysis and applications, often with a considerable combinatorial flavor. It has proven beneficial that students, not working together, can chat and appreciate problems in the same general area. The collection of mentors has never been the same from one summer to the next. Several faculty members have gradually drifted away from the activity. It became clear that some were not very good at it, while others are not so committed to students or had difficulty committing the necessary block of time in view of other activities. I am now the only one who has done it every summer. We have often had visitors help as

---

Received by the editor December 1, 2006.

mentors. These range from senior and junior collaborators to postdocs, other junior visitors, and Ph.D. students who are "graduates" of the REU program. This has worked very well for all parties (and has been very nice for me to have collaborators around for a sustained period). For example, in 2005, I had 3 junior colleagues from Poland come for the summer (two visiting the US for the first time).

The supported students have been admitted (primarily) competitively from a national applicant pool that is generally around 160. A very large percentage of these would benefit from an REU and be able to do good work. We try to achieve some gender balance, which is easier in some summers than others. When female candidates have declined our offers, it is usually to choose a topic more to their liking or to attend a program dedicated to females. It is important to attract able females to mathematics and to give them a realistic view of mathematical research (just as it is to attract talent from any other quarter). It may be that the time for female-dedicated programs has past. The only other consideration sometimes taken into account in admission is that a student who seems very strong, but has not had the opportunity for exposure to serious mathematics in their undergraduate experience thus far, may be given extra consideration. This entails some risk, but can also be very beneficial to the student, meeting an REU objective.

In addition, we have frequently had students not supported by the NSF involved in the program; these range from high school students to foreign nationals (from Portugal, Ireland, Korea, etc.) to qualified "walk-ons". The resulting vertical integration (e.g. a high school student, working with an undergraduate, working with a graduate student, working with me) has worked well and been beneficial to all parties. In fact, I strongly feel that carefully selected high school students should be involved in REU, and that, at rather low cost, this would serve all goals of the REU concept. In a like manner cooperative agreements with foreign countries could be helpful. Ireland is already implementing an REU program that would allow participation by foreign students. It would be of value if, at least, cooperative programs with other countries would allow US students to be involved in REU-like research abroad, while students from those countries came here.

3. Were do Problems Come From?

Perhaps the most important feature of an activity designed to expose pre-mathematicians to research is the availability of appropriate problems. If the activity is of some significant scale, such as a recurring summer program involving several students each summer, recurring availability of many good problems is both essential and difficult. Good problems for this purpose should be (*i*) accessible, (*ii*) unsolved, and (*iii*) important; (*iv*) it should be likely that some valuable progress can be made, but (*v*) the area should be large and open-ended enough that an exhaustively complete solution is unlikely before the end of a summer program. Importance is a judgment but the problem area should relate to other things or be an interesting part of a bigger picture and should be viewed as intriguing by the pre-mathematicians. We have had many large problems progress in parts over many summers. Accessibility depends, to some extent, on participant background.

Our REU activity has been centered around matrix analysis/linear algebra and its applications, broadly defined. A solid beginning linear algebra course is required for admission and a true second course is preferred. Each participant is given a copy of [2] at the beginning of the program, and copies of [3] are readily available. The participants are rapid learners, and learning that helps solve their problem is a very compelling motivation. With on-line and library resources and guidance from someone who best knows the area, the topic of matrix analysis has proved ideal as a problem source. (In addition, because of its connections with all parts of mathematics and most of its applications, learning about the area serves the participants well.)

As should be the case with any guided research activity, the ultimate source of problems is spin-off from an active research program of a top researcher. In my case, I have very broad interests in matrix analysis and combinatorics and their relationship to other parts of mathematics. I collaborate with several dozen mathematicians in the US and around the world. This naturally suggests many related problems and subproblems that can be very useful to the collaborations. In addition, I frequently receive e-mail queries, some of which the writer would like help with or suggest yet other questions. Occasionally, good problems are simply suggested by colleagues. Since prior REU work and ongoing work raise many fresh questions as well, this altogether allows accumulation in a year's time of many "good" problems, many more than I can pursue by myself.

Over the years, we have had many continuing themes for REU problems. Each has been a very rich source of specific problems and nice results, many of which have been published. These include: (i) matrix completion problems; (ii) the long standing conjecture from statistical physics that the coefficients of $p(t) = Tr[(A+tB)^m]$ are positive whenever A and B are positive definite matrices; (iii) possible multiplicities of the eigenvalues among the Hermitian matrices with a given graph; (iv) minimum rank among positive semidefinite matrices with a given graph; and (v) factorization of matrix and operator functions.

4. The Benefits of Undergraduate Research/REU

If the purpose of REU programs (in mathematics) is to attract strong students to mathematics by giving them a realistic view of mathematical research, that seems to be working well. (But, recent statistics about a return to low percentages of US students among those finishing Ph.D.'s in the US are a cause for concern.) To be sure, many successful REU students would have gone on to mathematics Ph.D. study anyway, but it is likely that there are many others who would not have gotten "into" mathematics were it not for the summer opportunities. However, any program has consequences beyond those intended and I want to mention here some benefits that I perceive besides those intended or often cited.

(i). It was for many years a wonderful tradition in Russia and parts of Eastern Europe that important and established mathematicians would go out of their way to nurture talented pre-mathematicians. This tradition not only improved the discipline and helped new entrants feel a part of the "community," but it also helped establish famous mathematical traditions and establish mathematics as an enduring cultural tradition that transcended politics that came and went. As a symbol, the commitment to REU and research with pre-mathematicians is a modern version of that tradition that helps to accomplish similar objectives here.

(ii). The actual research that results from REUs and the like should not be ignored. In fact it seems to me one of the most important tangible products of the activity. In my experience, I am able to pursue interesting questions that I would not have been able to otherwise and, though I have altruistic motives as well, I would not be as enthusiastic were it not for the actual results. (As a mathematician I am rather social and enjoy working with others as well.) It also provides a nice way to establish pieces of work that can be assembled into a bigger picture. See the last section for some examples. In many cases, the actual work done is important and is published in very credible journals, and we should be assembling and publicizing the examples. The research is also very inexpensive to the NSF.

(iii). What may seem to be a failure may also be a success. If a student finds through exposure to research, that it really wasn't for him, that is much better (and cheaper) than finding it out from 2 years of graduate school. Fortunately, this is an unusual outcome.

(iv). The reinforcement of interest in and commitment to mathematics that comes from a group working together (even if not on the same problem) should not be underestimated. I had not anticipated it, but realized very early that one of the biggest benefits to the students (especially those who were "one-of-a-kind" at their institutions) is being with others of similar interest and outlook. This may result in friendships and acquaintances that persist through a career, and in fact I have long term collaborators who were once my REU students. We also have an example of a marriage of two REU students who are now professors at Bucknell!

Let me indulge in closing this section with a story of my first pre-mathematician research experience as a mentor. (I had none as a student, though I did publish a paper without a mentor.) I was an NRC-NAS postdoc at the National Bureau of Standards (now NIST), which, at the time, had an Eastern European-like tradition of hiring Westinghouse talent research winners in the summer. I ended up working with one, Tom Leighton, which resulted in two very nice papers, a classic on possible sign patterns of inverse positive matrices and a very practical one on graph isomorphism and eigenvalues. Tom did well academically and ended up founding Akamei Corporation (along with a junior colleague, who, by chance, had been a student with my collaborator Raphy Loewy, and who tragically died on 9-11-01), which survived the bursting of the "tech bubble" and is now a very successful company. The value added to the economy by Akamei alone likely dwarfs the low cost of REU programs (and I wish that I had gotten agreement from REU-type students for only, say 0.1% of their life-time earnings!)

5. Some Examples of Recent Research

I have now published about 40 of my well over 300 papers with pre-mathematicians (many papers involving several co-authors), and quite a few others are in some stage of progress. In addition, there is a comparable total amount of publication by colleagues with undergraduates. In my case, I would not have had time to pursue much of this work were it not for the interaction with students. Yet, I am extremely happy with and proud of much of this work, which has demonstrably advanced the subject.

We mention a few specific examples here, but, for variety, do not include examples from the continuing themes already mentioned in Section 3. This should help to make clearer some comments made in other sections. Nothing else guides the choice of these few examples.

5.1 An n-by-n matrix A over a field has an LU factorization if there exist a lower triangular matrix L and an upper triangular matrix U (wlog over the same field) such that $A = LU$. Such a factorization is important both in theory and in many applications and is calculated, under certain circumstances, as early as in the first course in linear algebra. Though some sufficient and some necessary conditions are known for its existence, I realized that no characterization of matrices A for which an LU factorization exists was known. It definitely need not exist, and the sufficient conditions were not generally necessary, while the necessary conditions were not sufficient. In the REU program one summer Pavel Okunev took this up with me. I had noticed that it was necessary that the rank of the upper left k-by-k principal submatrix of A plus k should be at least the rank of the first k rows of A plus the rank of the first k columns.

We wondered if this were sufficient, which would require understanding how to arrive at an L and a U. Eventually, Pavel came up with a complicated proof of sufficiency which worked for an algebraically closed field. But, we felt that the answer should not depend upon the field. Finally, we found a field independent proof of sufficiency, one of the more fundamental results of a summer program and the first ever to reach its original objective before the end (a risk in a narrowly focused problem). But, there is always more to do. The answer raised the question, for a matrix A without an LU factorization, of how many and where entries above the diagonal of L and/or below the diagonal of U might be needed for a factorization. We gave partial answers, and fuller answers have been given with Maribel Bueno.

5.2 The natural partial order on n-by-n Hermitian matrices A and B is the positive semidefinite partial order: $A \geq B$ iff $A - B$ is PSD. It has long been known that if A and B are positive (semi-)definite, then $A \geq B$ implies $A^t \geq B^t$ for all $0 \leq t \leq 1$. In a conversation with a Polish statistician, Czeslaw Stepniak, the question arose of for which pairs of positive (semi-)definite matrices A and B should we have $A^t \geq B^t$ for all $t \geq 1$? It is clearly necessary that $A \geq B$, and this is sufficient if A and B commute. Another obvious sufficient condition is that all eigenvalues of A are at least any eigenvalue of B. It was not clear what a characterization should be, but in summer 2006, I took this up with undergraduate Becky Hoai and interested colleague, Ilya Spitkovsky (who often advises REU students). Another student also had an interest. After some thought we came up with and proved two characterizations, one very pretty and one of likely practical value. The former is that k matrices $C_1 \geq \ldots \geq C_k$ that pair-wise commute (i.e. C_i communites with $C_{i+1}, i = 1, \ldots, k-1$ and A commutes with C_1 and B with C_k) may be inserted between A and B: $A \geq C_1 \geq \ldots \geq C_k \geq B$. The number of matrices k may have to be as high as $n-1$, but not higher. Several related results were given.

5.3 If the zero/nonzero pattern A of an m-by-n matrix over a field is known, but not the values of the nonzero entries, the question of what the rank might be often arises. The maximum possible rank has long been understood, and all ranks between the minimum and maximum occur, but the minimum is quite difficult to characterize. If there is a k-by-k subpattern of A (in general position) that is permutation equivalent to a triangular pattern with nonzero diagonal (call this a "k-triangle") then the min rank of $A \geq k$. Further, min rank $A \geq T(A)$, the maximum of such k or the "triangle size" of A.

It was known from prior work of the author that min rank $A > T(A)$ can occur, the smallest known example being 7-by-7. This raises a natural question of for which m, n, r must an m-by-n pattern of minimum rank r (over the real field) have an r-triangle? Rafael Cauto (Spain) and I had obtained some important partial results about this, and I had suggested the question to many bright pre-mathematicians who had given up after a brief look. In 2005, undergraduate Josh Link took up the question quite seriously, initially with a combination of clever, ad hoc arguments and computing. Eventually, he and I found a clever (but still involved) complete solution, which I would never have found without his collaboration. This settled a long-standing natural question, but raises many more. For which m, n, r is the "first" instance of an m-by-n pattern A for which min rank $A = r = T(A) + 2$?

5.4 A square matrix is Toeplitz if its entries are constant along diagonals parallel to the main one. As a natural case in a progression of determinantal inequality questions, we raised the following question in summer 2006: which ratios of products of principal minors are bounded among all positive definite Toeplitz matrices? Two pre-mathematicians, Hyo-min Choi from Korea and Alex Porush, a bright high school student from the area, took up this problem. I had addressed such questions before for M-matrices (and inverse M-matrices), totally positive matrices, positive definite matrices and certain structured P-matrices with co-authors Shaun Fallat and Tracy Hall (ex REU). The general positive definite case, especially, is still very unresolved (important partial results) and presents some remarkable difficulties, but a cone theoretic approach evolved from that work. Choi and Porush used this approach but had much creative work to do. Positive semi-definite Toeplitz matrices with special distributions of rank among the principal submatrices had to be constructed (or their existence ruled out) and inequalities suggested by the method had to be proven. They pushed things through $n = 6$, a remarkable and fresh piece of work in a very classial area.

5.5 Ron Smith and I had pioneered linear interpolation problems for special classes of matrices. Example: for which pairs x and y of real n-vectors does there exist a P-matrix (positive principal minors) A such that $Ax = y$. We had found informative characterizations for many familiar classes.

This raised a natural (further) question, what about replacing x, y by n-by-k matrices X and Y (wlog of full rank k). This proves to be enormously challenging for virtually every class. I had done the positive definite case, which had a nice answer, but we hadn't done any other. We were concentrating on the P-matrix case and had a natural conjecture, but could not even prove it in the case $n = 3, k = 2$. Christian Sykes took up the problem and focused upon the 3,2 case. This is very geometric and analytic, which he liked. Ron and I had been trying to find an elegant 3,2 proof that might generalize. I encouraged Christian not to worry about how many cases he might have to consider, but that resolution of the 3,2 case would affect everyone's thinking. In the end, he found a rather nice algebraic proof of our conjecture via many cases. He has already spoken about this at a meeting in Portugal and won a prize for the talk.

References

[1] C.R. Johnson and D.J. Lutzer, A Decade of REU at William and Mary, pp.19-29, in Proceedings of the Conference on Summer Undergraduate Research Programs, J. Gallian, ed, American Mathematical Society, Providence, 2000.

[2] R. Horn and C.R. Johnson, Matrix Analysis, Cambridge University Press, NY 1985.

[3] R. Horn and C.R. Johnson, Topics in Matrix Analysis, CUP, NY, 1990.

DEPARTMENT OF MATHEMATICS, THE COLLEGE OF WILLIAM & MARY, WILLIAMSBURG VA
E-mail address: crjohnso@math.wm.edu

Traditional Roots, New Beginnings: Transitions in Undergraduate Research in Mathematics at ETSU

Anant P. Godbole

1. Background and Transitions

I have run an REU program more or less continuously since the Summer of 1991. Starting at Michigan Tech (1991-1999), the program moved with me to Tennessee (2000–2004; 2006–2008) when I became department chair at ETSU. Various incarnations of the program have been run under monikers such as *Discrete Random Structures, Probabilistic Methods in Graph Theory, Combinatorics, and Number Theory, Discrete Probability and Associated Limit Theorems*, etc., indicating quite clearly that this was a probability REU with inspiration being derived from the AMS 05 (Combinatorics) and 11 (Number Theory) classifications. This phase of the program has been well described in the First PURM Conference article "The Michigan Tech REU Program in Probability," Proceedings of the Conference on Summer Undergraduate Mathematics Programs, Joe Gallian, ed., American Mathematical Society, pp. 93–104, 2000.

During the Summers of 2002 and 2003, however, I began to notice that I personally had some talent in the area of discrete mathematics. I could ask the right questions and guide students in undergraduate research in these non-stochastic areas of inquiry. My latest REU site award is accordingly titled *Probability and Discrete Mathematics*, and at least half of the projects do not involve probability in any form.

Transition Point No. 1: *The program is no longer rooted as firmly in Stochastics as before.*

In the Fall of 2004, we extensively revised our curriculum at ETSU in response to a 120-hour curriculum conversion. A new track in Quantitative Biology was added to the existing tracks in Mathematics, Statistics and Mathematics Education. However, *completion of a three hour undergraduate research requirement became mandatory* regardless of the chosen track, i.e., undergraduate research is now a core requirement. Work at the level of an REU student was not expected, but in many cases, students were indeed able to perform at that level. In many other cases, the work was original but perhaps not as deep as that produced by my summer REU students.

Received by the editor December 1, 2006.

Transition Point No. 2: *I have begun to believe that undergraduate research is appropriate for all math majors, regardless of their ability, and that it should not be confined only to select students such as those in honors programs.*

After working on undergraduate research with students at all levels, I have come to realize that a C student feels as immense a sense of ownership of his/her achievements as does a straight A peer, and is, debatably, inspired to a larger extent. This realization has had a ripple effect. I am in the process of redesigning my REU webpage to send the message to applicants that a very important criterion for selection will be the "potential value added to the student's future endeavors."

Transition Point No. 3: *I expect that my short list will no longer contain the very best applicants in the traditional sense, but rather the students most likely to experience a long-term impact from the summer experience.*

Elsewhere in this volume I have written of my involvement with talent expansion in quantitative biology, and the NSF-STEP program in which a team of mathematicians and biologists will lead 60 students in a research intensive curriculum in quantitative and computational biology. I have already worked with REU and ETSU students on research in areas such as shotgun sequencing and random palindromic sites in genomes, and my deep involvement with the STEP program (I am PI) will undoubtedly create for me a parallel undergraduate research director persona.

Transition Point No. 4: *I am now as interested in stochastic models in genetics and molecular biology as I am in stochastics and discrete mathematics.*

In 2003, we launched at ETSU a new MS program in "Pre-Collegiate Mathematics" with tracks for in-service K–8 and 7–12 teachers. The first cohort of teachers from Southwest Virginia has now graduated with an MS in K-8 Pre-Collegiate mathematics. Their curriculum consisted of 5 math classes, 3 education classes, a three hour research thesis on classroom practice and innovation, and, last but not least, a *Research Experiences for Teachers* requirement, inspired by the following heart-felt quote from Al Cuoco.

"There are very few absolutes in education, but theres one thing of which I am absolutely certain: The best teachers are those who have a research-like experience in mathematics. Working for an extended period of time on a hard problem that has no apparent approach or solution has profound effects on how one perceives the nature of the enterprise. Teachers who have done this type of research are much less likely to think of mathematics as an established body of facts than are teachers who have simply taken a set of courses. They are more likely to stay engaged in teaching after they start teaching. And they are much more likely to organize their classes around large investigations rather than low-level exercises. An ideal teacher preparation program combines the kind of orchestrated assimilation of the main results in mathematics with the much messier unstructured explorations that come from working with a mentor and grappling with a research project."

The teachers have worked in teams on very creditable research projects (described elsewhere in this article), often with input from their own young students. Additionally, their work has inspired that of several REU students. I plan on applying for RET supplements to my Summer of 2007 and 2008 awards; if granted, these will be my fourth and fifth RET supplements in eighteen years.

Transition Point No. 5: *I have come to believe that research is appropriate for students and teachers of mathematics at all levels.*

I will now offer snapshots of what has transpired over the last three years. Readers will notice, I hope, the emergence of each of the above-mentioned five sets of transitions.

2. Activities

2.1. The Summer of 2004. This was a glorious summer for undergraduate math research at ETSU. Debbie Knisley ran her first MAA-SUMMA REU with four students working on chemical and biological graph theory. The students were Ary Clemons (ETSU), Francesca Duncan (Tuskegee College), Mariam Konate (Virginia Tech) and Jeremy Smith (UT-Knoxville).

The Math and Biology departments ran their first NSF-UBM summer program with participation mainly from ETSU (7 students) and Rice (1 student). The students were Mike Phillips (ETSU, advisors Jeff Knisley and Steve Karsai), Leanna Horton (ETSU, advisors Godbole and Karl Joplin), Particia Carey and Georganna Rosel (ETSU, advisors Hugh Miller, Joplin and Jeff Knisley), Emily Mullersman and Jennifer Whittington (Rice and ETSU, advisors Lev Yampolsky and Jeff Knisley) and Brad Wild and Daniel Lamb (ETSU, advisors Debbie Knisley and Celia McIntosh). Techniques used extensively were Poisson approximation, neural network modeling, game theory, and graph theory. Leanna Horton's work on palindromes in genomes was conducted jointly with REU students James Gardner (ETSU) and Anne Shiu (Chicago).

Overall, I consider the Summer of 2004 to be one of the more successful years of my REU program. Students were Gardner and Shiu, mentioned earlier and now grad students at Emory and UC Berkeley respectively, Nathaniel Watson (Washington University, now at Cal Berkeley), Carl Yerger (Harvey Mudd, now at Cambridge/Georgia Tech), Annalies Vuong (UC Santa Barbara, now at Dartmouth. Annalies won honorable mention for the Alice Schafer Prize the same year that former student Melody Chan won it, fresh from a very productive summer at Duluth), Nadia Heninger (UC Berkeley, now at Princeton), and Ian Wyckoff (Arizona State, now at Georgetown Law School). Among the highlighs of the summer were the resolution of the Stacking Conjecture in graph pebbling by Wyckoff and Vuong, a series of six papers written by Watson and Yerger (some in collaboration with Vuong and Gardner), the collaborative work between REU and UBM students mentioned earlier, a resolution of a conjecture on vertex neighbor integrity (made by former Duluthite Marci Gambrell) by Heninger, Vuong and Shiu, and a hearty attack on a very hard and still unsolved combinatorics problem by the same three students. Several papers have emerged out of these efforts.

Last but not least three K-8 teachers worked with me in teams on RET style research as part of their MS requirements. Donette Carter, Mary Jo Neal, and Crystal Hall did number theoretic work on coincidences in multinomial coefficients, coming up with the $p \sim \{p+1, p+2 \ldots q-1\}$ conjecture (p, q are adjacent primes); Sherry Mullins, Beverly Owens and Laura Robinson studied the graph theoretic properties of "standardized planarizations of K_n"; and Robin Herndon, Rita McFaddin, Crystal Anderson, James Ray and Janeane Young worked on "625 generalizations of Karpekar's process." Each of these problems was further worked on by my 2006 REU students (see below).

2.2. The 2004-05 Academic Year. MATH 4010 (the required undergraduate research class) was offered for the first time with me as instructor. This is a so-called writing intensive (WI) and oral intensive (OI) class operated under the scrutiny of ETSU WI and OI committees. Students worked in teams and were required to give four oral presentations as they turned in 5, 10, 15 and 20 page reports – each a refinement of the previous. This first year was rather successful:

- Patty Carey, Josh Fair and Jason Jones began work on "A Generalization of the Erdős-Ko-Rado theorem". They did creditable work but only Patty chose to continue beyond the Fall semester. Her paper is now under review by *Journal of Combinatorial Designs* and was presented (by me) at the Fields Institute Workshop on Covering Arrays, Ottawa, June 2006.
- Tom Intardonato and Garth Fine's work "Birthday Problem with Dependence" was presented by me at a Workshop on deBruijn Cycles held at the Banff International Research Station in December 2004. A paper might emerge with possible co-authors being experts in compound Poisson approximation.
- Mark Dula and Cristian Madar continued the work of 2003 REU students Alexa Mater and Deanna Turk, which I presented at the IMA in October 2003. The work extends random genome reconstruction to include the case of Markov nucleotide generation. A joint paper with four students is in preparation.

2.3. The Summer of 2005.

- I had not applied for an REU renewal in the Fall of 2004. Consequently, I had a year off from the REU;
- I directed the RET research of two teams of teachers. The projects were "Generalizations of the $M\&m$ problem," and "Variations of the $3X + 1$ problem";
- Debbie Knisley ran her second MAA-SUMMA REU program, with four students working with her on chemical and biological graph theory. The students were Tywanna Anderson (ETSU), Huda Hussein (ETSU) Veranda Moffet (Mississippi Valley) and Glenda Span (Mississippi Valley);
- The NSF-UBM collaboration between Math and Biology offered its second summer REU program. I did not take part in activities but my student Patty Carey was a participant. Some participants won a $100 award at the Undergraduate Poster Session at the Joint Math Meetings. There were some repeat students from the summer of 2004; new recruits were Holly Hicks and Shannon McConnell who worked with the protein folding group of McIntosh and Debbie Knisley; Dmitri Yampolsky who joined Karsai and Jeff Knisley's complexity group, and Erin Ashton and Jennifer Cooke who were part of the microarray group led by Joplin, Yampolsky, Edith Seier, and Jeff Knisley

2.4. The 2005-06 Academic Year. During this academic year students worked independently on undergraduate research projects with Edith Seier. Patty Carey worked independently with me on "Standard Planarization of K_n and $K_{n,n}$, which has proven to be an extremely hard problem that has stymied even my best summer REU students.

In the formal offering of MATH 4010, the following projects were completed:

- Brad Wild and Stephanie Goins worked on several novel aspects of the longest monotone sequence problem. I hope soon to see their small case studies appear on the *Integer Sequences* website;
- Tracy Holt and Jeff Worley's project "Domination cover pebbling of certain families of graphs" was presented at the 2006 Southeastern Combinatorics conference and we hope to submit a paper to the *Proceedings of the 2007 Conference*;
- Brett Kindle and Torey Burton worked on "The appearance of the first 123 pattern," which I presented at the Permutation Patterns conference in Reykjavik, June 2006.

2.5. The Summer of 2006. Much of the work from this summer is still in progress. Highlights included results on universal cycles for multisets (Josh Zahl, Caltech, and Toby Johnson, Yale), random tree compression (Zahl), random compression of random graphs (Joe Marincel, Washington University), random domino tilings (Katie Bennedetto, William and Mary), planarizations of Sierpiński type graphs (Benedetto and John Symms, University of Utah), first occurrences of 123 patterns (Symms), planarizations of complete graphs and circulants (Benedetto, Johnson, Helen Hauser, Ohio University, and Michael Wijaya, University of Rochester), pebbling thresholds for diameter two graphs (Ariel Levavi, CMU, and Hauser), coincidences in multinomial coefficients (Igor Konfisakher, Washington University and Wijaya), chordal permutation graphs (Levavi and Marincel), and a complete resolution of the Kaprekar problem by Zahl.

2.6. The 2006-07 Academic Year. Students in MATH 4010 this year have worked with me in the formally "taught" Fall class, but others have worked with collegues Jeff Knisley, Michel Helfgott, and George Poole on a one-on-one basis. My students are getting ready for their final presentations as I write this. The projects are as follows:

- Bijan Khairollahi is working on "Equitable division of pizza," and I am confident that when completed it will combine nicely with work of RET teacher Floyd Brown to make a publishable paper (the original version was rejected by *Math Magazine*);
- Beth Haun and Courtney Sanders' work on "Star of David Graphs" will be publishable in an "entry level" combinatorics journal, as will
- Glenn Quarles, Eric George, Joe Barwick, and Brooks Dearstone's paper "U-cycles of DNA and non-bijective functions;
- David Simpson and Nicole Holder's work on "3-good permutations" will likely lead to a submission to Neil Sloane's website of integer sequences.

In addition to these projects, Elizabeth Harris is making good progress on "Square induced Sierpiński graphs" as part of a Research Discovery Federal Work Study position, while Ryen Lapham's work on "2-balanced permutations" will be submitted to the integer sequence website as well.

3. Looking ahead

I am fairly certain that my NSF-REU days will end after the Summer of 2008. I simply do not have the time to spend 6 or more hours a day "in the trenches." Eighteen years of such work has been the defining part of my career, and I will miss my Mudders and Arizona State honors students and Harvard Math Concentrators, but I do look forward to other challenges. My UGR direction will continue through the academic year – with ETSU Math majors and ETSU-STEP students, and over the summers with ETSU-STEP students (one project per year). Over the summers too, I hope to play a secondary role in Debbie Knisley's NSF-MAA-NSA-SUMMA REU program for underrepresented minorities, in which she guides students to use graph invariants in the pursuit of certain kinds of secondary RNA structures (I have not been successful in attracting many underrepresented students to my REU program).

     Thanks, Joe, Aparna, Ive, and Frank for putting the conference together!!

East Tennessee State University, Department of Mathematics, Johnson City, TN 37614.

E-mail address: godbolea@etsu.edu

Undergraduate Research in Mathematics at Grand Valley State University

Steven Schlicker

Introduction

The Department of Mathematics at Grand Valley State University (GVSU) has a long and successful history of research with undergraduate students. Since 1994, faculty in the department of mathematics have mentored 65 students in funded research projects in mathematics. These projects have resulted in many publications in refereed journals and award winning presentations at the national level. In this article, we describe the undergraduate research programs at Grand Valley and why we believe these programs have been so successful.

Grand Valley State University

Grand Valley State University is a mid-sized, comprehensive regional university with an emphasis on providing the feel of a small college. GVSU enrolled 22,565 students in the 2005-06 academic year. The Department of Mathematics at GVSU is home to 34 tenure-line faculty. Our emphasis is on undergraduate education. Within that context, our faculty is active in research, with faculty members regularly publishing books and papers. This department- and university-wide focus on undergraduate education combined with active research and scholarship provides a unique environment for undergraduate intellectual growth and achievement. We see undergraduates as active participants in the scholarly life of our department and university.

Undergraduate Research at GVSU

GVSU has a history of successful undergraduate research. Formal research programs with undergraduate students in mathematics began at Grand Valley State University in 1994 when the Division of Science and Mathematics established the Summer Undergraduate Research Program (SURP). This program was created to foster and support collaborative faculty-student research projects in science and mathematics. The SURP program started out small, funding 5 proposals (1 in mathematics) in its first year. In 2003, the SURP was expanded to the Summer

Received by the editor January 20, 2007.

Student Scholars (S^3) program, a campus-wide program supporting scholarly activities in all disciplines. Between 1994 and 2006 the SURP and S^3 programs funded 17 student-faculty collaborative research experiences in mathematics. Examples of past SURP and S^3 mathematics research topics include Sierpinski polyhedra, the differentiability of certain area functions, differential games, polynomial root-dragging, investigation of middle school students' understanding of step and linear functions, spherical geometry, separable voting preferences, bioinformatics algorithm development, dynamics of the dual billiard map in the hyperbolic plane, hypergeometric summation and the Wilf-Zeilberger (WZ) algorithm, and circle packings.

The Grand Valley State University REU Program

The SURP program was intended to serve as "seed" money for the development of undergraduate research programs. With that in mind, we submitted a Research Experiences for Undergraduates (REU) proposal to the National Science Foundation (NSF) to support our research activities with students. Our first REU grant allowed us to run an REU program in the summer of 2000. Since then we have offered REU programs in every summer except 2001. We just completed (in 2006) the second summer of a five-year REU grant.

Overview: The goal of our REU program is to provide a complete research experience for student participants, including literature searches, background reading, focused work on one or more research problems, and both oral and written communication of results. Developing and honing students' communication is a critical component of our REU. We provide a broad range of research choices in various areas of pure and applied mathematics that reflect the diverse interests of the many department faculty who serve as mentors.

Research Groups: The GVSU REU program is an eight-week program that provides support for eight students. We recruit nationally and give priority in our recruiting to students for whom this REU will be their first research experience, and we actively recruit students from groups that are traditionally underrepresented in the mathematical sciences. The eight students work with four faculty mentors and form research teams of two students and one faculty member. Examples of research projects in past REUs include polynomial root-dragging and the location of critical numbers of polynomials, Lanczos' derivatives, wavelets, spans of derivatives of polynomials, Hausdorff metric geometry, bifurcations in dynamical systems, San Gaku problems in spherical and hyperbolic geometry, Lie theory, and Mandlebrot sets for ternary number systems.

Format of the Program: We begin the summer with an opening picnic on a Sunday afternoon. This provides an opportunity for both the students and faculty to get acquainted. After a group breakfast (supplied by the faculty) the following Monday morning, each faculty member gives a presentation on his/her research area. That same evening, each student sends his/her preference rankings of the four research areas to the faculty who use the rankings to determine the research groups. The research groups are announced on Tuesday morning, and the students immediately get to work on background reading, literature searching, "warm-up problems", and further preparation for research. All students are actively pursuing research problems in the first week of the program. During the summer, each research group meets regularly (almost every week day) to discuss progress on the research, brainstorm, and plan for the future. In addition to their own research, through a weekly program of outside speakers our REU students are exposed to a variety of mathematical topics and learn a bit about what life is like as an academic mathematician. We also engage our students in many events in addition to their research (e.g. workshops, student talks, group breakfasts, group social activities) to create a student-faculty team environment.

Communication is an important component of our program. During the first week the faculty mentors provide a workshop for the students on using LaTeX. Throughout the program, students are required to write and submit weekly reports on their research progress in LaTeX. As these weekly reports are updated, they provide a solid foundation for the final technical report each group will produce. When appropriate, these technical reports are used by the research teams to prepare manuscripts to submit for publication. In addition, students are required to give short bi-weekly presentations to the entire group. Not only does the frequent writing and speaking hone their communication skills, it also helps clarify and solidify in students' minds the work they have been doing. In addition, the presentations help our students prepare presentations for two conferences: the Michigan REU conference and MathFest.

We initiated the first annual Michigan REU conference in the summer of 2000 to expose our REU students to the research being conducted at the nearby Hope College REU site. We expanded the conference in 2002 and served as hosts to the Hope College REU and the new REU program at Central Michigan University (CMU). During this day-long event, all of the REU students (as well other students engaged in mathematical research) give presentations of their work in parallel sessions. The Michigan REU Conference now rotates among the three institutions (CMU hosted the last conference in 2006). As a culminating experience for our REU group, we plan our program so that it ends at MathFest, the annual summer meeting of the Mathematical Association of America (MAA). We take all of our REU students to this conference at which they give presentations in the MAA or Pi Mu Epsilon student sessions.

Philosophy: Our philosophy of research with undergraduate students is founded on four basic principles. The first is Intellectual Independence. We understand that research is much different than a typical classroom experience. Our program is designed to help students make the transition from the more structured environment provided by classroom instruction to the open, independent work required in research. In the beginning of our program, the faculty mentor takes the lead and provides a framework for the research experience. As the summer progresses, the student researchers are gradually led to more independence and the students and faculty work as research collaborators and colleagues. The second principle is Student-Student Communication. An important component of the research experience is communicating one's results. This is a new experience for most students, so we create a non-threatening environment in which students regularly prepare and practice presenting their work to their peers. Through the presentations, students receive feedback on their work and their communication skills from both students and faculty. These presentations also expose our students to the work of their colleagues and areas of mathematics other than their own research area. One advantage to the regular student presentations is that the presentations increase in formality as the summer progresses, and by the end of the summer our students are well-prepared to give presentations at the Michigan REU conference and Math-Fest. The third principle on which our program is based is Community. To build community, and to foster informal communication among the students, we house our REU group together in 4 townhouses on campus (2 students per townhouse). Living together encourages cooperation among the students and helps them form close bonds. We also foster community through regularly scheduled social events, seminars by outside speakers, group breakfasts (provided by the faculty each Monday), and other informal activities. We feel that providing community is important for students for many reasons. In particular, most of our REU students are new to the area and it is vitally important to have support when students are struggling to learn new mathematics. Our fourth principle is Student-Faculty Interaction. Our faculty mentors are co-participants in all of the REU activities. We format our program to foster independence in our students and strive to create an environment in which students ultimately see themselves as colleagues and collaborators with the faculty.

Research Projects: Selecting appropriate research projects for undergraduate students is a challenge. Many faculty mentors in our REU have used this challenge as an opportunity to become students themselves, reading books and journal articles to learn new mathematics outside their own areas of expertise and find problems that are mathematically interesting and substantial but still accessible to talented undergraduates. Since the faculty are not experts in these new research areas, they are able to convince their students that the research the students and faculty do in the program is truly collaborative. We search for problems that are significant enough mathematically so that a full solution will be publishable in a refereed journal, but at the same time are accessible enough to allow for substantial progress to be made during the summer. Ideally, a good research problem will come in layers - for example, early progress can be made on specific cases while the problem proves more challenging as the students progress in their mathematical sophistication. We want our students to be able to take gradual ownership of their research and be confident at the end of the program that their contributions have been substantial and real.

Post-Program: We expect our REU students to continue their work after the program has formally ended. Faculty work with their students to prepare manuscripts for submission to journals, and encourage students to give presentations on their work at their home departments and at local, regional, and national conferences.

Evidence of Research Success

Students and faculty in the Grand Valley State University REU, SUPR, and S^3 programs have produced publications in a variety of journals. A list of publications includes:

- "Sierpinski Polyhedra," *Pi Mu Epsilon Journal*

- "Integral Functions Whose Right Derivatives Are Average Values of Periodic Functions," *Pi Mu Epsilon Journal*
- "On the Differentiability of Certain Integral Functions," *Mathematics Magazine*
- "Period-3 Orbits via Sylvester's Theorem, and Resultants," *Mathematics Magazine*
- "When Lines Go Bad in Hyperspace," *Demonstratio Mathematica*
- "Breaking the Holiday Inn Priority Club CAPTCHA," *College Mathematics Journal*
- "Symmetry in Bifurcation Diagrams," *Pi Mu Epsilon Journal*
- "Quadratic Dynamics in Matrix Rings: Tales of Ternary Number Systems," *Fractals*
- "A Singular Introduction to the Hausdorff Metric Geometry," *Pi Mu Epsilon Journal*
- "Wavelet-Based Steganography," *Cryptologia*
- "Lisa's Lemonade Stand: Exploring Algebraic Ideas," *Mathematics Teaching in the Middle School*
- "A Note on PIPCIRs," *Pi Mu Epsilon Journal*
- "On the Ratio Vectors of Chebyshev and Equispaced Polynomials", *Missouri Journal of Mathematical Sciences*

Our students have also been recognized for their presentations:

- Christy Heideger and Amanda Taylor (REU 2006) - best student talk award in an MAA session at MathFest 2006.
- Samuel Kolins (REU 2005) - best student talk award in an MAA session at MathFest 2005.
- Chantel Blackburn (REU 2005) - Council on Undergraduate Research (CUR) award for best student research talk at the Pi Mu Epsilon sessions at MathFest 2005.
- Alex Zupan (REU 2005) - CUR award for best student research talk at the MAA sessions at MathFest 05.
- Matthew Katschke (REU 2004) - best student award in an MAA session at MathFest 04.
- Christopher Bay (REU 2003)
  - Best student talk award in an MAA session at MathFest 03.
  - Best poster award in the Undergraduate Poster Session at the 2004 Joint Mathematics Meetings.
- Kris Lund (S^3 2003) - best student talk award in an MAA session at MathFest 03.
- Micah TerHaar (S^3 2003) - presented a poster at the prestigious Undergraduate Research Posters on the Hill session sponsored by CUR in Washington D.C.
- Lisa Driskell (REU 2000) - won the Annual Greg Mellen Memorial Cryptology Scholarship Prize from Cryptologia for her article "Wavelet-Based Steganography".

Reasons for Success. There are many factors that we feel have made our undergraduate research programs so successful. One is institutional support. From the original Summer Undergraduate Research Program to the current Student Summer Scholars program, Grand Valley for years has encouraged faculty and undergraduate students to be collaboratively involved in research. Students are expected to communicate their work to others during Student Scholarship Day (SSD), a day-long campus wide event at which students present results of their scholarly activities to a university-wide audience. GVSU also provides financial support for students to give presentations at conferences. All of this support creates a climate on campus that acknowledges this type of scholarship as important for both faculty and students and builds an expectation that many students will be involved in research or scholarly activities beyond the classroom.

Another factor that contributes to the success of our undergraduate research programs is the emphasis the mathematics department places on student-faculty interaction. Almost half (15 of 34) of the regular faculty in our department have participated in senior theses or research projects with our students. We encourage faculty who have not had research experiences with undergraduates to participate in the S^3 program and, at the same time, to engage in activities with the REU program and to consult with the REU faculty. In this way the department mentors new faculty in this area. We recognize this type of scholarship as vital to the department and university mission and encourage faculty and students to be involved. We expect students to share the results with their peers and others at our department seminar, at SSD, and at outside conferences. We also hang student posters through our building to publicize their work.

Other than our senior thesis course, the mathematics department doesn't have any specific courses directed toward undergraduate research. Rather, in all of our courses we emphasize communication skills, both verbal and written. Our courses include writing assessments, from short essay questions to extensive projects, and student presentations, even as simple as having students present solutions to homework problems. As a result, GVSU students who participate in research programs have already developed the skills to allow them to effectively communicate the mathematics they learn. In addition, most mathematics faculty at Grand Valley expect students to work cooperative in small groups on class projects and activities. In this way our students come to understand how to work effectively in groups and can then be productively contributors to any research group. In a similar vein, we work in depth with our REU students who are not from Grand Valley to develop these communication skills during our REU program. REU students write weekly reports on their progress, give bi-weekly presentations, and speak at two major conferences. Throughout the program, the faculty work with the students, coaching them on their presentation skills and providing specific, directed feedback and advice on their writing. As the results indicate, our students give well-received talks at major conferences and their final reports at the end of the program often become published papers.

One other factor that we believe contributes to the success of our research with undergraduates is the nature of the research problems themselves. Our research problems are specifically chosen to be substantial and accessible to undergraduates. This has required most of our faculty to learn about new fields of mathematics, different from their thesis areas. An interesting consequence of this is that the research projects we conduct with our students are truly collaborative in nature. The projects are just as much research for the faculty as they are for the students. While the faculty of course have broader and deeper mathematics backgrounds than the students, the fact that we are often new learners in the research areas makes it possible, as often happens, for our students to truly discover new mathematics for themselves instead of just being led to the mathematics.

DEPARTMENT OF MATHEMATICS, GRAND VALLEY STATE UNIVERSITY, ALLENDALE, MI 49401
E-mail address: schlicks@gvsu.edu

The Hope College REU Program

Tim Pennings

Overview

Undergraduate student research has long been the hallmark of the Hope College Science Division with over 150 students involved in research each summer. This research is supported by a number of grants including a divisional HHMI grant and NSF-REU awards in Biology, Geology, Chemistry, Physics, Mathematics, and Computer Science. Excepting one year, the Department of Mathematics has been supported by NSF-REU funds each year since 1991 resulting in twenty papers published or submitted.

The objective of our program is to give students the opportunity to "get their mathematical wings." That is, over the course of the summer we aim to guide them through the entire research process - selection of a problem, literature research, reading and understanding mathematical articles and texts, grappling with and solving a problem, effective oral explanation of mathematical ideas, the writing of a mathematical paper suitable for publication, and presentation of the paper. That is much to do in eight weeks. It may or may not result in a published paper, but from our experience it nearly always results in students who have a new understanding and appreciation of – and confidence in their ability to do – mathematical research. This is the first step towards a successful career as a scientist/mathematician.

Significantly, such an experience is not typically a part of the undergraduate education of most mathematics students. Our objective is to give this opportunity to students who 1) have the potential to benefit from it (which means that they have requisite ability, are self motivated, and can work well with others), and 2) do not otherwise have the opportunity for such an experience. In particular, we look for students who have not yet participated in an REU and who are from schools where research experiences are not available.

Along with advanced undergraduates who have chosen to major in mathematics, we hope to attract and encourage some of the students who have finished just their first year of higher education and have not yet committed themselves to a particular discipline. We have found from past experience that some of the brightest and most talented students spend time exploring various interests until they are generally obliged to choose a major in their sophomore year. Providing them with a positive research experience at the appropriate level can serve to attract some of

Received by the editor January 20, 2007.

these students by showing them an appealing side of mathematics they would not otherwise see. Thus some of our projects will be tailored for students who have not yet taken advanced undergraduate courses.

Summer research projects supported by the NSF and Hope College are offered to 6-10 students in probability (involving pebbling and geometry), analysis (involving dynamical systems and computational mathematics), and algebra. Students work in pairs with faculty mentors for eight weeks beginning the first Monday of June. They receive a stipend of $3200 and are provided with free housing and transportation.

Two intentional trends occur throughout the summer. The first is a trend from dependence to independence in reading literature and solving problems. The second is a trend from giving informal presentations to the writing and formal presentation of a paper suitable for submission for publication. As such, the student has experienced a microcosm of graduate research. The ability and confidence so gained is transformational as students feel the flush of accomplishment at successfully investigating, solving, and presenting their research problems.

Nature of Student Activities

The goal of the summer research program at Hope College is to help talented and motivated students develop ability and confidence as mathematical researchers by achieving significant mathematical results in partnership with faculty mentors. In particular, students will 1) learn new mathematics, 2) gain mathematical independence by learning to read math texts and journals, 3) learn how to do library/electronic research, 4) solve an original mathematical problem, 5) learn to communicate mathematics verbally through informal and formal oral presentations, 6) write a mathematical paper suitable for submission to a journal and present it at a conference, and 7) learn the use of LaTeX, Maple, PowerPoint, and MathSciNet.

Students will initially learn about their own projects and those of the other students through introductory lectures given by the faculty researchers and/or by reading requisite background material provided by their mentors. Then, for the bulk of the summer, the students will be given their own project or research problem to solve - typically working in pairs. (We find from experience that pairs offer the best dynamic. Working with another student is generally more fun and enhances the communication skills. Groups larger than two allow for one to be left behind.) During this time students work on their problems with regular visits and help from their mentors as needed. Since students live and work together, they share ideas and strategies throughout the summer continually on an informal basis as well.

Our particular REU model has students working on projects in distinct areas of mathematics. This seems to work well; students do their own research in one topic and are exposed to a couple other areas via a weekly research seminar in which student researchers present their ongoing work. These presentations, which begin as informal progress reports and evolve over the course of the summer into more polished and professional presentations, hone the oral communication skills of presenters and critical listening skills of students in the audience. **Learning to effectively communicate mathematical ideas both formally and informally is a primary goal of the research experience.** In particular, students should learn how to be critical and creative in constructing and presenting a well-honed formal presentation. But they must also learn how to give informal, impromptu expositions. In both cases, the key is not to give a slick, flawless delivery, but to effectively communicate mathematical ideas. This involves focusing on the listener rather than on the material/presentation. It involves asking the question, "Did they understand?" rather than "Did I make any mistakes?" The difference is subtle but important, and we stress the distinction. Through formal and informal critique, presenters are evaluated on the clarity of their exposition and their ability to engage the attention of the audience. Students assess their own growth in effectively communicating mathematical ideas in an end-of-the-summer questionnaire.

Finally, for the last 1-2 weeks of the project students concentrate on writing their results in a paper suitable for publication. **Developing skills in writing mathematics is another primary goal of the research program** at Hope for several reasons. First, writing contributes to clear, precise thinking. Second, the ability to communicate well is an essential component of higher education. Third, since many students have no opportunity to write an extended technical paper as part of their undergraduate experience, it is one of the best preparations we can provide them for graduate school and/or a professional career.

Generally speaking, research projects have been inspired by problems in the faculty members' own research programs and will ideally generate original results. Past experience has indicated that it is best to find projects with a range of partial outcomes, and the projects have been chosen with this in mind.

The student workday is from 9 a.m.– 4:30 p.m. We encourage them to spend their evenings and weekends recreating. This helps avoid two possible pitfalls. First, sometimes even disciplined, motivated students who are working independently for the first time on a long-term problem can fall into a relaxed summer slump. Setting regularly scheduled work hours helps avoid that and allows students to work together most effectively. On the other extreme, conscientious students sometimes need to be encouraged to balance their research with leisure activities. Students should leave the program fully satisfied and proud with what they have accomplished, but also refreshed and energized for a new school year.

During the final week of the project, students give formal presentations of their research at a Michigan Mathematics REU Conference which we put on jointly with two other math REU in the west Michigan area (GVSU and CMU). Each summer ends with a Michigan Mathematics REU Mini-Conference (rotating between the campuses) which provides opportunity (and audience) for all of the REU students to present their research results. In addition to the student presentations, the conference includes a guest speaker or a mathematics game competition.

We also encourage our participants to present their results at regional or national mathematics meetings in order to gain a sense of the wider mathematical community, and our budget includes funds to support such travel. Over the years, many of our REU students have presented papers and posters at the Joint AMS/MAA Meetings, MathFest, various regional conferences, and at their home institutions.

Since an overarching goal is to promote mathematics research as a career, at some point in the project we often involve an outside research mathematician to give talks on his/her own research in particular and the nature of graduate mathematical research in general. Alternatively, we sometimes take the students to Chicago to visit the departments at the U of Chicago and Northwestern University.

The departments with REUs also take turns hosting seminars to which all research students are invited. Recent examples include: "Preparing for the GRE", "The ethics of scientific research", "Do dogs know calculus? - an example of the scientific method", and "Research similarities and differences across the disciplines."

Although our goals are centered around research, the success of the research experience depends heavily on the social aspects of the summer. First of all, such activities help foster good student relationships and friendships. Since students spend considerable time each day studying together, the increased intensity and enjoyment that comes from working with friends is an essential component to a productive summer of research. Secondly, providing the students with a wide variety and constant stream of activities for their free time helps ensure an enjoyable and memorable summer.

The first week is especially important since it sets the tone for the summer. Also, since students are coming to a new and unfamiliar dwelling for a relatively brief time, we want them quickly to get to know each other and the area's recreational opportunities (and to have little opportunity to feel lonely or homesick). So on the Sunday evening before research begins, we meet for pizza and a trip to the beach and sand dunes of Lake Michigan. We have a grill-out on the first Monday, and spend the first Saturday afternoon at another beach and visiting area attractions.

We find that if we keep them fully engaged the first week, by then they are comfortable and excited to go off on their own adventures. Our present group of students just returned from a weekend together in Chicago and they have already organized several bonfires at the Lake Michigan beaches and several bowling and movie nights. We plan events as well for the July 4th holiday weekend, some faculty mentors host board game nights, and the PI, being a classic movie buff, invites the students to his home for *Casablanca*, *The Sting* and other great movies.

Departments take turns hosting Thursday afternoon ice cream socials and Tuesday night beach volleyball picnics. Ultimate frisbee is available on Wednesdays and Sundays. On Wednesdays the math REU has lunch together. Students organize scavenger hunts and late night capture the flag. Between planned and spontaneous events, almost every day offers the opportunity for a social/recreational activity.

Student Recruitment and Selection

We aim to recruit and select students who i) have the requisite talent and interest, ii) are self motivated and disciplined, and iii) can work well with others. We identify such students through application materials which includes their college grade transcript, two letters of recommendation from professors, and their own statement of interest in mathematical research. We also note whether they have taken full advantage of extra-curricular mathematical opportunities available to them - such as competitions and participation in departmental responsibilities such as tutoring.

We target students who have not yet participated in an REU and who are from schools where research experiences are not available. We realize that some students enjoy and benefit from multiple REU experiences, but with the demand greater than the number of available positions, our site operates under the philosophy that the NSF-REU program should give as many students as possible an experience in mathematical research. For those wanting more, graduate school awaits. We also give lower priority to very good candidates who have already had substantial research experiences from their home institution. Some submit published papers with their applications. These students are to be congratulated; they are very deserving, but we give others higher priority. The ideal student is the one (often from a small college or regional university) who is described as being the best student in many years, who has taken advantage of everything the department has to offer, and who is still hungry for more.

Increasingly, students learn about our program through the NSF and MAA web pages. However, we still target certain schools, sending them a full information packet describing the Hope Undergraduate Research Program, with program posters, application forms, and project descriptions. These include regional and peer institutions as well as a smattering of colleges and universities throughout the United States in order to attract students from various institutions and different parts of the country.

Also included in the target group are a number of women's colleges and institutions with significant minority enrollment (such as Morehouse College and Spelman College in Atlanta) which have strong mathematics programs and/or an interest in undergraduate research.

Prospective participants can apply via an online application form. Deadline for submission is February 28, and we typically begin making offers within the second week of March.

Project Evaluation and Reporting

At the end of the project, students are given an evaluation form which asks the following questions:

(1) **Local Arrangements**
  - Were housing accommodations satisfactory?
  - Were office, computing, and library facilities sufficient?
  - Were the number and nature of organized social and recreational activities satisfactory?

(2) **Individual Projects**
  - Was the topic at a level appropriate for your background?
  - Was the amount of guidance you received from your advisor too little, too much, or about right?
  - Are you satisfied with what you learned both about mathematics and the nature of mathematical research?
  - Did you learn to effectively communicate mathematical ideas - both orally and in written form?
  - Were the opportunities to learn about the work of others adequate and helpful?
  - Are you more/less likely to pursue a career involving mathematics research after this research experience?
  - If you were directing the research/program, what would you do differently?
  - What advice would you give to a friend coming next year?

(3) **Any other comments?**

These evaluations have provided valuable feedback over the years, and we have implemented many of their ideas and suggestions. Long term evaluation occurs informally through continued contact with students as we revise papers for publication and prepare talks for major meetings. Students also write to ask for letters of recommendation, and still others stay in close contact through the friendships which have developed.

Feedback from student participants has been very positive. Many respondents indicated that they were more likely to pursue graduate study in mathematics because of the program. Some examples:

- "The REU was a good experience for me. I had already considered graduate school in mathematics, but the program gave me more confidence and direction with that decision. I learned skills which helped me greatly during my first year of studies here at the University of Texas at Austin. I also made friends whom I still keep up with to this day."

- "The REU program was the one experience that solidified my goals to pursue a higher degree. This was my first exposure to academic research. The experience was so enjoyable and challenging that I knew I wanted to seek a position that would keep me close to the study of mathematics."

- "It was one of the most interesting and most fun summers I have ever had . . . This REU was conducive to thought, creativity and play - which is an excellent combination." (Ph.D. from U of Michigan, post doc at U of Oklahoma.)

- "This experience convinced me that I will definitely follow a career in research mathematics. This was a tremendously great experience. I cannot imagine a better way for a mathematics undergraduate to spend a summer. The program was excellently run and the advisors did a great job of advising us."

- "The Hope College REU was my first crack at mathematics research, and got me started in the right direction. If I hadn't gone to it, I believe I would not be where I am today. It was truly helpful to get an idea of how math research goes at such an early stage in my career. (Ph.D. from Stanford, post doc at U of Arizona.)

DEPARTMENT OF MATHEMATICS, HOPE COLLEGE, HOLLAND, MI 49422-9000
E-mail address: pennings@hope.edu

Proceedings of the Conference on
Promoting Undergraduate Research in Mathematics

The REU Experience at Iowa State University

Leslie Hogben

1. Overview

Undergraduate research is a high priority at Iowa State University. The ISU Mathematics Department has hosted summer REUs for seven of the past nine years. In addition, mathematics faculty regularly work with ISU undergraduates on research projects, primarily through the ISU Honors Program. This report covers only the ISU Math REU that took place during the summers 2004-2006, a program which involved research in a wide range of areas, and whose largest source of funding was an NSF REU-site grant. The ISU Math REU was directed by Justin Peters (Mathematics Department Chair) and Leslie Hogben, and was managed by Hogben. It is anticipated that this program will serve as a model for future summer mathematics REUs at ISU.

FIGURE 1. The 2005 REU. Photo by Melanie Erickson.

Received by the editor October 6, 2006. The ISU Math REU was supported by the National Science Foundation through grant DMS 0353880 and other grants, and by Iowa State University. The opinions expressed are those of the author.

With up to sixteen undergraduates and a total of about thirty people involved each summer, the ISU Math REU was one of the larger NSF mathematics REU sites and one of the largest undergraduate summer research programs on the ISU campus. The program was specifically designed to exploit the strengths of a large research university (ready access to a large number of faculty and graduate students doing research), while at the same time giving students the kind of individual attention and mentoring that is often found only at smaller colleges. The research topics varied from year to year, depending on which faculty members served as mentors, although projects have been offered every year in dynamical systems, linear algebra, and mathematical biology.

2. Research Projects

The research projects were selected by the faculty mentors. During the fall, the mentors for the following summer were identified and general descriptions of the areas to be offered were posted on the ISU Math REU web-site, http://orion.math.iastate.edu/reu/homepage.html. The applicants were instructed to identify up to three projects and explain their interest. The mentors and REU directors jointly selected the participants. In late spring the mentors selected the specific research problems, based on their current research and student interests.

A typical research team consisted of two undergraduates, one graduate student and one faculty member; an average of eight projects were offered each summer. Research groups met daily for at least an hour. Initially, faculty taught the necessary background to students; later students reported progress and discussed obstacles. Each project team produced a final paper (typically 20 pages) and presented an hour-long report at the symposium held in the eighth (final) week of the program.

FIGURE 2. The 2006 REU students ask, "What do you call a baby eigensheep?" (A lamb, duh.) Photo by Haseena Ahmed.

As all mentors are active researchers who publish regularly, it was intended that many of the projects result in papers in professional journals, and this in fact occurred. Since 2004, five ISU Math REU papers have appeared in journals such as *Linear Algebra and Its Applications*, *Journal of Mathematical Analysis and Applications*, and *Mathematical Biosciences and Engineering*. Four additional papers are under review, and several from summer 2006 are in preparation.

Approximately ten students have made research presentations at undergraduate conferences (primarily the Young Mathematicians Conference at Ohio State).

The mathematical diversity of the ISU Math REU can be seen from list of projects and the number of faculty involved. Faculty mentors have supervised the research projects listed below:

- **Leslie Hogben** Matrix theory: Matrix completion problems; Rational realization of eigenvalues of tree sign patterns; Minimum rank of symmetric matrices described by a graph; Matrix D-stability.
- **Wolfgang Kliemann, Justin Peters, Jiyeon Suh** Dynamical systems: Dynamical systems in projective space; Morse decompositions, attractors and chain recurrence; Dynamical systems and group theory; Mathematics of the heart beat cycle; Dynamically coupled linear ODEs and Markov chains.
- **Khalid Boushaba, Howard Levine, Michael Smiley** Modeling of tumor angeogenisis: Negative feedback systems; Fibroblast growth factor competition; Tumor dormancy; SELEX against multiple targets; Regulation of secondary metastases by plasmin.
- **Roger Alexander** Numerical Analysis: Analysis of software for stiff ODEs; Runge-Kutta Design and Optimization.
- **Sung-Yell Song** Combinatorics and graph theory: Polygonal designs; Strongly regular graphs.
- **Zhijun Wu** Biomolecular modeling: Thermodynamic fluctuations of proteins; Optimizing protein structural alignment.
- **Dan Ashlock** Evolution of hybrid grid robots.
- **Cliff Bergman, Jennifer Davidson** Methods of steganalysis.
- **Jonathan D. H. Smith** Partial semigroups and binomial coefficients to prime square modulus.
- **Eric Weber** Attack on hiding messages in oversampled Fourier coefficients.

3. Student Recruiting and Funding

The primary funding source for the ISU Math REU was the NSF REU-site grant DMS 0353880, which supported twelve undergraduates annually; these students were recruited primarily through the ISU Math REU web-site and the NSF REU web-site. As is typical of REUs listed on the NSF web-site, there was no difficulty recruiting far more well-qualified applicants than could be accepted. The students came from all over the country, attending school in more than 20 states. Due to student self-selection and regional recruiting, approximately 45% of the students lived or attended college in Iowa or adjacent states. The students came from a variety of undergraduate institutions, including about 45% from liberal arts colleges.

The ISU Math REU also recruited students through the Alliance for the Production of African American Ph.D.s in the Mathematical Sciences. The Alliance is an NSF-supported partnership between the Iowa Regents universities (Iowa State, University of Iowa, and University of Northern Iowa) and several historically black colleges and universities, including Alabama A&M University, Florida A&M University and Jackson State University. The Alliance has supported up to four participants annually. Due to a miscommunication, the ISU Math REU was not listed on the 2006 Alliance web-site and had no 2006 Alliance participants, but plans have been made to host Alliance students in 2007.

Additional participants have been funded through faculty member grants from various agencies and through the NASA/Pipelines Sciencebound program.

The ISU Math REU had no funding targeted specifically at women, but actively recruited women and collaborated with the ISU Program for Women in Science and Engineering, both in recruitment and to offer additional programming through PWSE to female participants. More than 40% of the undergraduates who participated are female, one third of the graduates students are female, and several female faculty were involved (including one of the directors).

Since the research teams functioned better when students had similar mathematical preparation, the majority of the research teams paired students from the same funding program, but when feasible, teams mixed students across programs.

All ISU Math REU students focused their main efforts on their research projects, and participated in additional common academic activities. All were invited to the social activities, regardless of funding source. However, in practice, students funded by another program were sometimes required to participate in activities for that program and/or housed separately from the students funded by the REU-site grant. This sometimes restricted their involvement with the ISU Math REU.

4. Graduate Students and the REU

Recognizing the many benefits of graduate student involvement in the REU, the ISU Mathematics Department funded seven or eight research assistantships each of the three summers for graduate students who participated in the REU. These graduate students did not replace faculty involvement; normally the entire project group met together, although the graduate student led the project when the faculty member was away for a short time at a research conference.

The graduate students were encouraged to talk to the undergraduates about the graduate school application process and graduate student life, in addition to serving as research mentors. Evaluations show that the undergraduate students found the participation of graduate students to be very valuable to their experience, and the graduate students found the experience beneficial also. During the summers of 2005 and 2006, a senior graduate student served as the assistant REU manager, coordinating student activities. These students gained leadership experience and were a great help in managing the program.

As many ISU graduate students accept positions at liberal arts colleges where they will be expected to advise undergraduate research, participation in the REU is valuable training for the graduate students. (Although unrelated to the current ISU Math REU, the ISU Mathematics Department is one of the leading producers of REU site directors: Both Timothy Pennings of Hope College and J. D. Phillips of Wabash College received their Ph.D.s from Iowa State.)

5. Activities

In addition to the research projects described in Section 2, the ISU Math REU had an extensive program of activities, both academic and social. During the first two weeks, the students attended several classes in the computer lab about Matlab and LaTeX. Students have heard lectures on topics varying from "The mathematics of secrecy: an introduction to public key cryptography" to "Mathematical modeling of phytoplankton", as well as presentations such as "Graduate school: What is it *really* like?"

Social activities played an important role in enhancing the REU experience and were supported by the REU-site grant. Once a week undergraduates, graduate students and some faculty had lunch together, where extensive discussions about life as a graduate student and as a faculty member occurred. Every Tuesday and Thursday afternoon participants gathered for "pop and cookies" and conversation; once a week this was followed by a lecture or presentation. There were two major social activities each summer: a picnic at Big Creek Lake including boating, and a trip to an amusement/water park. Other social activities have varied with the summer and included participation in an intramural soccer team, ice skating, bowling, math movie night (*Fermat's Last Tango*), a concert by a local folk singer, etc.

FIGURE 3. Racing on Big Creek Lake, 2004. Photo by Leslie Hogben.

The undergraduates were housed together in Frederiksen Court student apartments. This has worked well; by living together they developed strong friendships. The apartments offer the opportunity to cook, and during 2006, when all ISU REUs collaborated, there were regular student pot-lucks.

6. Cooperation with other ISU REUs

During 2006, there was a campus-wide group of ISU REU directors who met regularly. This resulted in many positive outcomes, including a mentoring workshop for graduate students, a campus-wide REU picnic funded by the Dean of the Graduate College at the beginning of the summer, a presentation to all REU students about graduate study by the ISU Vice Provost for Research, an end of the summer campus-wide poster session, etc. The undergraduates in the Math REU enjoyed interacting with those in other programs and this cooperation gave all the REUs greater internal visibility. It is expected that this collaboration will continue.

7. Goals and Program Evaluation

The goals of the ISU Math REU are:

- Provide a supportive environment where students can discover the joys of mathematical research.

- Increase the number of U. S. citizens receiving Ph.D.s in the mathematical sciences.
- Produce high quality, publishable research.

These goals have been met as measured by student surveys, data on subsequent student placement, and publications (see Section 2 for information on the latter).

Due to the need to report data separately to each funding agency, analysis of student responses was separated by funding source. The overall response from NSF REU-site grant funded students was extremely favorable. One 2005 student commented, "It was the finest summer of my life."

Table 1 shows the average responses of all 36 NSF REU-site grant funded undergraduates to some of the evaluation questions, on a scale of 1 to 5, with 5 being "strongly agree" and 1 being "strongly disagree."

TABLE 1. Average responses of 2004-2006 NSF REU-site grant undergraduate students

Statement	2006	2005	2004
I learned a lot from my project.	4.5	4.3	4.5
I enjoyed the REU.	4.7	4.4	4.5
I am seriously considering going to graduate school in mathematics.	4.1	4.6	3.9

We are still in the process of gathering data from the students concerning their placement after completing their undergraduate educations. Table 2 contains the information gathered so far.

TABLE 2. Fall 2006 plans of 2004-2005 NSF REU-site grant undergraduate students

Outcome	2005	2004
PhD math, applied math or statistics	3	2
Ph. D. other NSF discipline	1	1
Ph. D. other discipline	0	1
M.S. math sciences	0	1
Undergraduate	5	3
Other	2	0
No response	1	4

The 36 NSF REU-site grant funded students included 13 women. Of the five students from the first two years who are currently enrolled in doctoral programs in the mathematical sciences, four are female.

While complete data from Alliance students is not available, it is known that of the four Alliance students who participated in the 2005 ISU Math REU, one is currently enrolled in a mathematics doctoral program and one in an engineering doctoral program.

DEPARTMENT OF MATHEMATICS, IOWA STATE UNIVERSITY, AMES, IA 50011
E-mail address: LHogben@iastate.edu

Lafayette College's REU

Gary Gordon

Lafayette has been an NSF REU site in math since the early 1990s, and is currently running a program that will be funded through the summer of 2010. Each year, we fund 10 - 12 students who are partitioned into three research teams for an eight-week research experience. Eight of the students are funded by the NSF from a national search, and the remaining students are undergraduates at Lafayette who are funded through the college's EXCEL program. Typically, students work in groups of 2, 3 or 4 with a single Lafayette faculty member (although we occasionally select married faculty mentors). Here are some of the specifics for our program:

- How do the mentors decide whom to choose for their programs?

We get between 150 and 200 applications for eight positions. Student applications contains three important pieces of information: a transcript, an essay from the student and 2 letters of recommendation. These three pieces all contribute, but different mentors weigh them differently. Some mentors look for students who might not have a research opportunity at their home institution, and we are all interested in students with lots of enthusiasm for the project who have written a positive essay. Grades are important, but so is the breadth and depth of the courses taken. The essay should not include statements like:

> I have trouble working with other students.
> You will be very happy to have me in your program.
> I have been fascinated with research mathematics since 7th grade, and have developed my own approach to the Riemann Hypothesis.
> For the reasons I have outlined in this essay, I would love to spend the summer at _____ studying _____ with _____.

(with the blanks appearing in the letter).

Finally, good letters that speak to the student's personality are invaluable. Letters that read

> He got an A- in Real Analysis and asked several probing questions. I believe he would really benefit from a research experience at your institution.

are of limited value. Students from traditionally underrepresented groups are especially encouraged to apply. Gender balance is important to us, and we have had some success attracting minorities to the program.

Received by the editor January 20, 2007.

In sum, the criteria for selection includes the student's academic credentials (transcripts and letters of recommendation), the level of interest the student displays in the project, and the likely benefit the REU experience will have on the student. We consider this last condition one of our key criteria for admission.

- How do mentors decide on project topics?

This is hard. Different mentors use different strategies, including combing the literature for interesting problems. In general, we shy away from choosing open problems that are well known, for obvious reasons. Most undergraduates will have a research experience that is long on frustration and short on success in such a situation. Instead, we try to find areas of our own research that can be smallerized to fit an eight-week time frame and an undergraduate background. For example, a general question on matroids might be simplified to graphs, or a question about geometric groups could be analyzed for a specific class of groups. In the best of situations, this can be very fruitful, but it requires the mentor to be flexible during the summer, willing to shift the problem significantly.

- How do students get selected for individual projects?

On our web site, we advertise three projects, complete with abstracts and pretty pictures. Students read the project descriptions and apply to whichever one they like. This happens in the spring, well before they arrive on campus (usually during the first full week in June). Although individual mentors occasionally share applications with each other, the typical application is only read thoroughly by the mentor.

- What should students gain from an REU experience?

We want to give the students a sense of what it means to undertake mathematical research in a meaningful way. This will be accomplished over the summer by fostering interaction among the students within a team, between the students and the faculty advisors and between the different teams. This environment helps to build a mathematical community in which the students are integral members. In each of our previous programs, we have seen students experience the excitement of discovery, the frustration of trying to solve a difficult problem, the hope engendered from the infusion of new ideas and the satisfaction that is only obtained from a deep understanding of a problem that was mysterious or intractable a few weeks earlier. While we strongly believe that the creation of this mathematical community is the central goal of the REU, the publication of papers resulting from these projects can be an extremely valuable component of the program. Final publication gives the students a finished product of which they can be proud. We have had a strong record of publications resulting from previous REUs; in fact, nearly all of our projects have resulted in one or more publications, as is noted in Publications from REU Programs at Lafayette College. A complete list can be found at http://ww2.lafayette.edu/~math/reu/Pubs.html.

- Do students give talks?

Well, this wouldn't be a bullet point if they didn't. In addition to sharing their ideas with other members of their team, the students will present their work to the entire group at regularly scheduled seminars. Each group gives three presentations during the summer; one presentation introducing the problem to the audience (during week 1 or 2), a second presentation midway through the summer to update the other groups on their progress (during week 4 or 5), and a final wrap-up presentation (during the last week). At each meeting, one research team will present material. These meetings will have several benefits. The presentations help prepare the students for speaking at conferences. These meetings also help foster a sense of camaraderie among the different teams. Students who are not presenting can add their insight to the problems of the team that is presenting, and listening to presentations by other teams may inspire a non-presenting team with its own research. And finally, as a simple matter of fact, students often make significant progress under the pressure of preparing their talks.

- What other mathematical activities do we have for our lucky REUers?

There are activities planned throughout the summer. The main activity is a trip to a research conference. This is a tricky enterprise; the conference should meet all of the following criteria:

- Conference must take place during the eight-week REU program (preferably within weeks 2-6);
- Conference should be accessible to undergraduates;
- Conference should last no longer than three or four days; and
- Conference must be within a one-day drive from Easton.

The advantages of attending a research conference are manifold; students experience research at the highest level, not watered down for undergraduates. In addition, they get a chance to interact with a large group of professional mathematicians this interaction has a valuable mentoring component. Finally, the students can bond as a group in the conference setting. Last summer we went to the CBMS conference at North Carolina State University for a series of lectures on Cluster Algebras.

In addition, we frequently attend a topology conference in at Lehigh University and we also visit another REU site (usually Rutgers University). Each of these activities takes one or two days.

Finally, we support travel to a conference after the summer, usually the annual joint math meetings in January, where we encourage the students to speak and/or give poster presentations on their work from the previous summer.

- What about outside speakers?

We bring 2 or 3 speakers to campus each summer to give an undergraduate research talk. Last summer we were fortunate to have Frank Morgan, Maria Chudnovsky and Eric Gottlieb speak to the group. We also have Lafayette math faculty speak to the group on occasion. In the past, we have taken advantage of our location to attract speakers from Penn (Herb Wilf), Princeton (John Conway and Maria Chudnovsky), Binghamton (Matthias Beck, Thomas Zaslavsky), Cornell (Bob Connelly) and Williams college (Colin Adams and Frank Morgan).

- What about fun and games?

Students are spending their summers here we want to make sure they have fun and plan lots of activities that have no mathematical goal. Last summer we took a trip to Philadelphia (to see the Franklin Institute and do some sightseeing), we took a day-long hike to Sunfish Pond, a glacial lake in the Pocono mountains, we took a day trip to Ringing Rocks Park, and we went rock climbing at Lafayette's indoor rock wall. We've seen baseball games in Philadelphia, gone rafting down the Delaware River, seen a play in Manhattan.

On a more regular basis, we have several get-togethers at faculty houses (opening night cookout, cookouts for various guest speakers, weekly dvd nights, etc.) and the students organize activities themselves, too. In the past, they have taken trips to New York, organized a weekly bowling night, and many more similar activities. Student reaction (see below) indicates this is a strength of the program.

- What about the logistics?

Students get a stipend of \$3200 for their work during the summer and free housing in a new dorm at Lafayette. They get free access to Lafayette's newly renovated library and fantastic gym (with a challenging rock wall), and they have many food contacts provided for by the REU (when at conferences, at cookouts, when entertaining guest speakers). They also get some travel money for one post-summer math conference.

- How can we tell if we're doing a good job?

We run a survey immediately after the program and another one about 9 months after the program. Here are some student responses to past surveys.

(1) Did the program meet your expectations mathematically? How cohesive was your group?

- The math was exactly what I'd hoped it would be, with the right amount of guidance from our leader (not so little that I had no clue what was going on, not so much that I knew what was going on and didn't like it) and the right range of difficulty (since the problem was a bit open-ended, I could pick and choose an area that was interesting to me and also just hard enough to be fun). My group was very cohesive, in that we all liked each other and were interested in each other's work.

- The program met my expectations. I enjoyed working with my group. We worked together and individually and it was good meeting as a group every once in awhile to find out how everyone else was doing. I learned a lot about doing math research. The Rutgers trip was a good idea and I really liked the speakers. It was also good to present to other groups and hear updates about what they were doing.

- My group met all of my expectations mathematically. The math was challenging, but it was fulfilling when we came up with results that worked. Our group was extremely cohesive. I felt comfortable with every person in my group, and we all worked together well, allowing us to challenge one another and help each other in our research.

- Yes. The program definitely met my expectations mathematically. I had talked to some other people who had done REUs that failed to meet these expectations, so I paid special attention when applying to attempt to insure that I would have a good experience in that regard, and I did.

(1) How did the Lafayette REU influence your future plans?

- The Lafayette REU made it clear to me that I would enjoy life as a theoretical mathematician. While I look forward to getting a different experience this summer in another field, my experience at the Lafayette REU will be invaluable in figuring out if theoretical math is what I want to do with my life.

- The REU program definitely made me realize that I should work towards a PhD and stay in the field of academia. I had other 9-5 type jobs in previous summers and could not stand them. I enjoyed the fact that we were able to set our own hours while working on research and not forced to be working prescribed hours. I also found my job experience in previous summers mind-numbing, while research kept me stimulated and excited about meeting with my group. It definitely made me begin to consider grad school and ultimately led me to my decision to pursue a PhD.

- I became more confident that I can actually do research, and that I actually like doing research. I also got a whole lot out of the conferences I could attend because of the REU (meeting cool people, hearing about new math, whitewater rafting). So I guess it influenced my future plans by making them the same as they always were (I've been planning to be a mathematician for a while now), but even more appealing.

- The Lafayette REU was a huge influence on my plans. I learned of the Penn State MASS program and of the Budapest Semester program. I applied and was accepted to both and made the decision to attend the Budapest Program. The program also helped me gain exposure to thinking like a mathematician. I was able to attend conferences about topics of advanced mathematics. I could meet many wonderful people in the field of mathematics. I am still in contact with many people who I worked with during the summer. I could practice presenting my work. The whole experience of the REU including the Boston trip and the Joint conference helped me realize I want to do mathematics as my career. The faculty members at Lafayette helped steer me further into mathematics. I cannot thank everyone at Lafayette for helping me so much.

- The Lafayette REU introduced me to the vast world of research in mathematics and the impact that pursuing such research could have on others, which is why I am convinced that pursuing my doctorate in Mathematics education will allow me to make a difference in the lives of future students with a background similar to my own.

- Lafayette was basically my first experience with mathematical research, and the fact that I enjoyed it so much has led me to go to grad school and get my Ph.D. The Lafayette REU was an amazing introduction to research. My best memories are sitting around doing math and really making a breakthrough, finally getting a result that we had been trying to get for most of the summer.

A somewhat facetious summary of our experiences with the REU program can be found at http://ww2.lafayette.edu/~math/reu/REUpaper.htm in an article published in the November, 2004 issue of Math Horizons.

- Do students publish their work?

Another silly question. Of course they do. A complete list can be found on our web site.

- So is there some sort of conclusion, or what?

Running an REU program is an enormous amount of work, from lining up good projects and project leaders to planning all of the mathematical and non-mathematical activities for the summer. We do all of the logistical work ourselves; Lafayette does not have a graduate program, so there are no graduate students to help out. Being flexible is important (last summer we had two trips to the emergency room, one after the hike to Sunfish Pond and another after an intense capture-the-flag game; both students made complete and quick recoveries).

Our hope is that the students look back on their time at Lafayette with some satisfaction and a real sense of what it means to do research in mathematics. If we have influenced the students in a positive way, then we can take some pride in our efforts.

DEPT. OF MATHEMATICS, LAFAYETTE COLLEGE, EASTON, PA 18042-1781
E-mail address: gordong@lafayette.edu

LSU REU: Graphs, Knots & Dessins in Topology, Number Theory & Geometry

Neal W. Stoltzfus, Robert V. Perlis, J. William Hoffman

1. Overview

The LSU Math REU has run continuously since 1993, with funding from both the Louisiana Board of Regents and from the National Science Foundation. The targeted participants will normally be applying for their first REU experience and will have done well in an undergraduate abstract algebra course. A diversely balanced group of participants, especially in terms of gender balance and type of home institution (university versus four-year college) is selected from the national pool of applicants.

Our mathematical focus *Graphs, Knots & Dessins* lies in the overlap of Geometry, Topology, and Number Theory where many fertile interconnections can be explored. We seek out problems of current research interest that can be approached with the tools of group actions, group presentations, group representations, and various types of counting functions (such as a zeta function) where explicit computations can be made by undergraduates. We make a special effort to choose sets of problems that center around a common technique or theme so that students can interact with each other.

The Mathematics REU at LSU has these goals: to involve the next generation of young people in high level mathematical research in areas that are currently active, and to foster clear communication of mathematical results.

The basic structure of the REU is this: After a week of lectures by the mentors providing background, motivation, examples, general research directions, and specific entry points, each student selects a project on which to work. A cluster is formed consisting of a faculty mentor, a graduate assistant, and four students working on related problems. Each student then works primarily within the cluster and directly with the mentor for the rest of the summer. However, the entire group (twelve students and faculty and graduate mentors) meet each afternoon for mathematical conversations, refreshments and announcements relevant to the entire group. The clusters help develop strong collegial student-mentor and student-student relations, and the daily larger meetings help build a group identity.

On week-ends there are parties for the students at faculty homes, or field trips to nearby sites of interest. When the participants leave the program they should view

Received by the editor November 30, 2006.

Louisiana and LSU as intellectually stimulating places, they will have made some life-long friendships, and they will have made meaningful progress on challenging problems. As a result, many will decide to pursue graduate study in mathematics.

The students give three oral presentations of their work: a ten-minute statement of their chosen problem in the second week, a fifteen-minute progress report in the middle, and a twenty-minute final project summary at the end. Each student turns in a written project report, prepared in LaTeX in the form of a research paper.

2. Nature of Student Activity

The student projects center around problems in graphs, knots and *dessins d'enfants* in topology, number theory and geometry. This quickly leads to sophisticated mathematics. Nonetheless there are entry points suitable for getting students to work successfully in these areas. Often this involves computation and judicious construction of examples.

Significant features of the program include:

- The PIs lecture during the first week on interesting project directions, focusing on accessible entry points. Sometimes a professor from outside LSU is invited to present material. We believe that it is important that students have some idea of the larger context in which their chosen problem sits.
- The students select their own problems on which to work within the areas proposed in the lectures. In the next days, each student gives a short lecture explaining the problem in simple terms. This process gives the student a sense of commitment, ownership, and motivation.
- Student research starts with examples and computation. By the end of the program students formulate conjectures and prove theorems.
- Students work within clusters. They have individual projects, but interact with other participants. There are clusters of related projects and the students in a cluster are encouraged to help each other.
- Interaction with graduate students. This component has been possible through additional funding from the Louisiana Board of Regents.
- Communication of the results through oral and written presentation. Final projects are written in LaTeX and put on the REU website. The students give three oral presentations: a short project description at the beginning, a midterm progress report and a final report.
- Publication in undergraduate journals, and in regular mathematics research journals when appropriate. We encourage and financially support former REU participants to attend AMS poster sessions and other math conferences where the students can present their research.

We have also featured lectures by visiting experts. In 2002 Kiyoshi Igusa came from Brandeis to give a series of lectures on *pictures* related to free resolutions of modules over certain kinds of groups, and in 2006 Sergei Chmutov came from Ohio State to give lectures on knot and link invariants. In the weeks after Chmutov's lectures there were two videoconferences between the Ohio State University VIGRE and the LSU Math REU with talks by students at each institution.

3. Research Environment

Our projects center around our theme of *Graphs, Knots & Dessins in Topology, Number Theory & Geometry*. This continues the general direction of our past summer REUs (low dimensional topology, graphs, zeta functions), but with a new emphasis, that of *dessin*, a combinatorial objects which, informally, is a graph together with a cyclic ordering of the edges meeting at a vertex. This additional information provides an embedding of the graph into an oriented surface. There are other names for this concept (ribbon graph, fat graph, *dessins d'enfants*) and they occur in multiple, diverse contexts. We will use the term *dessin*.

Twenty years ago Grothendieck (who coined the term *dessins d'enfants*) discovered unexpected connections to algebraic geometry and Galois theory. In fact, the absolute Galois group of the rational field $\mathbb{Q}$ acts faithfully on the set of dessins, but this action is far from understood. The action of Galois groups on the fundamental group of $\mathbf{P}^1 - \{0, 1, \infty\}$ has led to profound studies and conjectures (by Deligne, Drinfeld, Goncharov, Ihara and others), some of which can be attacked by advanced undergraduates. This is an exciting domain whose problems will not be exhausted for many years.

What makes dessins especially suitable for our REU is that there are entry points that are completely elementary. Start with an ordered pair (σ_0, σ_1) of permutations in a symmetric group S_n. This is all the information required for the specification of a *bipartite dessin*. The set of edges is $E = \{1, 2, \ldots, n\}$, the bipartition of the vertices is given by the orbits of the two permutations and the endpoints of an edge are the orbits in which it lies. The faces of the dessin are the orbits of σ_2 determined by the identity: $\sigma_0 \sigma_1 \sigma_2 = \text{Id}$. This determines a bipartite graph (which we picture with hollow vertices for the orbits of σ_0 and dark vertices for the orbits of σ_1. Attaching k-gons to the edge cycle determined by the k-cycles of σ_2, we construct an (isotopy class of an) embedding of the graph into a surface. The genus of the surface is determined by the usual Euler characteristic formula, $2 - 2g = v - e + f$, where v is the total number of vertices; e, the number of edges and f, the number of faces, respectively.

For an example, let $n = 4$ and consider the genus 0 dessin $\mathbb{D}_1 = (\sigma_0 = (1234), \sigma_1 = (12)(34))$ and the genus one dessin $\mathbb{D}_2 = (\sigma_0 = (1234), \sigma_1 = (13)(24))$. The dessins are:

FIGURE 1. Two Dessins

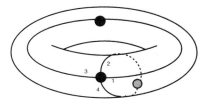

The following theorem concerning bipartite dessins summarizes work of Grothendieck and Belyi:

THEOREM (Grothendieck-Belyi). *The following are in a canonical one to one correspondence:*

(1) *Isomorphism classes of ordered pairs (σ_0, σ_1) where the group $\langle \sigma_0, \sigma_1 \rangle$ is a transitive subgroup of S_n.*

(2) *Connected graphs Γ with a bicoloring of its vertices, and for each vertex of the graph, a fixed cyclic (counterclockwise) order of the edges incident on that vertex.*

(3) *Isotopy classes of (X, Γ) where X is a compact oriented surface, $\Gamma \subset X$ is a connected graph, with a bicoloring of its vertices, and such that $X - \Gamma$ is a union of contractible 2-cells.*

(4) *Isomorphism classes of (X, β) where X is a nonsingular irreducible projective algebraic curve defined over an algebraic number field, and $\beta : X \to \mathbf{P}^1$ is a morphism, also defined over a number field, such that β ramifies only over $0, 1, \infty \in \mathbf{P}^1$.*

Moreover, every nonsingular irreducible projective algebraic curve defined over a number field admits a morphism β as in the fourth item above.

The following gives a selection of problems related to dessins suitable for REU students.

Gassmann triples: A Gassmann triple (G, H, H') of groups together with a choice of two elements $g_0, g_1 \in G$ can be used to create two different pairs of graphs: Their dessins, and their Cayley-Schreier graphs. For the dessins, the coset spaces $G/H, G/H'$ label edges while for the Cayley-Schreier graphs the cosets spaces label vertices. Find an algorithm to relate the dessins to the Cayley-Schreier graphs. Try to use that algorithm to relate the Ihara zeta function of the graph underlying a dessin to the Ihara zeta function of the Cayley-Schreier graph.

Discrete Jacobians of Dessins: For each dessin $\mathbb{D}$, Shabat and Voevodsky, [**SV90**], define an abelian variety $J_{\mathbb{D}}$ which we will call the *discrete Jacobian* of the dessin. Their fundamental result is that, under a subdivision process, these Jacobians converge in a suitable sense to the usual Jacobian of the Riemann surface X in which the dessin is embedded, [**SV89**], [**Cav99**]. In contrast to the Jacobian of a Riemann surface, whose definition involves computation of periods of integrals, the discrete Jacobians of dessins have an elementary combinatorial definition, and so far have not been studied in detail. Investigate these discrete Jacobians.

Bollabas-Riordan-Tutte polynomial in link theory: Some of these problems involve an important polynomial invariant of dessins, called the BRT polynomial. It was defined by Bollobas and Riordan [**BR02**] in a manner generalizing the spanning tree construction of the Tutte polynomial of a graph. Very recently, new connections between the theory of dessins and knot/link theory was discovered by two groups Chmutov & Pak [**CP**], and Dasbach, Futer, Kalfagianni, Lin & Stoltzfus [**DFK**⁺], by relating the BRT polynomial of certain dessins constructed from link diagrams to the Jones polynomial of the link. Find a formula for the BRT polynomial of the 2-parallel (more generally: n-parallel) of a knot in terms of the original knot. In particular, develop an understanding of this construction from the relationships among the permutations of the associated dessins.

Past graduates of the LSU REU program continue to make mathematical news. Shelley Harvey, now Assistant Professor at Rice University has just received a Sloan Fellowship. Her description of her experience at the LSU (and Cornell) REU was published in the Notices of the AMS [**Har98**]. The semester before participating in our REU she was majoring in engineering. A second graduate, Dorothy Buck, of the Imperial College, London, will give an MAA Invited Address at MathFest this year, August 10-12, in Knoxville, TN. She is an expert in the developing field of mathematical biology. Her talk is entitled "The Circle (and Knot and Link) of Life: How Topology Untangles Knotty DNA Questions".

We also had three of our past participants, Yaim Cooper (2005), Michele Lastrina (2005) and Stacey Goff (2003), participate in the 2006 Women and Mathematics Program of the Institute for Advanced Study and Princeton University entitled *Zeta functions all the way*. Michele and Yaim also presented posters in the AMS 2006 National meeting in San Antonio. Yaim is preparing a paper, *Properties determined by the Ihara zeta function of a graph*, for publication. Selected additional publications are listed in the references.

4. Student Recruitment and Selection

Participants are recruited nationally, and are drawn from all major regions of the country. Over the years we have worked to recruit students from underrepresented groups into our program. In particular we have been successful in achieving gender balance. We also strive to provide places for students from four-year colleges in addition to those we recruit from major universities.

From the participants during the years 2001-2004, 33 (out of 48) subsequently attended graduate school. For the same years, 18 of the participants were women and 16 were from four-year college programs.

The initial contact with applicants has been through recommendation from previous participants, announcements in professional notices, or via the Internet. Our web page (`http://www.math.lsu.edu/REU.html`) includes detailed information on the program and provides an email address and electronic application forms. We also request that links to our web page are present at mathematics portal sites throughout the Internet.

We receive between one and two hundred email inquiries about our program each year. Each inquiry generates an electronic information pack that includes a description of the program. We request that applicants supply an academic record supplemented by descriptions of advanced courses they have taken, two letters of reference, and a personal statement explaining their interest in the REU.

Our website is set up so that applications can be made completely electronically, except for official transcripts. Reference letters can be emailed. We select students based on evidence of their ability in mathematics, willingness to commit to completing a project, and ability to interact well with others. We normally require a participant in our program to have successfully completed a course in abstract algebra; this is necessary for working at the mathematical level of our program. Many students come to the program with advanced courses such as algebraic topology, or algebraic geometry. Finally, we give preference to students seeking their first REU experience.

References

[BR02] Béla Bollobás and Oliver Riordan, *A polynomial of graphs on surfaces*, Math. Ann. **323** (2002), no. 1, 81–96. MR **MR1906909 (2003b:**05052)

[Cav99] Renzo Cavalieri, *Curve algebriche definite su* $\mathbb{Q}$, Master's thesis, Universit'a degli studi di Milano, 1999, http://www.math.lsa.umich.edu/~ crenzo/professional.html.

[CP] Sergei Chmutov and Igor Pak, *The Kauffman bracket and the Bollobas-Riordan polynomial of ribbon graphs*, arXiv:math.GT/0404475.

[Cza05] Debra Czarneski, *Zeta functions of finite graphs*, Ph.D. thesis, Louisiana State University, 2005, http://etd.lsu.edu/docs/available/etd-07072005-121013/.

[DFK+]   Oliver T. Dasbach, David Futer, Efstratia Kalfagianni, Xiao-Song Lin, and Neal W. Stoltzfus, *The Jones polynomial and dessins d'enfant*, arXiv:math.GT/0605571.

[GS]   Josh Genauer and Neal W. Stoltzfus, *Explicit diagonalization of the Markov form on the Temperley-Lieb algebra*, arXiv:math.QA/0511003, accepted for publication in Math. Proc. Cambridge Phil. Soc.

[Har98] Andrea Harvey, Shelley & Ritter, *A research experience for undergraduates*, Notices of the American Mathematical Society (1998), no. 2, 267–68.

[Kha06] Zuhair Khandker, *Defining a zeta function for cell products of graphs*, R. H. U.G. Math. Journal **7** (2006), no. 1, http://www.rose-hulman.edu/mathjournal/.

[Pra06] Neeraj Pradhan, *Hyperplane arrangements and the Bernstein-Gelfand-Gelfand correspondence*, CUSJ **1** (2006), no. 1, 28–34, http://cusj.columbia.edu/.

[SV89]   G. B. Shabat and V. A. Voevodsky, *Piecewise Euclidean approximations of Jacobians of algebraic curves*, preprint, 1989.

[SV90]   _____, *Drawing curves over number fields*, The Grothendieck Festschrift, Vol. III, Progr. Math., vol. 88, Birkhäuser Boston, Boston, MA, 1990, pp. 199–227. MR **MR1106916 (92f:11083)**

DEPARTMENT OF MATHEMATICS, LOUISIANA STATE UNIVERSITY, BATON ROUGE, LA 70803-4918

E-mail address: stoltz@math.lsu.edu, perlis@math.lsu.edu, hoffman@math.lsu.edu

Mount Holyoke College Mathematics Summer Research Institute

Margaret M. Robinson

1. Introduction

Mount Holyoke's undergraduate summer research site for undergraduates was established in 1988 and has run for all but two summers since then. It is an eight-week program that consists of two research groups of five or six students each. Each group works with a faculty member on a significant mathematical problem or small set of related problems. Computation and examples play important roles in these problems and give the students a way to get started. The problems are never so straightforward that computation alone suffices. Instead they require a constant interplay between computation and mathematical theory. The role of examples to illuminate theoretical concepts is a main focus of the program. Each student finishes the program with a carefully written preprint of the summer's work. We have always published our students' preprints on our web site, worked to help students do undergraduate honor's theses extending REU results, and published in regular mathematics research journals.

Because of the traditions and priorities of Mount Holyoke, at least half of the participating students have been, and will continue to be, women. We draw students from all parts of the country and from all different sorts of undergraduate institutions. We make sure that a good portion of our students come from institutions without summer research opportunities.

Mount Holyoke College is the oldest undergraduate women's college in the country. Although small (2100 students, 200 faculty), it has a history of preparing women for graduate school in the sciences. In 1976, A. Tidball and V. Kistiakowsky wrote in "Baccalaureate origins of American scientists and scholars" (*Science* , Aug. 20 (1976)) that Mount Holyoke College had produced a larger number of women baccalaureates who eventually received doctorates in science and engineering than any other American institution in that century. A key factor in this record has been the science faculty's success in engaging undergraduates in joint research projects. As further evidence of institutional commitment, the Mount Holyoke Dean of Faculty's office and the mathematics department have provided and will continue to provide substantial matching funds for the program. These funds are used for

Received by the editor October 10, 2006.

student travel, for student housing in Mount Holyoke dorm rooms, for the visiting seminar speakers' honoraria, and for celebratory program meals and afternoon coffee breaks.

Primary documentation on our activities can be found on our web site at www.mtholyoke.edu/acad/math/reu

2. General nature of student activities and the research environment

The program runs for eight weeks, starting in early June. Each student receives $3,000 for the eight weeks of work. Their housing is paid for by the program but they must buy and cook their own food, although the program does arrange at least one dinner (after the student seminar) and two lunches (one with invited speakers and one after the weekly reporting seminar) each week. We try to feed the students as much as possible. The student's all live in the same Mount Holyoke dorm with other Mount Holyoke summer research students from different disciplines. They share a kitchen with the other research students. In the mathematics department, each research group is assigned a large room equipped with a conference table, a blackboard, several comfortable chairs, desks, a library of relevant texts and papers, and dual-boot Windows/Linux computers. (Apple computers are available as well.) Other rooms are available for study and quiet. A refrigerator, microwave and coffee-making facilities are available in the department office. The entire area is air-conditioned.

Each group begins the morning by meeting with the faculty advisor to plan the day's activity. A group's daily schedule might begin with a presentation by the faculty member of new material, a presentation by students of their own progress, a discussion summarizing what has been done, or a restatement of the project's short- or long-term goals. The faculty advisor remains in touch throughout the day. The day ends with afternoon tea in the common room.

Once a week each group gives a formal progress report. The inexperience of undergraduates is never more painfully apparent than in their first presentations. We have learned over the last few years how valuable it is for them to have repeated opportunities to say what they are doing, and in the process to clarify for themselves and their friends in the other groups what their problem is about and how they approach it. In the course of the summer everyone speaks regularly and each group becomes familiar with the other group's problems.

There are also visiting speakers, most of whom are paid for with Mount Holyoke funds. In the summer of 2005, for example, there were six such visitors: David Cox on Origami Constructions (Amherst College), Tom Weston on Fermat's Last Theorem (University of Massachusetts at Amherst), Jean Steiner on Geometric Analysis (then a Post-doc at NYU), Thomas Wright on the ABC Conjecture (Graduate student at Johns Hopkins, REU 2002), Seth Sullivant on Algebraic Statistics (Harvard University), and Jason Starr on Diophantine Equations (MIT). The visitors do not just give a talk, but also spend the day with the students discussing the projects, graduate school, and mathematics in general. Some summers the visitors will bring their own summer research students with them.

The students also run and speak in their own weekly seminar series with a pizza dinner afterwards. One student volunteers to run the seminar and that student lines up the others to give talks on some unusual topic they have learned about from courses, independent work, or other summer REUs. In 2005 we had six such talks and some of the topics were: Tilings of the plane, Ramsay numbers, Galois theory, the fundamental group and the universal covering space, representation theory, and singular homotopy.

Visits with other undergraduate research sites are often arranged during the course of the summer and almost every summer the students arrange to borrow the Mount Holyoke van for a two or three day trip to visit graduate schools. They have found that visiting graduate schools in a group is very valuable. In the past the whole REU has visited other REUs like the groups at Williams, Worcester Polytechnic Institute, Amherst College, the University of Massachusetts, and Boston University. In 2006, for example, the whole REU attended and spoke at the Young Investigator Conference at Ohio State in August. We encourage the students to present their results in the undergraduate sessions at mathematics meetings during the following year. We ask for a modest amount in our NSF grant to provide travel support so that the students can attend these meetings as well as travel during the summer to visit other REU groups. Since our travel funds are very limited, we ask the students' home institution to pay travel to the January conference and use our NSF funds only when the home institution does not have the funds available.

Five students in a group seems to work best. This size allows for diverse interactions and division of labor, yet is small enough so that no one is lost. All of our faculty except for one new faculty member have been involved in the program before. In fact most of us have led a group five or six times.

Each summer the administrative tasks are divided among the instructors. These are what one might expect: advertising the program, selecting students, finding extra funds, arranging housing, preparing the work areas and computer facilities, arranging payment of student stipends with the financial office, finding outside speakers, arranging exchanges with other REU sites, planning afternoon tea, planning social events, and so forth. Our administrative assistant organizes the incoming applications each summer (a nontrivial job, since in 2006 we had about two hundred applications for our 10 spots).

3. Student recruitment and selection

We are generally looking for students between their junior and senior year in college; we expect them to have taken the calculus sequence, linear algebra, and at least two courses beyond these such as abstract algebra and real analysis. Often one faculty member will have a more specific course requirement in mind for his or her group. The program is advertised in January. In the past, we have sent flyers to all New England college mathematics departments, and also advertised on various email listings. Recently we have not done mailings; our long history of activity, our web site, and the NSF and AMS listings continue to draw many applicants.

The applicants are selected based on the courses they have taken, their grades, recommendations from faculty members, and the applicant's statement of interest. Each faculty advisor chooses his or her own work group in consultation with the other advisors. We try to form each group so that it has a variety of talents. In particular, we make sure that each group has one or more students with extensive computer experience.

Lastly, and very important, we try to balance the number of men and women in each group (professor included). Most summers we have a Mount Holyoke student in the program but almost always she is an international student who is funded with Mount Holyoke summer fellowship money. We encourage our students who are U.S citizens to participate in REUs at other institutions, for we consider the breadth of experience excellent for them.

4. Project evaluation

The evaluation part of our program has several parts to it. First, we collect questionnaires from each participant at the end of the program. Second, we contact them one year after the REU has ended to find out how their summer experience affected their career choice. Third, we contact them after three years to ask the same questions. Janice Gifford, in our department, is a Statistician whose primary area of expertise is educational measurement and she designed the questionnaires we use. In addition, one member of our department is in charge of the evaluation and this person writes an evaluative summary of the data from each questionnaire we distribute. A copy of our Program Evaluation is on the MHC REU website at http: \\ www.mtholyoke.edu \ acad \ math \ reu \sample_eval.pdf

All but the last four questions questions have a rating attached to them and we report the number of students who give each response along with the written answers to the last four more personal questions in our annual reports. We also investigate and respond to each negative response. We consider our program very successful to quote from the 2005 Summary: "Notably, the mean judgment of their improvement in their ability to tackle research with confidence was 3.8 [out of 4], to organize and write mathematical papers was 3.9 [out of 4], and to organize and give mathematical talks was 4.0 [out of 4]." We are happy to share any of this data and the details on our 1-year and 3-year questionnaires but, for brevity, have not included them in this paper. We use our data to make improvements in the program. In fact many of our innovations like the student seminar series and the trip to graduate schools come from student suggestions. The one reappearing suggestion that we have not been able to respond to until now is the suggestion that we air-condition the dorms and/or arrange better kitchen equipment. Although the mathematics department area and all student work areas are air-conditioned, we just had fans in the dorms. We bought our group their own refrigerator, toaster, and microwave and our students often set them up in a separate location from the dorm kitchen. Most years this works fairly well but some years the kitchen can get extremely crowded. However, this is about to change. There is a new dorm currently under construction and it will be air-conditioned. We are also in discussions with other science faculty and the administration about arranging for a summer student food plan. We recently had a much younger summer program on campus and their food plan worked well. The administration is now considering opening this option up to us too.

Our whole department participates in the REU and this is a real strength of our program. However it does mean that responsibility for the program rotates between department members every year. Each research group leader is responsible for maintaining contact with their own students, documenting the summer's work for the annual report, and for updating their own entry on the REU web site after the summer is over, as well as managing the grant finances and the program events during the summer. This decentralized control has made it hard to ensure that everyone remembers what they are expected to do as far as evaluation goes.

Having one person in charge of evaluation has ensured that each of the three pieces gets done every year in time for the annual report.

The faculty advisors of each REU group keep in touch with their students, help them continue the summer work in their honors theses, and advise and recommend them for graduate school and for prizes.

5. MHC REU Papers published since 2000

1) Alexander, Balasubramanian, Martin, Monahan, Pollatsek and Sen, "Ruling Out (160, 54, 18) Difference Sets in Some Nonabelian Groups," Journal of Combinatorial Designs, 8 (2000), no. 4, 221–231. (H. Pollatsk's 1994 REU)

2) G. Cobb and Y. Chen, "An Application of Markov Chain Monte Carlo to Community Ecology, American Mathematical Monthly, 110 (2002), no. 4, 265-288. Available from http: \\ www.mtholyoke.edu \ courses \ gcobb \ REU __MCMC \. (G. Cobb's 2001 REU)

3) P. Kim, L. Stemkoski, and C. Yuen, "Polynomial knots of degree five," MIT Undergraduate Jounal of Mathematics, v. 3 (2001) p. 125-135. (A. Durfee's and D. O'Shea's 2000 REU)

4) D. Meuser and M. Robinson, "Igusa Local Zeta Functions of Elliptic Curves", Mathematics of Computation, 71 (2001), no. 238, 815-823. Student work from 1999 REU is summarized and cited in this paper. Available from http:\\ www.mtholyoke.edu\ robinson\reu\ reu99\ reu99.htm. (M. Robinson's 1999 REU)

5) M. Peterson and Y. Rubinstein, "Turbulence on a Desktop," Computers in Science and Engineering, May/June, 2001, pp. 86-93. Available at http:\\ www.mtholyoke.edu \ mpeterso \reu\ 99\ reu99.html. (M. Peterson's 1999 REU)

6) B.D. Marko and J.M. Riedl, "Igusa local zeta function of the polynomial $f(x) = x_1^m + x_2^m + \cdots + x_n^m$ ", Yokohama Mathematical Journal, 51, no. 2 (2005). (Paper from student honors thesis following M. Robinson's 2002 REU)

MOUNT HOLYOKE COLLEGE, SOUTH HADLEY, MA 01075
E-mail address: robinson@mtholyoke.edu

The Director's Summer Program at the NSA

Tad White

The Director's Summer Program at the National Security Agency is a research experience aimed at the very best undergraduate mathematics majors in the country. Each summer, we invite a number of exceptional students to participate in a 12-week program in which they collaborate with each other and with Agency researchers to solve classified, mission-critical problems. The inaugural DSP in 1990 hosted eight students; it was successful even beyond the high hopes of the mathematicians who organized it, in ways that would not become apparent for years. The program has grown steadily with its success, and we now invite about two dozen students each year.

The DSP is not intended to be a recruiting program, though dozens of alumni have since joined the NSA as permanent employees. Nor is it intended to be an educational program, though the participants learn a great deal about mathematics and life at the NSA. Rather, the DSP was born of the recognition that the health of mathematics at NSA depends on both the health of the external mathematics community, and on a robust connection between NSA and that community. Obviously, as the nation's largest employer of mathematicians, NSA relies upon the mathematics community to provide a technically strong workforce. However, no matter how skilled our people are, we cannot hope to keep up with the research frontiers in all fields of mathematics. Often, advances which seem unrelated to our current work turn out to provide essential clues to the solutions of our most difficult problems. We rely upon the outside community to help us identify and apply these advances, and we can be successful only if the top mathematicians have a deep understanding of NSA's problem set and mathematical culture.

In the late 1980s, the NSA mathematics community explicitly recognized the importance of fostering a close relationship with the nation's academic mathematics community, and undertook a number of initiatives to reinvigorate this relationship. Richard Shaker, then head of mathematics research at NSA, described a few of these initiatives in an address at the 1992 Joint Mathematics Meetings [1]. In 1987, we invited a hundred mathematicians to come hear ten unclassified talks on research being done at the Agency. We set aside a few million dollars annually to support academic research proposals. (While this figure represents a small fraction of the U. S. government's support to pure mathematics research, it amounts to a large portion of our technology budget.) We established a sabbatical program to allow

Received by the editor February 7, 2007.

mathematicians to visit us while retaining their academic affiliation, and another program to enable NSA mathematicians to visit universities or industry. We have also initiated a number of programs which support mathematics education at both the undergraduate and K-12 levels.

The Director's Summer Program is a cornerstone of our strategy to engage the academic mathematics community. While some of the other NSA summer mathematics programs are targeted primarily at recruitment, the aims of the DSP are higher:

- to introduce the future leaders of the U. S. math community to the Agency's mission, and share with them the excitement of working on mathematics problems of national importance,
- to provide a deep understanding of the vital role that mathematics plays in enabling the Agency to tackle a diverse set of technical challenges,
- to encourage bright undergraduates to continue their study of mathematics and pursue careers in the mathematical sciences,
- to advance science at the Agency, thereby setting the stage for future successes, and of course
- to provide solutions to current operational problems.

Despite containing the term "summer" in its name, the Director's Summer Program is the result of a year-round commitment by some of the Agency's top mathematicians, and by a dedicated support staff. What follows is a rundown of a typical year of the DSP.

Autumn. Since participants work on classified problems of operational interest, the application process for the DSP is more complicated than that for an REU at a university. While many REUs have application deadlines as late as April, applications for the DSP must be received by October 15 of the previous year in order to accommodate the lengthy security clearance process. Application is open to any U. S. citizen, and the top U. S. citizen scorers on the Putnam examination receive invitations to apply. The most distinguishing characteristic of a potential DSP participant is not a concentration in any particular field of mathematics, but rather an interest and ability in collaborating to solve hard problems. A certain comfort level with computer programming is also helpful, as is experience with some of the common mathematical computing packages, but these are not absolute requirements.

Each year, the program is coordinated by two or three technical directors. The TDs rotate every few years, but they are always top NSA scientists who take time away from their own work to mentor the DSP students. In late October, the TDs review several hundred applications. The NSA personnel office mails a package, including security forms, to the top ten percent or so of the applicants. Once the package is returned, our Human Resources office arranges an interview, which includes a polygraph exam.

Winter. By the time the students come to Maryland for applicant processing, the leaves have long since fallen off the trees. As a respite from the HR interviews and the polygraph exam, the students spend some time talking about math with the technical directors. Unfortunately, the TDs can't say much about what specific problems the students will be working on; not only are the problems classified, but they won't be determined for months!

Spring. By the time the leaves return in April, the participant list is firming up, and the technical directors begin to settle on a set of problems. Like good thesis advisors, the TDs apply their wisdom and background to select an array of problems balancing risk (difficulty) against reward (payoff). The importance of problem selection is one reason why we ask our most experienced mathematicians to serve as TDs. The problem selection is usually done as late as possible; while this may seem like procrastination, the real purpose is to ensure that the problems are current and important. The opportunity to do mathematics for which the payoff is clear and of great value is one of the unique features of the DSP – it stimulates extremely hard work and dedication from both the students and the Agency staff.

The TDs canvass the entire Agency for problems, and recruit sponsors and subject matter experts for each one. In the early years of the DSP most of the problems focused on cryptography, but the range of problems has since expanded to include such topics as signals analysis, image processing, and algorithms for analyzing large data sets. Some problems are concrete and applied; others are more theoretical and connected to long-term research programs. The problem supporters work closely with the TDs to prepare background material for the students and put together introductory talks.

Summer. The program itself runs for about twelve weeks, starting right after Memorial Day and continuing through mid-August. The Agency facilitates housing, either at a nearby apartment complex or university, so that the students can be co-located to the maximum extent possible. In addition to increasing the social opportunities for the participants, this also helps to solve the problem that most of the students don't have cars.

During the first two weeks of the summer, the students receive a crash course which provides an introduction to NSA mathematics and presents the specific problems to be tackled. The course is tailored to the problem set, so it differs from year to year. The students are not assigned to problems; once the problems have been presented, the students are free to work on whichever they wish. They quickly form themselves into a number of overlapping groups, each participant finding his or her own unique way of contributing to the group effort.

For many of the students who are alumni of various REUs, the highly collaborative environment in the DSP is a new and invigorating experience. The math problems at the Agency are difficult enough that the great advances are typically achieved through cooperation, not only among mathematicians but across disciplines. Indeed, many of our problems defy categorization by discipline – but we need to solve them nonetheless.

During the summer, the students gradually become an integral part of the NSA mathematics community. An annual day-long classified math conference, called Mathfest, and an annual awards banquet, are timed so that the students can attend and learn more about the community, its history, and its luminaries. (In fact, two DSPers were recently invited to give Mathfest lectures on their work from previous summers – both gave great talks, despite having to wait to return to secure spaces to prepare them!) The students take interesting tours, both at NSA and at other agencies in the intelligence community. They participate in office outings and organize their own activities. As the students and Agency researchers become friends and colleagues, they share both the frustration of failed approaches and the elation of sweet successes. The importance and urgency of the problems leads to an even greater emotional commitment among the participants, and an incredible bond often forms.

In the end, not all of the problems are completely solved, though most see partial solutions or progress. In each year, the DSP has contributed at least one or two remarkable solutions. The students spend the last few weeks documenting their work in classified technical papers. In recent years, the students have had the opportunity to present summaries of their work to the Director of NSA in person. In addition, the Director has usually taken the time for a leisurely question-and-answer session with the students. This provides a unique opportunity for participants to learn about the Agency, and truly understand why what they do is so vital to our nation's security.

Once the students have left, the TDs and problem supporters put finishing touches on the research papers from the summer, and compile a comprehensive report. The results of the summer go back to the sponsoring offices, and follow-on research often begins while the seats in the DSP room are still warm. No sooner do the TDs catch their breath in September, than the next "summer" is upon them.

As of this writing, the DSP has just completed its seventeenth incarnation. While some of the over three hundred alumni have joined the Agency, many have gone on to illustrious research careers, and continue their involvement with the intelligence community in other ways. Some work full time at research centers dedicated to solving NSA problems. Some are now academic mathematicians who regularly spend summers with us working on our problems. One former DSPer recently wrote us to recommend one of his students and stated that his own DSP experience was what convinced him to become a professional mathematician. In the past few years, our alumni have collected four Rhodes scholarships, three Marshalls, two Schafer Prizes (and several close finishers) and a Salem Prize. So our technical directors have had some success identifying the future leaders of the community.

Seven years ago, when I agreed to serve as a DSP technical director, I could not have known that the DSP would afford me some of the most rewarding experiences of my career. It is no exaggeration to say that lives have been saved by the successes of the Director's Summer Program. The job satisfaction that comes from working together to contribute directly to the nation's security is difficult to describe to someone who has not experienced it firsthand. But we believe our alumni try, for when we ask applicants where they heard about the DSP, very often the answer is that they heard about it from their friends.

Reference

[1.] R. J. Shaker, The Agency That Came In Out of the Cold, Notices Amer. Math. Soc., 1992, May/June, 39, 1992, 408-411.

National Security Agency
E-mail address: tpwhit1@nsa.gov

REU in Mathematical Biology at Penn State Erie, The Behrend College

Joseph P. Previte, Michael A. Rutter and Scott A. Stevens

1. Development and History of the Program

Mathematical biology is one of the fastest growing fields of mathematics, and many believe that biology will dominate the twenty-first century as physics dominated much of the twentieth century [4, 5]. An NSF sponsored REU Program in Mathematical Biology has been conducted every summer at Penn State Erie since 1998. As a primarily undergraduate institution, Penn State Erie made a conscious effort in the 1990's to establish a research group in mathematical biology, an area of mathematics that lends itself to involvement of undergraduates in research. The REU program was started by Drs. Joseph Paullet, Richard Bertram (now at Florida State) and J. Carl Panetta (now at St. Jude Childrens Research Hospital and University of Tennessee). The original intent of the program was to promote undergraduate interest in the area of mathematical biology as well as equipping the participants with the tools needed to pursue research in this field. The program was one of the first REU programs dedicated to the emerging field of mathematical biology.

In 1998, the program consisted of six students participating in a four week program, with each of the participating faculty taking one week to describe his particular research in mathematical biology. One final week was allotted for independent research. The research topics included modelling electrical activity in the pancreas, studying spiral waves on the heart (responsible for heart arrhythmias), and modelling cancer response to various chemical treatments. In the beginning of the program, more emphasis was placed on exposing students to current research in mathematical biology rather than on actively involving them in research. Part of the reason for this was an initial skepticism that undergraduates could be actively involved in meaningful research beyond "scratching the surface" of a project.

As the program developed, the emphasis began to shift from exposing the students to mathematical biology toward involving them directly in active research. The length of the program expanded to five, then six weeks. In keeping with this shift, the structure of the program was inverted, with one week of faculty instruction followed by five weeks of independent research. The faculty mentors are now Drs. Joseph Previte, Michael Rutter, and Scott Stevens. Current research topics

Received by the editor 12/28/06.

include population dynamics and modelling, biomedical fluid dynamics, genomics, and bioinformatics.

2. Philosophy and Goals of the Program

A primary goal of the program since its inception is to expose the participants to mathematical biology and to the mathematical tools that are frequently used in this field. From the beginning, the creators of the program felt that mathematical biology lends itself to involvement of undergraduates in research. Indeed, the only prerequisite for participation in our program is that the student complete a course in ordinary differential equations. The number of students involved in the program is intentionally small to allow students to work closely with one another and with the faculty mentors.

Another goal of the program is to give our students an honest experience of mathematical research in this fast growing field. The hope for each project is to produce research at a level that contributes to the field and is publishable in a peer-reviewed journal in mathematical biology. Although this objective is not always met, students receive a true taste of the research endeavor. Regardless of the outcome of their work during the program, all students present their findings in a mini conference setting at the close of the program and are encouraged to speak about their findings at other undergraduate research venues. Students have also used their REU work to satisfy senior projects and honors theses at their home institutions. Those students who are interested in continuing the research that was started during the six week program remain in contact with their REU advisor well beyond the program. This is usually the case for those projects that result in a journal article. With no graduate program in mathematics at Penn State Erie, the REU program allows participating faculty to augment their research programs, providing a graduate school-like setting.

3. Program Structure

Once the students arrive on campus, the first week is used to get the students up to speed on the possible research projects. Each faculty mentor typically takes a day to introduce the students to potential projects. Students are given a choice of at least four projects from which to choose. While a certain amount of background information is given, a majority of the time is spent introducing students to the mathematical techniques and tools that will be needed to conduct the research. This exposes students to a wide range of topics even before the research begins. If time allows, a guest lecturer completes the week's presentations. At the end of the week, students decide upon which research projects they would like to be involved. Students can choose to work on more than one project and are afforded the opportunity to switch projects midstream. Recently, several students have suggested their own projects, and we have accommodated them provided that the faculty mentors were comfortable overseeing these projects. The REU also has the ability to bring in additional faculty members, either from Penn State Erie or from other institutions, broadening the scope of available projects.

The students spend the remaining five weeks of the program conducting research. Participants are encouraged to work in small groups, but independent work is also allowed. Faculty mentors work closely with the students, but much of the research is conducted by the students independently of the faculty mentor. Students utilize such programs as MAPLE, MATLAB, and R to aide in their research. During the five week research period, students have two daily requirements: each morning, the entire group meets to discuss the research goals of each participant for the day, and each afternoon, a small presentation by each research group is made, describing what was accomplished that day. These presentations require the students to possess a level of understanding of their work sufficient to communicate to the entire group. Furthermore, the presentations involve each student in every project, thereby enabling students to move between groups.

At the end of the six week period, members of the Penn State community, as well as faculty from nearby colleges, are invited to attend a mini-research symposium in which the students present summaries of the research that they have conducted during the REU. Ultimately, we hope the students will give similar talks at regional, national, or international conferences. Students are also encouraged to pursue publication of their research if they choose. If a project is deemed publishable in an appropriate journal, the students will engage in the process of writing an article. The writing process usually takes place beyond the six week program, with significant direction from the REU mentor.

Students live on campus in university apartments and are given a food stipend. Penn State Erie treats the program participants as summer students, and allows them to utilize the same campus facilities (e.g., the gym) and organized group activities for the other students. In addition, the coordinating faculty plan a number of social activities, including picnics, baseball games, and boating trips, to encourage camaraderie among the participants and faculty mentors as well as provide a break from research. At least one day is taken to discuss graduate programs in mathematical biology. Additionally, we have invited past REU participants to discuss their graduate experiences with our students and have taken our students to visit a university with a program in mathematical biology.

4. Student Recruitment and Selection

Recruiting student applicants to the REU occurs in a number of ways. Information about all REUs funded by the National Science Foundation are aggregated by a number of web sites, and many students discover our REU through these services. We also mail a poster advertising the REU to a number of colleges and universities throughout the United States, concentrating on smaller schools that do not have graduate programs in mathematics. We also find that applicants learn about our program from former REU students presenting their research at regional and national meetings.

The average applicant pool is approximately 60 students, of which a majority are juniors and seniors. We do receive, and have accepted, some applications from freshman who have completed differential equations in their first year of study. The program has attracted students from all types of colleges and universities, from large research institutions to small, liberal arts schools. The participants from the 2005 REU are a typical representative group. They hailed from Hiram College, UCLA, Keene State University, University of Alaska at Anchorage, Hillsdale College, and Concordia College of St. Paul. Past participants have come from many parts of the country: Alaska, Washington, California, New Jersey, New Hampshire, New York, Ohio, Oklahoma, Pennsylvania, North Carolina, Mississippi, Idaho, Arkansas,

Texas, Michigan, Wyoming, Georgia, Iowa, Virginia, Minnesota, Tennessee, and Guam. Over the span of the program, only two student participants have come from the host institution, Penn State Erie, as a goal of the program is to afford a research opportunity to those who do not have such an opportunity at their home institution.

From the applicant pool we select six students. An application consists of a letter of interest from the participant, a transcript, and a letter of reference. In judging an applicant, we try to assess the student's interest in mathematical biology, prior coursework, GPA, whether the student has opportunity at their home institution to engage in research in mathematical biology, and whether the student comes from an underrepresented group. Approximately two thirds of the applicant pool is female. We have had great success in attracting both female and minority participants.

5. Examples of Research and Outcomes

In earlier years of the REU, publications with undergraduate authors did not occur. In recent years, after the program structure evolved from primarily instruction to primarily research, it became evident that undergraduates could be fully involved in the research activities of the coordinating faculty. Recently, the REU has averaged about one publication per year in a peer reviewed journal. These include [1, 2, 3] as well as two submitted articles and one in preparation.

Students have also had success presenting their research at national and international conferences. Ben Nolting (U. Alaska at Anchorage) won an award for outstanding undergraduate poster at the 2006 joint mathematical meetings in San Antonio. Brittany Parker, a senior at Mercyhurst College (Erie, PA), presented her REU research with fellow participant Tamar Wilson (Mt. Holyoke) at the Thirteenth International Conference of Forum for Interdisciplinary Mathematics on Interdisciplinary Mathematical and Statistical Techniques (2006) in Tomar, Portugal and won an award for outstanding undergraduate oral presentation. Two students from 2006 plan to present their findings in a poster session at the AMS/MAA joint meetings in New Orleans.

More importantly, around two-thirds of all participants have gone on to graduate school with about one-half pursuing research in mathematical biology or a related field (e.g., medical school). Equally important, as a result of participation in the REU, several participants have made conscious decisions not to pursue careers involving mathematical research, opting for careers that are less open ended and having more structure.

6. Feedback and Evaluation

Many of the changes to the program have taken place as a result of feedback in the form of correspondence with past participants and through proposal reviews by the NSF. The inversion in the time of instruction and the time of actual research took place as a result of this feedback. The creation of a final presentation at our mini symposium was also a suggestion that has been implemented.

All participants are formally contacted at least a year after the REU to determine the path that they chose after the REU and to solicit feedback. Students are also afforded a time of feedback at the close of the summer program in an informal brainstorming session. Participant feedback has been unanimous in assigning value to the REU, with at least half mentioning that their participation in the REU had made a significant impact in career and graduate school decisions.

On average, three of the six participants per year continue regular contact with the participating faculty beyond the formal termination of the program. This interaction ranges from direct involvement in writing research articles and continued research, to help for preparing research presentations or papers generated at the participant's home institution. A higher percentage of students solicit advice and consultation concerning future career decisions. Several students have made or plan to make return visits to Erie to continue research that they began in the program.

References

[1] Chauvet, E., Paullet, J., Previte, J. P., and Walls, Z., A Lotka-Volterra Three Species Food Chain, Math. Magazine 75, 243-255 (2002).
[2] Ellison, R., Gardner, V., Lepak, J., O'Malley, M., Paullet, J., Previte, J., Reid, B., and Vizzard, K., Pattern Formation in Small Arrays of Locally Coupled Oscillators, Int. J. of Bifurcation and Chaos 15 2283-2293. (2005)
[3] Merrill, K., Beauchesne, M., Previte, J. Paullet , J., and Weidman, P., Final Steady Flow near a Stagnation Point on a Vertical Surface in a Porous Medium, Int. J. Heat and Mass Transfer 49 4681-4686 (2006).
[4] Steen, L. A., editor., Math and Bio 2010 Linking Undergraduate Disciplines, MAA, (2005).
[5] Reed, M., Why Is Mathematical Biology So Hard?, Notices AMS, Volume 51, No 3, 338-342 (2003).

PENN STATE ERIE, THE BEHREND COLLEGE, ERIE, PA 16563
E-mail address: jpp4@psu.edu, mar36@psu.edu, sas56@psu.edu

The Rice University Summer Institute of Statistics (RUSIS)

Javier Rojo

The Rice University Summer Institute of Statistics (RUSIS) has been generously supported by The National Science Foundation (NSF) and The National Security Agency (NSA) for five years. A total of 17 undergraduate students per summer spend 10 weeks at Rice University. Twelve students are supported by NSF and five students are supported by NSA.

The main goal of RUSIS is to encourage undergraduate students to pursue Ph. D. work in the statistical sciences. Selected junior and senior underrepresented minority students and students with no easy access to a career experience at their institutions are recruited. Requirements for admission to the program include: three semesters of calculus, one semester of linear algebra or matrix analysis, a minimum 3.0 GPA, and at least two letters of recommendation. However, students with a GPA below the 3.0 threshold, or students who have not met the mathematical requirements, are not automatically rejected. Rather, when the letters of reference and statement of purpose give an indication that the student is capable of pursuing graduate work, telephone interviews are conducted to better ascertain the students potential. Whenever feasible (e.g., local students) applicants are invited to Rice for an interview. RUSIS believes that grades constitute but one indicator of the potential for a successful career in graduate school. Other important dimensions that impact success in graduate school derive from personal qualities such as perseverance and creativity. The program also enrolls the help of the advisory board in the search for qualified candidates.

Through intensive short courses in areas of current research interest, intensive seminars in computation, close mentoring and supervision by faculty, students develop skills that help them during their graduate careers. In addition, attendance to a series of lectures by top scientists keep their interest and focus on science and graduate school. By the end of the summer students have participated in at least one research project analyzing data, running computer simulations, developing algorithms, and, when appropriate, engaging in theoretical work. Every student is encouraged and expected to prepare their research findings in close collaboration with their faculty mentors and peers for presentation at student sessions at national

Received by the editor December 18, 2006.

meetings and subsequent submission for publication.

At the end of the summer, all students meet with the advisory committee, in the absence of any personnel associated with the RUSIS, to provide feedback on ways to improve the program.

1. Institute Activities. The first day of the Institute is reserved for introductions of students, mentors, and support staff; tours of the campus and Library, and applying for Library cards. Students register, at no charge, as Rice students and earn one hour of credit for Independent Study. Students fill out questionnaires related to their backgrounds, and expectations for the program. A pre-test and a post-test are given, and the information is utilized as part of a set of indicators to measure the impact of the program.

As most students come without the needed background in probability and statistics, an intensive course in probability, stochastic processes, statistical inference, and survival analysis, is taught for the first three weeks of the summer. In addition, an afternoon short course in computation is taught during the first three to four weeks of the institute. Beginning in the fourth week students start working on their research projects. Mentors, with the support of postdoctoral and graduate students, work closely with students to provide them the background material specific to their group projects and the needed research direction as the projects evolve.

Throughout the summer, invited speakers from MD Anderson Cancer Center, the Michael E. DeBakey Department of Surgery at the Baylor College of Medicine, the University of Texas Health Sciences Center at Houston, lecture on survival analysis applications in cancer research, liver transplant, and other health-related and environmental applications. In addition, Sallie Keller-McNulty, Dean of the School of Engineering at Rice, and President of the American Statistical Association, visits with the students. For example, during the summer of 2005, Dean Keller-McNulty discussed opportunities for graduate work in Statistics and the job market. Students commented in the exit questionnaires that they felt that RUSIS really cared about them as a result of the visits by the Dean and the amiable conversations with the advisory committee members. Visits by NSA have been motivational and elicit very positive reactions by the students.

Friday afternoons are dedicated to meet as a group to discuss improvements of the program and, more importantly, discuss progress of the research projects as well as requirements and strategies for graduate school. Statistical/mathematical videos (Fermat's last theorem, If Copernicus had a computer, MSRI on-line lectures) are presented to spark the interest of students in pursuing graduate work.

At least two field trips are scheduled during the summer. Students spend a day at NASA facilities where the group is allowed to operate the flight simulators and tour the facilities including Mission Control. David Mains and Daniel Adamo have provided excellent educational tours. Students also visit The MD Anderson Cancer Center. Gary Rosner organizes a session where various researchers present their work on cutting-edge cancer biostatistical research.

In addition, a consultant from the Cain project www.owlnet.rice.edu/~cainproj helps the students improve their writing skills. As part of their final projects, students write letters to their Congresspersons describing their summer experience and encouraging them to continue supporting the federal funding for these activities.

2. Teaching and computational facilities. All teaching, computational, and mentoring activities take place at Rice University. RUSIS focuses explicitly on Statistics and its applications. Rice University is committed to facilitate the development of nontraditional groups in undergraduate science and engineering programs. Several well-equipped classrooms exist in Duncan Hall, the home of the School of Engineering and the Statistics Department.

Classrooms are equipped with the latest technology: A 32-Button Wireless Remote Control, One Video Projector (1024 x 768 native, 1280 x 1024 max.), one VCR (VHS / S-VHS) with Cable TV; PowerMac G4 w/ MacOS 10.4 (with floppy, CD / DVD); Pentium w/ Windows 2000 (Floppy, CD/DVD); Laptop connection with video, audio, network (DHCP); and a Microphone with Distributed speaker system.

The Computer Laboratoy has room for 22 students. The laboratory is equipped with one VCR (VHS), a DVD Player, a Document Camera, a video projector, and a Distributed speaker system. Instructors work from two Compaq Pentium PC with Windows 2000, and the students have access to twenty-two active network ports (DHCP) and twenty Compaq Pentium PC with windows 2000. A wireless network is available, and students are provided access to all the library electronic holdings in addition to their Rice students library privileges. Laser printing facilities are available in the room.

The Statistics Department has access to the Rice Virtual Laboratory in Statistics (http://www.ruf.rice.edu/~lane/rvls.html) to support teaching. The site contains: HyperStat Online – An online statistics book with links to other statistics resources; Simulations/Demonstrations: Java applets that demonstrate various statistical concepts; Case Studies: Examples of real data with analyses and interpretation; Analysis Lab: Some basic statistical analysis tools. RUSIS students use Mathematica, R, SAS, Gauss, and Matlab to support their research projects. The School of Engineering provides network, hardware, and software support through the Information Technology Department.

3. Examples of Projects. It is understood that not all students come with the same background and training. Our goal is to provide them with a valuable experience that is helpful during their graduate school career. Therefore, several projects are presented in class through the first 3 or 4 weeks and students are allowed to choose their project(s). The current need to understand, develop, and assess the merit of new methodologies in the areas of multivariate survival analysis, multivariate extreme value theory, analysis of microarray data, and analysis of massive data sets presents a great opportunity for the training of selected undergraduate students. Problems of current interest are used to motivate the students and serve as a point of departure for the research projects. Some examples of projects follow.

Extreme Value Theory: Extreme events have large impact on various areas of engineering, science and economics. Events such as extreme waves, rainfall, and floods are of fundamental importance, as are high wind speeds and extreme temperatures, and extreme value theory provides a foundation to study corrosion and metal fatigue. Health hazards develop as a result of high concentrations of pollutants, and damages to the economy develop from extreme changes in the market. The development of new statistical and probabilistic methods for extremes is an active area of research and one that has potential to impact risk management, pollution and weather forecasting. Bivariate extreme value distributions can always be transformed so that their marginal distributions are exponentially distributed. Once this is done, the bivariate distribution is characterized by the dependence function which must satisfy certain constraints. Rojo et al. (2001), proposed a nonparametric estimator for the dependence function. The students have the opportunity to work on various computational aspects of these problems. Ozone level data from various National Parks covering over a period of fifteen years is available and provides an interesting set of theoretical and applied challenges. Questions of interest include: Are extreme ozone levels, as measured by exceedances over the 12 ppm threshold level, decreasing in size and/or in frequency? Are there any significant differences among the parks? Can the impact of the Clean Air Act be observed in the ozone level trends? These issues are explored through a mixture of statistical computer modeling and computer graphics exploration.

Multiple comparisons: The renewed impetus in this area is due in part to the arrival of microarray data. The statistical interest in microarray experiments derives (See, e.g., http://www.sbm.temple.edu/ ~sanat/cbmsconf/lectures.html, and http://elib.zib.de/mailing-lists/public/st-net/2002/msg00045.html) from its much larger data sets and coincides with the arrival of computer technology which allows for computer-intensive re-sampling based methods. (e.g., Westfall, P. H. and Young, S. S. (1993)). Genomics and bioinformatics have spawned challenging problems. Modern biotechnology allows researchers to collect high-throughput genetic data which leads to thousands of significance tests. Advanced undergraduate students, in the context of differential gene expression, learn multiple comparison issues and methods relatively quickly, and they compare the various testing procedures that are in current practice using computer simulations. Substantive problems and (real) data is available through various publicly accessible websites.

Dimension Reduction: One major challenge in the analysis of survival microarray data derives from the number of replicates being very small, while the number of genes is usually very large (10,000-20,000). There are at least two procedures used for dimension reduction before fitting a survival analysis model. Principal Component Analysis (PCA), and Partial Least Squares (PLS). The latter has been heralded by their proponents as a clear winner over PCA. Students in the last two RUSIS have investigated the properties of both methods in terms of Prediction Mean Squared Error, and the conclusion has been that in the modest simulation studies performed by the students, there is no clear distinction between the two. Both tend to select the same genes, and both tend to have similar Mean Squared Error properties. This area continues to be explored.

Other areas of interest have included: multivariate survival analysis with censored data; ROC curves as a way to identify differentially expressed genes; Linkage analysis with sib-pair data; and inverse problems in geophysics.

4. Program assessment. An advisory committee provides an annual assessment of RUSIS. The purpose of the committee is two-fold: **(i)** Provide advice for evaluation and improvement of the program and, **(ii)** To serve as ambassadors of the program at their own institutions and help in the recruitment of students.

The following currently serve on the advisory board: Arturo Bronson, University of Texas at El Paso; Willie Pearson, Georgia Institute of Technology, William Velez, University of Arizona; Anna Baron, University of Colorado at Denver; Cristina Villalobos, University of Texas Pan American; Maria Acosta, Texas State University San Marcos. The advisory board convenes during the tenth week of the program to assess the merits of the program and the progress of the students, and to provide insights and feedback on ways to improve the Institute. In addition, board members provide an invited lecture on a statistical aspect of their discipline. These lectures have received a lot of praise from the students who think that lectures are inspirational. These lectures also provide a different perspective on science and engineering and on opportunities for career development and employment.

Student Recruitment and selection Program information and application packets are mailed to mathematics and statistics departments throughout the country. A website in the Statistics Department provides detailed information on application deadlines and requirements for admission. World Wide Web sites that specialize in mathematics and statistics are utilized to post announcements about RUSIS. Similar electronic announcements are sent to most of the universities located in predominantly underrepresented minority areas in the United States. In addition, SACNAS conferences offer excellent opportunities for recruitment, and various committees of the American Statistical Association are contacted seeking their help in making the information available to undergraduate advisors.

Project evaluation and reporting The most important goal of the RUSIS is to motivate students to pursue graduate work in the Statistical sciences, through a series of activities designed to entice them and excite them to engage in research. As such, the real impact of the program is difficult to gauge, and the outcomes of interest will be observed only until after a few years have passed. The program keeps track of all students after their participation in RUSIS. Students are asked to provide permanent contact information, and faculty and staff reciprocate to encourage the students to stay in touch. During their visit, each member of the advisory committee presents a fifteen-minute talk that will highlight how statistics play a role in their discipline. In addition, the students present their research projects and interact with members of the board. These interactions are valuable for the board in formulating their recommendations for improvement of the program. These recommendations are communicated to the RUSIS Director during an exit interview, and a formal site visit report is written. Through student questionnaires and exit interviews, students provide feedback on various aspects of the program.

5. Summary of RUSIS accomplishments. The number of applicants has been stable at 51-55 per year. Most of the students so far have come from the southwestern and southern states, although several students from the Midwest and New England have participated. Roughly 56% of the students have graduated. (The program accepts sophomores-seniors). Of these, 85% have expressed a desire to pursue graduate work, and 75% are currently in Ph. D. programs or are applying for admission to Ph. D. programs. For example, four students are applying for admission to Rice University, while three students are already members of the statistics department at Rice University. Che Smith (Spelman) has begun her Ph. D. in biostatistics at the University of North Carolina, Chapel Hill. Sarah Williams is a Ph. D. student in environmetrics at Colorado State University. Tahira Saleem was accepted to the Ph. D. program at Rice but opted for a job with NSA. Daisy Wang is a Ph. D. statistics student at UC Berkeley. These are examples of the many success stories of former RUSIS participants. There is, however, an example of a student who has blossomed as a result of his RUSIS participation. Juan Gallegos, the first student in his family to attend college received a bachelors degree from the downtown campus of the University of Houston. Juan took advantage of the opportunities offered by RUSIS and presented his work at several student conferences. His self-confidence grew and he was accepted and offered a fellowship by the University of Texas at Houston to pursue Ph. D. work in Epidemiology.

Students present their work at national conferences, and many have participated in SACNAS, American Mathematical Society meetings, and the Young Mathematicians Conference. Students have won recognition for their work. Jean Kongpinda and Venessa Tavares received an award at the 2004 AMS meeting; Stacey Ackerman, Israel Cabello, Cyrus Aghili, and John Ratana won second place at the 2005 SACNAS meeting; and David Kahle and Darren Homrighausen won a poster award at the 2006 AMS/MAA meeting. Tollie Thigpen writes in a recent email: "I have had the opportunity of presenting it (RUSIS project) at conferences in Mississippi and North Carolina. I was awarded 2nd and 3rd places at two of the conferences in which I was unaware of being judged."

The RUSIS program has been featured in the Rice News University paper, (http://senews.rice.edu/hotnews.cfm?mode=details&status=Archived#1058), and in a short article that appeared in Math Horizons in September 2005. Student photos and abstracts, as well as the students' home institutions, may be found at the RUSIS webpage – http://www.stat.rice.edu/RUSIS03.

References

[1] Rojo, J., Villa, E., and Flores, M. (2001). Nonparametric Estimation of the Dependence Function in Bivariate Extreme Value Distributions, *Journal of Multivariate Analysis*, pp 159-191.

[2] Rojo, J. (2005). REU Spotlight: Rice University Summer Institute of Statistics, *Math Horizons*, Vol 13, pp 30.

[3] Westfall, P. H. and Young, S. S. (1993). *Resampling-Based Multiple Testing: Examples and Methods for P-value Adjustment*, Wiley, New York.

STATISTICS DEPARTMENT, RICE UNIVERSITY, 6100 MAIN STREET, HOUSTON, TX 77005
E-mail address: jrojo@rice.edu

The Rose-Hulman REU in Mathematics

Kurt M. Bryan

1. Overview

In the past 18 years 126 undergraduates have participated in Rose-Hulman's NSF-REU program in mathematics. The program began in 1989 under the direction of Gary Sherman, who mentored six students in computational group theory each summer. In 1996 Allen Broughton began co-directing students, and subsequent renewals from 1997 to 2003 under Dr. Broughton expanded the program to eight students per year, working in teams of four with two faculty, including John Rickert and Kurt Bryan on a rotating basis. Kurt Bryan took over as director in 2004, and faculty members Tom Langley and David Finn began mentoring students. The list of research topics was also enlarged based on the new faculty research areas, to include hyperbolic geometry, number theory, inverse problems, and geometric analysis. The program has evolved from a single mentor to one in which faculty rotate in and out from summer to summer, offering a wider range of problems for undergraduates to explore.

Although the areas of research we offer are quite diverse, we do have a common approach: the significant use of computational tools like Magma, Matlab, or Maple to help students understand the problems at hand, begin making conjectures or designing algorithms, and then follow up with rigorous analysis. The problems are carefully designed so that students can start work immediately, without extensive background beyond standard undergraduate course work in the area of interest. However, we do not give them "canned" or "toy" problems, and students always find that they need to master new mathematical tools as the summer progresses.

We emphasize collaboration and communication throughout the summer. Students typically work in pairs, and the four students in each research group share a common work area. The students make regular presentations of their progress to the other REU participants and faculty, and research results are also published in the Rose-Hulman Mathematics Technical Report series. Many papers from our REU have appeared in professional journals.

Students are selected based on having adequate preparation to succeed in the program, and we particularly seek students who express interest in a career involving research, but who have not had the opportunity to engage in research at their home institution.

Received by the editor October 20, 2006.

2. Goals, Philosophy and Problem Selection

2.1. Goals. The goal of our program is simple: to give undergraduates the chance to play the game of mathematics at the professional level. Specifically, students

- Tackle research problems which start out a bit nebulous, and for which no "answer key" exists. Emphasis is placed on clarifying and simplifying the difficult initial problem as a means of gaining understanding, analyzing specific examples, making and proving conjectures, and adding additional complication once simpler versions are thoroughly understood. Computational tools are frequently useful here.
- Improve their collaborative skills by working intensively with other students and professional mathematicians.
- Develop their speaking skills by giving frequent formal presentations of their results during the program, as well as at conferences after the program ends.
- Improve their written communication skills, by preparing a report in the Rose-Hulman Mathematics Technical Report series, and when appropriate, refining and submitting the work to a professional journal.
- Develop their ability to be independent learners, by mastering relevant material in professional texts or journals (including those written by previous REU participants).
- Develop a sense of identity as mathematicians, as being capable of original insight and contribution to mathematics.
- Develop a sense of belonging in the mathematics community, by forming relationships with professional mathematicians, in and out of our REU program.
- Deepen and broaden their knowledge of one or more areas of mathematics.

2.2. Student Recruitment. Each year we mail notices of the program to several hundred schools nationwide, and in particular to schools in which students may lack access to the kind of research program we can offer, perhaps because of a lack of sufficient faculty involved in research. The mailing list is updated regularly to reflect any good leads we get on appropriate students, and of course our large web of prior year REU participants and their institutions can be a good source of applicants. Faculty also use professional contacts at other institutions in order to identify appropriate applicants.

The students we seek will generally have completed their junior year, though well-qualified sophomores or even freshmen may be accepted. The desirable student will have completed a year of differential equations (ordinary and/or partial), or a year of abstract algebra, depending on the research area of interest to the student (which they specify on the REU application form). They should also have experience in some computer programming language, though not necessarily any of those we use, an ability to work with others, and of course a drive to achieve in mathematics. Of particular interest to us are students who have not had a significant research experience, but wish one, as a basis for making a career decision. To demonstrate these qualities students must submit a letter of interest, academic resume, transcripts, and two letters of recommendation.

2.3. Program Approach and Philosophy. Students are selected for the program by mid-March and arrive in early June. In that interval we make available to them, via email or the web, a small amount of preliminary material to look over, to familiarize themselves with the problems and past results, though we don't impose any great workload, since most students are still in class.

The problems we give the students are typically an integral part of the faculty mentor's research program. The problems are thus "real" research questions, of significant interest to mathematicians, but which can be initially simplified if necessary to gain understanding and begin making progress. Some involve extending techniques which worked in one setting to new situations. In all cases we emphasize experimentation with software to allow students to build insight and begin making conjectures or trying algorithms.

We put a strong emphasis on getting students to work on the problems very quickly, preferably within the first week. The first few days of the program are split between lectures from the faculty mentor followed by students working together on simple versions of the problems (workable in an afternoon or less), as a means of building the students' understanding of what the research question is, if not an approach to the solution. The students are expected to have settled on a specific problem early in the second week. At this point the students have generally decided with whom they would like to work, and are strongly encouraged to team up with at least one other person.

Each group of four students has exclusive use of one classroom, outfitted with tables, chairs, whiteboards, and computers, at least one workstation per student. All computers run up-to-date versions of TeX/LaTeX, Magma, Maple, Matlab, and Femlab (a powerful 3D finite element PDE solver). No fixed daily schedule is imposed, but the students have 24 hour access to the room. The faculty mentor has an office next to the classroom and wanders in and out during the day, talking with students, suggesting things to try, giving lectures when the students need to know something. Lunch is taken as a group almost every day. The close physical proximity of all eight students is a powerful factor in building collaboration and preventing anyone from feeling isolated.

Presentations are required from each problem team, about every two weeks starting in the third or fourth week. These presentations are made to the entire REU group of eight students and faculty, and are frequently attended by other Rose-Hulman faculty as well. In addition, during the final week of our program in the summers of 2002-2006, the faculty at nearby Indiana University staged a conference for all mathematics REU participants in Indiana (30 to 50 students, 8 to 12 faculty), in which all of our students presented their work, a tradition we intend to continue. These regular presentations and final presentation to an outside group encourages students to keep plugging away at the problems. Each student must also hand in a draft of a technical report by the final day of program. The faculty mentor reviews the report and students make revisions after the program ends (these are usually fairly minor).

The relationship we strive for with the students is one of junior and senior colleagues, working as a team to understand a difficult problem. Although we choose problems on which we can be reasonably certain students will make progress, we do not steal the problem from the students by telling them how to solve it. This approach, along with the polished technical report and presentations, gives the students a sense of pride and ownership of the results.

3. Project Evaluation and Reporting

Our central goals are to improve the participants' communication skills, both oral and written, to develop their collaborative skills, to help them develop intellectual independence, and get them excited about a career involving mathematical research.

The most immediate assessment of our success occurs by the end of the program itself: We expect each student, collaborating with others, to contribute to a first draft of a technical report which advances mathematical thought, however modestly. We expect every student to give a minimum of three formal presentations during the course of the program, incorporating feedback to iteratively improve each succeeding presentation. We also conduct exit interviews and have students fill out a questionnaire to obtain immediate feedback on the quality of that summer's program. The students have been overwhelmingly positive about their experience in our program, and most feel it helped them make a more informed career decision.

In the longer term, a primary measure of our success is that every participant complete a polished technical report and give at least one presentation at his or her home institution or at a conference, concerning the results obtained during the REU. Indeed, a significant portion of our budget is devoted to supporting students who travel to conferences after the program has ended; in recent years many students have presented the REU work at the national AMS/MAA joint meetings. The REU mentors also usually attend these meetings and hence have a chance to follow-up with students on research and career plans. We also gauge success in improving student mathematical and writing skills by the number of reports which prove suitable for publication in professional refereed journals.

After the program ends we maintain close contact via email with almost all of the students (for at least a couple years, since we serve as important letter writers and advisors for graduate schools). Because of this continued contact we have built up and maintain a database of student career choices and/or graduate schools, as evidence for the impact of our program in producing independent mathematical scholars. Of those 82 participants from 1995-2005, 64 went to graduate school, 7 went into industry, 4 are still undergraduates, and there are 7 on which we have no data.

DEPARTMENT OF MATHEMATICS, ROSE-HULMAN INSTITUTE OF TECHNOLOGY, TERRE HAUTE, IN 47803

E-mail address: Kurt.Bryan@rose-hulman.edu

The REU Program at DIMACS/Rutgers University

Brenda J. Latka and Fred S. Roberts

1. Introduction

DIMACS, the Center for Discrete Mathematics and Theoretical Computer Science, headquartered at Rutgers University, runs a unique and pioneering REU program with both domestic and international components. Traditional themes of the domestic program are discrete math and theoretical computer science with considerable emphasis on applications. However, the program also has an academic-industrial co-mentoring component, a component sponsored by a new Department of Homeland Security Center for Dynamic Data Analysis headquartered at DIMACS, and a shared component with the Rutgers Mathematics Department that encompasses a wide variety of other mathematical topics. DIMACS has run its domestic REU program since 1993 under NSF support. For seven years the DIMACS REU has been totally integrated with that of the Rutgers Mathematics Department, expanding the scientific scope and student experience of both programs.

Since 1999 the DIMACS REU has had an international element co-sponsored by DIMATIA, the Center for Discrete Mathematics, Theoretical Informatics, and Applications, a distinguished research center based at Charles University in Prague, Czech Republic. This was a pioneer in international REU programs. The joint international program includes outstanding Czech students with strong European training, gives U.S. students a global international perspective, and introduces them to the mathematical sciences as a global endeavor.

The REU program has almost 30 students per year. Students from across the U.S. participate in the 8-week domestic component at DIMACS or the Rutgers Mathematics Department, with academic mentors or academic and industrial co-mentors; or in the first 7 weeks of the domestic component with an additional 3 weeks at DIMATIA. Students from the Czech Republic participate in the first 7 weeks of the domestic component and then act as hosts at DIMATIA to the U.S. students. More details of our program can be found at the DIMACS REU website, http://dimacs.rutgers.edu/REU.

2. Overview

The goal of our program is to provide participants with an exciting research experience that will help them decide on future educational and career paths and

Received by the editor December 27, 2006.

give them the confidence to pursue their choices. All students, including those only in the domestic component, get a taste of the international scientific enterprise. The U.S. students going to Prague get a more direct international experience and benefit from the scientific atmosphere at an international center of research. The Czech students benefit similarly from exposure to their U.S. counterparts and contribute to providing the global perspective that we seek. All students are exposed to the industrial research environment, providing a broadened view of career possibilities.

Our REU program is unique because it is run in the context of two major research centers with many scientific activities and long-standing relationships with industrial partners. The richness of the intellectual community, the synergy of the academic and industrial collaborations, and the international flavor contributed by the many foreign scientists participating in DIMACS and DIMATIA activities at the same time as the REU students adds to the overall atmosphere.

DIMACS was founded in 1989 as an NSF "science and technology center" and is a consortium of Rutgers, Princeton, AT&T Labs-Research, Bell Labs, NEC Laboratories America and Telcordia Technologies, with partners at Georgia Tech, RPI, Stevens Institute of Technology, Avaya Labs, HP Labs, IBM Research, and Microsoft Research. Undergraduate research has been a major focus of DIMACS programs since its founding. For more about DIMACS, see http://dimacs.rutgers.edu.

3. Nature of Student Activities

3.1. The Program at DIMACS. The key to our REU program is the one-on-one research experience under the direction of a mentor. The domestic component begins when the students arrive at Rutgers in mid-June, move into campus housing, and receive offices and computer and library accounts at DIMACS. A graduate student coordinator introduces them to their mentors to begin a program of directed study and research, including regular student/mentor meetings.

There are regular lunches and evening social activities, to which the mentors are also invited, as well as a weekly REU Seminar Series featuring local speakers and renowned outside speakers, at least one devoted to a presentation about careers and graduate school and another to developing good technical communication skills, including writing abstracts, making presentations, and developing project websites.

Students make two presentations about their projects. Early on, each describes their research problem. These short talks encourage collaboration and discussion and provide students the opportunity to work on multiple projects or even switch projects. Near the end of the domestic component, students make second short presentations about their work to the DIMACS community. Those going to Prague repeat these presentations there and make further presentations before they depart. In addition, students are asked to prepare project websites.

We introduce the students to industrial research by making trips to our industrial partners for tours and technical presentations. The students in the academic-industrial component of our program meet regularly with their academic mentor and at least once a week at DIMACS with their industrial mentor. At the industrial locations, these students are introduced to other researchers and projects. The academic-industrial co-mentoring broadens all the students' views of the applications of computer science and mathematics.

REU students are invited to participate in the wide variety of summer programs at DIMACS. This includes tutorials and workshops in the DIMACS "Special Focus" programs such as Computational and Mathematical Epidemiology, Communication Security and Information Privacy, Computation and the Socio-Economic Sciences, Information Processing in Biology, and Discrete Random Systems.

Students benefit from discussions and collaborations with researchers, postdocs, and graduate students who are not their official mentors and are able to switch projects or engage in more than one. It is not unusual for a student to work on a problem under the direction of a visiting researcher from a foreign country or start on a theoretical problem in computer science and end up working on an applied problem in engineering or biology, or vice versa.

After the DIMACS summer, students are encouraged to stay in touch with their mentors and enhance their project websites. Some prepare a research paper or give a presentation at a scientific meeting. Student papers are published in the DIMACS and DIMATIA technical report series and in journals or conference proceedings. Many students present talks back at their home institutions and others make their REU project a major piece of their senior thesis or eventually of a master's thesis.

3.2. Research Areas. The REU program focuses on the common strengths of DIMACS and DIMATIA in discrete math and theoretical computer science and their applications, natural areas for undergraduate research, with topics from such areas as graph theory, computational geometry, logic and complexity theory, and combinatorial optimization. Specific topics include variants of graph coloring, evolution of massive graphs, extremal combinatorics, finite geometries, new complexity classes, algebraic models of computation, large-scale linear and integer programming, binomial proportions, and statistical models of uncertainty in data streams. We also build on the interdisciplinary strengths of DIMACS and DIMATIA faculty, to include topics in DNA topology, biomedical engineering, epidemiological modeling, economic applications involving e-commerce, sensor location, information management in massive data sets, and network analysis, to name just a few. The joint program with that of the Rutgers Mathematics Department allows inclusion of a wide variety of mathematical topics involving PDEs, ODEs, commutative algebra, numerical analysis, mathematical physics, and topology/geometry.

The new Homeland Security Center at DIMACS allows us to provide an REU experience in topics such as pattern detection from multiple information sources, bioterrorism event detection, and port of entry inspection algorithms. This phase of the program exposes all students in our REU to homeland security applications and provides them with an opportunity to interact with visitors from DHS and national labs such as Lawrence Livermore, Los Alamos, and Sandia.

3.3. Social and Community Building Activities. Throughout the summer the students are invited to participate in planned group social activities. There is an orientation dinner, and other informal meals such as picnics or barbecues during the first week of the program to encourage students to mingle and meet. Lunches and dinners involved with the seminars, workshops, and field trips give the students additional opportunities to interact. In Prague, the students join faculty at lunches, are hosted by the Czech students at dinners, and go on cultural excursions organized by the Czech students.

3.4. The Program at DIMATIA. The goals of the experience in Prague are different from those for the experience in the U.S., and center around introducing the participating students to a wealth of open problems and questions and appropriate problem-solving techniques and strategies.

The Czech REU students make informal presentations while at DIMACS, preparing the U.S. students for life in Prague, and then act as hosts for the 3 weeks the U.S. students are in Prague. The U.S. graduate student coordinator and sometimes a DIMACS faculty member accompany the group. The arrival in Prague is timed to occur the week before the annual Prague Midsummer Combinatorics Workshop.

During the first week, there are tutorial presentations by DIMATIA faculty as an introduction to their scientific interests, to prepare students for the topics of the Midsummer Workshop, and to present potential research problems. Background material for these presentations is based on lecture notes from the DIMATIA "Spring School on Combinatorics." There are presentations orienting the students to the culture and history of Prague and the country, with special emphasis on the rich Czech mathematical tradition. The U.S. students make presentations about their domestic REU research. The faculty members presenting tutorial lectures act as mentors to the students. We let the emphasis on problem solving lead to natural connections between students and mentors.

The REU students participate in the informal problem sessions of the one-week Prague Midsummer Workshop and some give research presentations at the workshop. The U.S. and Czech coordinators/mentors lead discussions on workshop topics, helping students focus on research problems. Group meetings emphasize approaches to unsolved problems, problem-solving strategies, and group attacks on problems. Students are encouraged to explore, in collaboration with Czech mentors and students, the research questions that arise during the visit to Prague and to pursue the research project begun at DIMACS with Czech mentors.

After the Midsummer Workshop, the program concludes with more intensive one-on-one sessions between students and mentors, group meetings for problem-solving, and presentations of research. While the short visit in Prague doesn't give students as much time as they had at DIMACS to get deeply into research, the experience teaches them how research questions are formulated and pursued and enables them to make a good start on research they can pursue after leaving Prague. They are encouraged to remain in contact with their Czech mentors.

3.5. Selection of Mentors. Mentors who are able and willing to inspire undergraduate students are a key element in the success of an REU program. Mentors are invited to participate early in the Fall semester based on previous success in the program, reputation for research with undergraduates, accessibility of their research, and enthusiasm for the program. Our mentors come from a wide variety of departments besides mathematics and computer science, including operations research, statistics, chemistry, industrial engineering, biomedical engineering, and library science. This leads to a variety of project topics, which we believe is one of the primary attractions of our program.

4. Student Recruitment and Selection

The REU website serves as a primary source of information about the program and application process. We also send out paper flyers. Our mailing targets "elite" universities, minority institutions, and those with little opportunity for undergraduate research. Surprisingly, students from "elite" schools often tell us they don't have opportunities for research. We make a special effort to recruit U.S. participants from members of groups historically under-represented in the sciences. Our mailing list has been compiled from many years of experience in running programs aimed at under-represented groups, and through our connections with professional organizations devoted to enhancing the involvement of minorities and women in Mathematics and Computer Science.

The application for U.S. participants is web-based. The first round of acceptances is usually sent out by February 15 in order to allow sufficient time for students in the international program to obtain passports and make other preparations for international travel. The program is highly selective. The main criteria we use in selection are readiness to undertake a research project and good match of interests with a mentor. After that, we seek a diversity of participants in terms of gender, ethnicity, geography, and types of institutions they represent. Opportunities to continue their project at their home institution are considered. Usually, we give priority to students entering their senior year. For the Prague program, we select students whose backgrounds and interests and U.S. project seem to make them good candidates for success in the Prague program. Projects proposed by mentors are described on our website, applicants rank their choice of projects, mentors select students, and we put the mentor and potential student in contact.

Our advertisements and recruitment build on things that students have reported as attracting them to our programs: DIMACS reputation as a major research center; careful advance presentation of projects; the quality of the faculty mentors; the opportunity to explore the potential for graduate study in numerous fields; and the interdisciplinary opportunities available at DIMACS. Of course, the international aspect of our program is also a major attraction.

5. Closing Remarks

The DIMACS REU program provides both a scientific and a cross-cultural experience. In addition to the scientific papers and talks by our students, there is networking among the students, domestic and foreign, that we have seen last far into the future. Based on our experience, the program greatly influences the choices about further education and future careers of the students involved, and gives them the confidence to pursue their choices. With a few years experience, students develop a better understanding of the impact of the REU on their careers. We recently completed a survey of students who went through our REU programs from 1993 to 2005. A question about their advanced degree plans or achievements provided a remarkable result: 100% of the respondents had either achieved, were currently in, or were planning to attend graduate school in the near future. In almost all cases this was for a Ph. D.

Students returning from our programs in the last four years have made over 40 presentations in a variety of venues, from high schools to international research conferences, and have published a variety of papers as technical reports or in journals or conference proceedings, some solving well-known open problems and others describing new approaches to applied questions. They have won recognition such as first place in the ACM student research competition, prizes for poster presentations at the MAA/AMS Joint Meetings, the Alice T. Schafer Mathematics Prize for excellence in research by the Association for Women in Mathematics, and "best student paper" awards at the MAA national meeting "Mathfest." The words of some of our prior program participants sum up the essence of the DIMACS program.

"REU has given me the chance to get to know people who really want to make a difference in math and science and I feel that has helped me see my capability to succeed in math research. The weekly seminars I attended offered an insight into the numerous applications of math. The lectures on grad school were very informative because I learned about fellowships I can apply for and strategies for applying. The faculty at Rutgers University is incredibly helpful. My mentor and other professors took the time to thoroughly answer my research and grad school questions. The DIMACS faculty made me feel like they truly care about helping me succeed. This program was a great experience for me because not only did my research project teach me a lot, I also made many good friendships with the other REU students. I think it will be very interesting to see where my fellow REU friends go in the future and know that this summer at Rutgers had a huge influence on our decisions. I am forever grateful for having the opportunity to participate in the REU program at Rutgers University this summer. Thank you!" (REU 2006)

"I really appreciated the fact that this REU included such a range of projects, from the very theoretical to the very applied. ... I also appreciated how large the REU was, giving me a chance to meet a lot more of the people I'll no doubt run into in grad school or at conferences in the future ... The opportunity to go to Prague was what made this REU my top choice, I like to travel, and it was great to meet and work with students from the Czech Republic, to get a new perspective on life in general and mathematics in particular. ... I thoroughly enjoyed my project and plan to keep working on it next year, my mentors were wonderful, and best of all–I got paid to do mathematical research!" (REU 2006)

"The Rutgers program was in many ways a fantastic experience for me. ... The environment was very relaxed, but people there loved math and wanted to do good research, and so the environment was mathematically very rich. The program made a life in math research seem really enjoyable. I made a lot of good friends that I stayed in touch with afterwards. ... I still remember the REU as making life in math research seem interesting and superfun, and I think I've definitely been influenced by the exposure to the bright mathematical minds (...) I encountered at Rutgers...I was selected for a NDSEG math fellowship and a Stanford SGF fellowship ... and I think the Rutgers program had a lot to do with that." (REU 2002, 2003)

"The REU program was a unique experience in my life. It helped (me) to realize what I want to do in my life as well as got (me) into contact with leading experts in the area. A fruitful and abundant cooperation between DIMACS and DIMATIA centers is a great opportunity for students both from the U.S. and the Czech Republic to meet and work with experts from both the institutions. I think the REU program in its current form allows both the U.S. and the Czech students to meet different cultural environments and leads to a better understanding of different styles of life in the U.S. and in Europe. In the current world, this becomes very important. ... In the current world, I think it is impossible to conduct a research in the area of mathematics or computer science without a close and intensive international cooperation." (Czech participant)

"At the REU I first learned about many fields of math and computer science that I had not seen in my undergraduate studies. And, I believe I may have never learned about had it not been for the REU. I had wonderful mentors, ..., who taught me many things including how to do research. I was excited about the chance to participate in an REU again, and last summer I taught a two week course at the University of Chicago REU. I advise many undergraduates to consider REU programs and I always suggest considering DIMACS." (REU 1998, 1999, and currently an L.E. Dickson Instructor at U. of Chicago)

DIMACS, RUTGERS UNIVERSITY, 96 FRELINGHUYSEN ROAD, PISCATAWAY, NJ 08854
E-mail address: latka@dimacs.rutgers.edu, froberts@dimacs.rutgers.edu

The SUNY Potsdam-Clarkson University REU Program

Joel Foisy

1. History and Structure of the Program

The SUNY Potsdam-Clarkson University REU program began, with support from NSF, in the summer of 1997 with Kazem Mahdavi (algebra, SUNY Potsdam), David Powers (graph theory, Clarkson University) and Joel Foisy (geometry/topology, SUNY Potsdam) involved as faculty advisors. Professor Mahdavi had for several summers before that led individual SUNY Potsdam students in various research projects, with financial support from SUNY Potsdam. In 1996, the author was a new PhD, but he had spent two very positive summers working as a student in the Williams College SMALL program. We were all delighted to begin the program. In 1998, we did not have a summer program, but for every summer since 1999, we have had a summer program, with financial support from NSF and NSA. We have had some changes in the faculty members involved each year, but the overall structure of the program has changed little. We have typically had between 3 and 5 different research groups each summer. This past summer, our faculty advisors were Joel Foisy (graph theory, SUNY Potsdam), Chrisino Tamon (applied graph theory, Clarkson) and Blair Madore (ergodic theory, SUNY Potsdam).

For eight weeks, students work daily in groups of 3 or 4 with an advisor on an original research problem in mathematics. We have ambitious goals for our students. We would like to help them build their confidence in their ability to do research independently, and to help them build their appreciation and understanding of the vast field of mathematics. We also want to help them to improve their oral and written communication skills. Each group presents their results to the other students, at least twice during the eight weeks. Students are encouraged to present their results at a national meeting. Each group also uses LaTeX to prepare a paper on their work. We also try to expand our students' mathematical horizons through hosting a weekly guest speaker, followed by lunch.

In order to encourage student-student interaction, we house the students in a cluster of on-campus apartments. We also organize a couple of outings for the students, including a day-trip to Ottawa, and a hike in the Adirondacks. We have also hosted barbeques at the beginning and endings of the program.

Students are assigned to their team before the program begins. We believe this allows them to start focusing in on a specific problem on the first day they

Received by the editor February 7, 2007.

arrive. Though there will be differences among the different teams, each day is structured fairly rigorously. A typical group might meet in the morning to discuss background, brain-storm on possible theorems and counter-examples, and to check proofs. Lunches are informal, except on days we have a guest speaker. In the afternoons, in a spirit of helping our participants become independent researchers, students generally work individually or in groups, without the faculty advisor. The faculty advisor would be available for consultation and help. If the situation demands, the entire group may arrange to meet.

The faculty members provide support for the students by helping them get started on a problem, and directing them to relevant background sources. The students and the faculty members check each other's work, and ideally we jointly develop publishable mathematics. At the later stages of the program, the faculty members help the students communicate their results both orally and in writing. In our experience, we have found that small groups of students working on a research problem is a successful paradigm. Brain-storming in a small team setting is useful for both experienced and less experienced participants. Explaining their ideas, both orally and in writing, to their peers is excellent for both the explainer and for the one who is trying to understand what the other is trying to say.

In choosing research problems for the students, we have several requirements in mind. First of all, problems should be approachable by good undergraduate math majors of varying backgrounds and talents. Second, they should require very little start-up time so those students can begin engaging in real research as soon as possible (eight weeks is a short amount of time!). Finally, it should be possible to obtain substantive results with enough time for writing up these results.

We feel we have been successful in finding such problems for our students in past summers. Having said that, finding good problems is probably the most difficult job for the faculty advisor. Where do we find good problems? We use a variety of approaches: looking through journals (MathSciNet is very useful for finding potential sources of inspiration), talking to research mathematicians, talking to mathematicians active in research but who primarily teach undergraduates, attending conferences, and attending MAA and Pi Mu Epsilon student talks at Mathfest are just some of the ways we find inspiration for problems. Often a good problem from one summer can be modified into another good problem for the next summer.

2. Recruitment

We recruit nationally, though we find comfort in knowing that a wide pool of qualified students can be found at numerous universities and colleges within three hundred miles of Potsdam. We recruit students in their junior (or in exceptional cases: sophomore) year, and we expect they will continue their undergraduate studies in the following Fall. All should have completed three semesters of calculus, one semester of linear algebra, and at least one more advanced mathematics course.

We will ask students to send the following material as part of their applications: (1) A letter describing mathematical background and interests, with an indication of topic preference. (2) A transcript. (3) Two letters of recommendation from mathematics faculty members, addressing mathematical aptitude, enthusiasm for the subject, and the ability of the student to work within a group. (4) A resume.

We feel that the two biggest recruitment tools are our web-page and word of mouth. In the fall of 2005, we improved our program's web-page, so that it now has links to descriptions of our various projects, as well as links to local attractions in the Potsdam area. We feel that this improvement has helped to increase the volume and quality of our applicants. In terms of word of mouth, we plan to continue to e-mail colleagues and the Project NExT list-serves to advertise our program, as we have done in the past. Our previous year's web page is located at:

http://www.clarkson.edu/mcs/reu.html

We do hope to attract students from under-represented groups to our program. The Principal Investigator has a history of working with SUNY Potsdam's CSTEP (Collegiate Science and Technology Entry Program) program. The purpose of CSTEP is to increase the number of historically underrepresented students who enroll in and complete undergraduate and graduate programs leading to professional licensure or to careers in mathematics, science, technology (MST), and health related fields. In our 2006 program, CSTEP and SUNY Potsdam provided funding for one Native American math major from SUNY Potsdam to take part in our program.

3. Project Evaluation and Results from Prior Support

In addition to giving numerous talks and posters at the local, regional and national level (several were award-winning), our groups have published several papers in journals ranging from the Pi Mu Epsilon Journal on up to standard professional journals.

We track our students after they leave the program. If we have not heard from a student in a while, it is quite easy to do a web search to determine if our former student is in graduate school. We administer pre and post-project surveys. Our surveys ask students questions that relate to their career plans, how well they liked their research topic, how well they interacted with other students and faculty in the program, the facilities, how they learned about our program, and what they perceived to be general strength and weaknesses of our program. Our surveying has shown our students to feel very positive about mathematics and the program. This was not surprising. We have learned other useful information, however, from the surveys. For example, in our 2006 surveys, we learned that several of the students were disappointed that we did not have any female guest speakers. This was an important piece of information that we will try to address in our 2007 program, should we be lucky enough to receive funding.

We have had an NSF supported REU program every summer since 1997, with the exception of 1998. According to our records, from 1997-2005, we have had 105 different students (31 women) participate in our REU program (some more than once), 9 of them participated in two different summers, and one student managed to participate for 3 summers. We have had 70 student positions from NSF funding, 18 from NSA, 15 from SUNY Potsdam, 8 from Clarkson University, and the remaining positions were funded from outside universities. We know about all but 15 of those 105 different participants. A total of 9 of these students have completed their Ph.D.s and are employed either in post-doc or tenure-track positions in mathematics. Forty-six of these students are currently in graduate programs in the mathematical sciences (one of these in physics, one in biostatistics, one in computer science and one in engineering). Nine of our former students are teaching or are preparing to teach at the K-12 level. Ten are still undergraduates. Thirteen are engaged in various occupations. One is working for the department of defense, and yet another is a peace activist. We also have a former student in a graduate program in music technology at McGill University. Yet another is teaching at a Community College. One recently completed her Ph.D. in statistics and is working in industry.

Our post-doctoral REU participants have been recently employed at the following institutions: Rice University, the University of Chicago, University of Arkansas Fort Smith, California State East Bay, University of Kansas, Cleveland State, Texas A & M, and University of Colorado at Boulder.

4. Conclusion

The most important aspect of our program is that the students and the faculty both enjoy immensely working on mathematical problems together. We are grateful for the support NSF, NSA, SUNY Potsdam and Clarkson have provided over the years, and we look forward to next summer's program.

DEPARTMENT OF MATHEMATICS, SUNY POTSDAM, POTSDAM, NY 13676
E-mail address: `foisyjs@potsdam.edu`

The Trinity University Research Experiences for Undergraduates in Mathematics Program

Scott Chapman

Trinity University is one of the Southwest's leading undergraduate institutions. For the last 14 years, U.S. News and World Report has ranked Trinity number 1 among comprehensive universities in the Western part of the United States. Undergraduate research is central to the overall goals and priorities of the academic program at Trinity University and Trinity's Chemistry, Computer Science, Biology and Mathematics Departments have a long history of substantial external funding for such activities. As evidence of this strong commitment to summer research, during May, June and July of 2006, 97 undergraduate students conducted research on Trinity's campus; 80 of them were Trinity students, 17 were from other institutions. Their research in Biology, Chemistry, Engineering Science, Mathematics, Biochemistry, Physics, Computer Science, Psychology, Neuroscience, Geosciences and Political Science was funded by 11 different external agencies[1].

The Mathematics Department at Trinity University recognizes and endorses the University's overall goals as described above. The Department has functioned as an NSF supported REU Site under three different grants (from 1997 to 1999, 2001 to 2003, and 2004 to 2006). The program has become an intricate part of the Department and has played a key role in strengthening and improving the mathematical education we offer our own students. While the accomplishments of our Program have been detailed in publications such as *Math Horizons*[2] and *Focus*[3], we operate our program on the principle that no matter how successful we are, there is always room for improvement. Our goals have remained consistent over our 9 REU Programs.

• To provide 12 student participants per year with an understanding of, an appreciation for, and an experience in the nature of mathematical research and the life of a mathematical researcher, to a degree that encourages them to pursue the study of the mathematical sciences on the graduate level.

Received by the editor December 1, 2006.

[1]Information in the first paragraph was provided by the Trinity University Office of Academic Affairs or taken from their summer research web site http://www.trinity.edu/org/student_summer_research/index.htm.

[2]S. T. Chapman, REU Spotlight: Trinity University, *Math Horizons* **11**, pp. 26–27, 2004.

[3]M. Martelli, The undergraduate student poster session, *Focus* **25**, p. 23, 2005.

FIGURE 1. The 2006 Trinity REU Participants

- To offer these experiences with an eye toward students who do not typically have these opportunities. Of particular interest are students who are either female or a member of an under-represented group.
- To develop in each participant superior skills in mathematical writing, oral mathematical presentation and poster design.
- To produce quality mathematical work appropriate for publication on our web site or in undergraduate research journals and, when possible, in higher level research journals.
- To extend the research experience beyond the 9 weeks at Trinity by motivating the students to present their REU research at a regional/national meeting.
- To gain experience in the use of computers and their interaction in mathematical research.
- To immerse the participants in the culture of mathematics and instill in each member a lifetime appreciation of the value of collegiality and group interaction.

While the primary focus of our Program is mathematical, the Trinity mathematics faculty believe deeply that a strong recreational component of the program is vital to its success. Our past programs have featured a strong schedule of activities that included a cultural tour of the downtown area, weekly flag football and ultimate frisbee contests, social gatherings at faculty member's homes, and a tubing trip down the Guadalupe River. One of the more popular parts of our previous programs has been the daily 3PM refreshment "Break," which gives all participants and faculty an opportunity for camaraderie and conversation.

We traditionally break our 12 students into 4 research groups, which vary (according to faculty availability) but usually consist of Algebra, Combinatorics, Discrete Dynamical Systems and Mathematical Biology. Students are selected for the Program by the Individual Project Directors and hence are aware of their assigned Projects when they accept their summer offer. Participants are housed in campus dormitories and roommates are assigned outside of each participant's research group in hopes of developing a higher level of interaction between students working on different projects. Once participants arrive on campus, they are assigned a work area in the Mathematics Department's computer laboratory, which includes for each participant a Dell PC dedicated to their personal summer use. We believe strongly in an intense opening week of the Program. The first morning includes a comprehensive orientation of the campus, the local area and the procedures of the University and the Program itself. The initial meeting of participants and their Project Directors takes place early the first afternoon, and is followed by a Panel Discussion of 4 to 5 Trinity REU Alumni. Group meetings continue every morning for the remainder of the first week as faculty review essential introductory material. From Tuesday to Friday, the remaining first week afternoons will consist of two colloquia. The first is a "Topic Colloquium" is delivered by one of the participating faculty and is intended to give all participants an introduction to that person's area of research. A second "Enrichment Colloquium" is directed toward broader issues that impact the participants' mathematical development,such as (a) how to give a successful mathematical presentation, (b) how to choose and apply to a graduate program in the mathematical sciences, and (c) getting started with the typesetting system LaTeX .

The remaining 8 weeks of the Program are dedicated to the completion of each group's research project. Participants will begin in early May to receive reading material and suggested problems from their Project Director. During the first week's meetings between the Project Directors and their research groups, the faculty will review this material and discuss the positive and negative aspects of each proposed research project. Early in the second week, each research group will make a decision on the specific problem/topic that the group will pursue. Our past experience indicates that this "choice" is an important aspect of the Program. After the first week, research groups traditionally meet daily with their Project Directors. At the end of the second week, each research group will give an oral presentation (no more than 30 minutes in length) explaining their selected problem to the entire group. A written document supporting the presentation will be submitted to the Project Director. Each Project Director evaluates their group's oral presentation and written description and communicate this evaluation in writing to the research group.

During the fifth week, each research group presents a written midterm progress report to their Project Director. The report will not merely be an outline of the progress to this point, but should contain arguments supporting the student's research. They will present an oral progress report (of no more than 30 minutes) to the entire REU group and again each Project Director will evaluate both the oral and written reports and communicate this finding in writing to their group. This process will culminate during the ninth week when the group gives their final 60 minute oral presentation and turns in their final written report. The written report becomes part of Trinity University's *Mathematical Technical Reports* series and be posted in the REU Project Archives on our web site.

We have put a substantial amount of time and energy into the development of our web resources for student recruitment. This has paid off as the number of applications we have received has risen drastically since 2001. Moreover, since we began tracking our web site in February of 2003, the site has had over 19,000 unique hits [4]. The web site is one of our main tools for recruiting. It contains an

---

[4]http://extremetracking.com/open;sum?login=tureu

indepth description of the Program, including descriptions of the available research projects, photographs from past programs, a summary of past participant current activities, a summary of written student evaluation remarks, an archive of past REU projects, a summary of published research, and the forms that prospects need to apply. We plan to continue our important emphasis on the recruitment and participation of female students and students from under-represented groups. Our goal is to *exceed* the figures of 44% participation by females and 17% participation by under represented groups which were achieved by both the 2001–2003 and 2004–2006 Programs. Some data concerning recruiting is contained in Table 1.

TABLE 1. Summary Data for Years 2001-2006

Years	# of Applicants	# of Offers	Male Participants	Female Participants	Participants from Under-Rep. Groups
2001–03	330	46	20	16	6
2004–06	381	45	20	16	8
Total	711	91	40	32	14

Our system of program evaluation is four tiered and includes 1) an anonymous written evaluation by the participants and Graduate Mentors, 2) an exit interview between each participant and Program Personnel, 3) comments and opinions solicited from Program Alumni, and 4) an external review by a mathematician of national stature. Our written evaluation consists of both numerically scored and free response questions. We set as a goal an average response from the participants of at least 5.0 on each scaled question. Should the average on such a question fall below this level, a written plan addressing corrective measures will be included in our following yearly progress report to NSF. We include below comments from our 2005 external review.

Here are the highlights of the program: a rigorous and fair selection process that ensures the presence of highly-motivated student participants, a group of dedicated faculty mentors with the experience and expertise to select challenging projects that match student interests and abilities, good physical facilities that encourage academic interactions and facilitate the exchange of intellectual ideas, the establishment and maintenance of a strong communications network with participants from prior years. ... I was particularly impressed by the strong sense of community that appears to have developed among the student participants. ... The development of such a group dynamic can clearly be attributed in my opinion to the enthusiasm and guidance provided by the faculty mentors. ... The Trinity REU program is to be commended for its serious efforts to encourage participation by a diverse group of students. There is clearly a healthy balance of student participants both in terms of gender and ethnicity. ... There is no question that the program is an unqualified success. - from 2005 External Review by **Efraim Armendariz, Professor and Chair, Department of Mathematics, University of Texas at Austin**

Since its inception, the Trinity REU has hosted 88 different participants. As of the submission of this article, 67 have received their undergraduate degrees and 51 of those students (or 76%) have enrolled in some type of graduate program.

Of the 51 REU participants who began graduate study, 36 (or 71%) enrolled in a program in pure or applied mathematics, 5 enrolled in computer science, and 2 or less enrolled in each of the fields of law, logic, engineering, mathematics education, physics, statistics and economics. Of the 36 who began in mathematics, 28 (or 78%) enrolled in programs which are ranked as Group I by the American Mathematical Society. We are particularly pleased with the following statistics: (a) of the 26 female REU participants who have received their undergraduate degrees, 22 (or 85%) have enrolled in graduate study, (b) of the 6 REU participants who identify themselves with under-represented groups and have received their undergraduate degrees, 4 (or 67%) have enrolled in graduate study. The group which has completed our program has received further distinction. Two of our participants (M. Holden (2002) and W. Meyerson (2003)) completed in 2005 the prestigious Tripos III Program at Cambridge University. Two more of our participants have received Homeland Security Fellowships to begin their graduate study (P. Baginski (2001) and K. Cervello (2004)). Two other students (D. Morris (2002) and G. Harrison (2005)) have spent time overseas supported by Fulbright Fellowships. In total, eight of our participants have completed the Budapest Semesters in Mathematics Program (T. Moore (2002/3), M. Bannister (2003), T. Landry (2003), J. Chaika (2003), B. Finklea (2003), P. Blain (2005), E. Treviño (2005), C. Vinzant (2005) and M. Gallant (2005)). Another participant (J. Bauman (2006)) has completed the Mathematics in Moscow Program.

During the 2003 program, we began to emphasize student oral presentation and poster design skills. We added improvement of these skills formally as a goal to our 2004–2006 proposal. We believe that this emphasis has paid off. Over the past 3 years, 12 of our participants have presented talks at the annual Young Mathematicians Conference (YMC) at Ohio State University. During 2003 and 2004, 7 more of our participants presented talks at the annual Big Sky Conference on Discrete Mathematics at the University of Montana-Missoula. Beginning with the January 2003 joint meeting, 12 participants have presented posters at the annual AMS-MAA Undergraduate Poster Session. Several of our former participants have won awards for presentations, including 6 Meritorious Awards of $100 each for posters at the AMS-MAA Undergraduate Poster Session.

Since our initial program in 1997, 16 papers have been published or accepted in regular research level journals which contain Trinity REU participants as co-authors. As of the submission of this article, 3 other papers have been submitted and numerous papers (including all those from work in the 2006 program) are in preparation. We are particularly pleased with the following statistics: (a) of the 75 REU participants prior to the 2006 program, 35 (or 47%) are a co-author of a published or accepted paper in a regular research level journal, (b) of the 30 female participants prior to 2006, 18 (or 60%) are a co-author of a published or accepted paper in a regular research level journal, (c) of the 11 participants from under-represented groups prior to 2006, 4 (or 36%) have a similar co-authored publication.

Publications Resulting From The Trinity REU Program
(Undergraduate co-authors marked with an asterisk-*)

(1) J. Amos*, E. Treviño*, I. Pascu*, V. Ponomarenko and Y. Zhang*, The multi-dimensional Frobenius Problem, to appear in *Advances Appl. Math.*

(2) P. Baginski, S. T. Chapman, C. Crutchfield*, K. G. Kennedy* and M. Wright*, Elastic properties and prime elements, to appear in *Results Math.*

(3) P. Baginski*, S. T. Chapman, K. McDonald* and L. Pudwell*, On cross numbers of minimal zero sequences in certain cyclic groups, *Ars Combin.* **70**, pp. 47–60, 2004.

(4) P. Baginski, S. T. Chapman, M. Holden* and T. Moore*, Asymptotic elasticity in atomic monoids, *Semigroup Forum* **72**, pp. 134–142, 2006.

(5) M. Banister*, J. Chaika*, S. T. Chapman and W. Meyerson*, On a result of James and Niven concerning unique factorization in congruence semigroups, to appear in *Elem. Math.*

(6) M. Banister*, J. Chaika*, S. T. Chapman and W. Meyerson*, On the Arithmetic of Arithmetical Congruence Monoids, to appear in *Colloq. Math.*

(7) C. Bowles*, S. T. Chapman, N. Kaplan* and D. Reiser*, On Delta Sets of Numerical Monoids, to appear in *J. Algebra Appl.*

(8) A. Brown*, A. Gedlaman*, A. Holder, and S. Martinez*, An Extension of the Fundamental Theorem of Linear Programming, *Oper. Res. Lett* **30**, pp. 281-288, 2002.

(9) L. Cayton*, R. Herring*, A. Holder, J. Holzer*, C. Nightingale*, and T. Stohs*, Asymptotic Sign Solvability and the Dynamic Nonsubstitution Theorem, to appear in *Math. Methods Oper. Res.*

(10) K. Cervello*, D. Terry*, V. Ponomarenko and L. Zhu*, The Extraction Degree of Cale Monoids, *Semigroup Forum* **72**, pp. 149–158, 2006.

(11) S. T. Chapman, V. DeLorenzo* and H. Swisher*, On the asymptotic behavior of irreducibles in block semigroups, *Semigroup Forum* **63**, pp. 34-48, 2001.

(12) S. T. Chapman, J. Herr* and N. Rooney*, A factorization formula for class number two, *J. Number Theory* **79**, pp. 58-66, 1999.

(13) S. T. Chapman, M. Holden* and T. Moore*, On full elasticity in atomic monoids and integral domains, to appear in *Rocky Mountain J. Math.*

(14) J. Cuomo*, N. Nwasokwa* and V. Ponomarenko, Jump Systems and Manhattan Polytopes, *Australas. J. Combin.* **31**, pp. 135–143, 2005.

(15) D. Dunn*, S. Graham* and G. Salazar, An Improvement of the Feng-Rao Bound on Minimum Distance, *Finite Fields Appl.* **12**, pp. 313–335, 2006.

(16) V. Lyubashevsky*, C. Newell* and V. Ponomarenko, Geometry of Jump Systems, *Rocky Mountain J. Math.* **35**, pp. 1675–1688, 2005.

TRINITY UNIVERSITY, DEPARTMENT OF MATHEMATICS, ONE TRINITY PLACE, SAN ANTONIO, TEXAS 78212-7200,

E-mail address: schapman@trinity.edu

Undergraduate Research in Mathematics at the University of Akron

Jeffrey D. Adler

1. The local setting

The University of Akron (UA) is a state-assisted institution. About 95% of our approximately 23,000 students come from Ohio, a state where relatively few adults have a college degree. The rest come from 41 other states and 68 foreign countries. Because UA has open enrollment, we will always fare poorly on many of the traditional measures that go into college rankings. Thus, to anyone who is too focused on bean-counting, we appear to be a far from ideal climate for conducting undergraduate research.

But despite what the averages say, we have many fine students. They come to us via a rapidly growing honors program, a large college of engineering, or other routes. For example, one of our top mathematics students of recent years came to Akron because of our dance program.

2. Early history

Intentionally or not, Judith Palagallo and Thomas Price founded our undergraduate research program. In each of 1998 and 1999, Palagallo invited one student to join her on a research project. The net results were an honors paper; presentations at the AMS/MAA Joint Meetings in 1999, the MAA MathFest in 2001, and at the Nebraska Conference for Undergraduate Women in Mathematics (2002); and two publications [4, 6].

The pace of activity accelerated in 2000, when Palagallo and Price conducted a summer undergraduate research program with local funding. This resulted in two MathFest presentations and one publication [9].

At the same time, we started encouraging some of our students to attend REU programs elsewhere. This was the start of a trend, and in recent years eight of our students (seven women, one man) have participated in eight different external REU programs, sometimes continuing their projects upon their return to Akron. So far, this effort has resulted in an article [7] and a technical report [2].

In the summer of 2002, Price established the Fibonacci Forum, which involved eight undergraduates over several years. One of the students was a McNair Scholar.

Received by the editor October 25, 2006.

When Palagallo and Price went on sabbatical, the Forum continued under the direction of one of the students. This activity led to several honors projects; three presentations at MAA meetings; and an article [8] that has been submitted for publication.

In the summer of 2003, our (still informal) program attracted a McNair Scholar from a nearby university. She worked with Palagallo, and presented her results at the 2003 MAA MathFest, winning a Best Presentation award.

Where are they now? Of the student participants in the activities described above, one has completed a Ph.D., and is now an assistant professor of mathematics. Another received an NSF Graduate Fellowship, and is expected to finish her Ph.D. in May. Our internal McNair Scholar is in graduate school in statistics. Five other students, including our external McNair Scholar, are presently in Ph.D. programs in mathematics or physics. (One of these was an invited graduate student participant at the 2006 Nebraska Conference for Undergraduate Women in Mathematics.) Two completed master's degrees in mathematics at UA, and now work as a mathematical modeler and a public school teacher. One spent a year working for the NSA, and recently completed a law degree. One is a college senior.

3. Formal REU program

Starting in the summer of 2005, we have operated an eight-week REU program with major support from the National Science Foundation, and additional support from our unit within UA, the Buchtel College of Arts and Sciences. Palagallo, Price, Jeffrey Riedl, and I run the show. In the summer of 2006, we used local funds to employ a graduate assistant. Not only could she help with clerical tasks, but her master's research was related to one of our summer research problems.

Each year, we recruit around eight students from around the country, including one local student. We received 131 applications the first year. The second year, partly in order to decrease this, we deliberately presented fewer sample projects on our web page. It worked: We received only about 100 applications.

Community. The students all live together in campus dorms, together with the students in UA's summer REU in polymer science. We organize a few social events and field trips, and the polymer science students join us for one or two of these. For example, during the first week, we attend a game of the Akron Aeros, our local AA baseball team. Later in the summer, we hear a concert at Blossom Music Center, the summer home of the Cleveland Orchestra. The students also tend to find plenty of social outlets on their own. It helps that one of the students is local.

We take a few field trips to local industrial concerns and research laboratories.

Since we want all of our students to understand each other's research problems, during the first week all attend all introductory lectures. Once a week, we require all students to give progress reports. This is followed by a communal lunch.

Visitors. We typically have four or five visiting faculty, who speak on a variety of topics in both pure and applied mathematics. (Examples: wavelets; coatings of nanofibers; Dirichlet's Theorem; working for the NSA; the Seven-Color Theorem; p-adic numbers; etc.) In addition to mathematical talks, we provide an introduction to LaTeX (sometimes in the form of Scientific Workplace), and discuss how to give a technical presentation.

Finances. In addition to free lodging, the students receive a stipend, food allowance, travel reimbursement, and a parking pass if necessary. When the travel budget allows (which so far it always has), we provide partial support to attend the Joint Meetings.

Problems.

- Gessel [5] proved the curious (to me) result that a natural number x is a Fibonacci number if and only if $x^2 \pm 4$ is a perfect square. More specifically, if F_n and L_n denote the nth Fibonacci and Lucas numbers, respectively, then the only integral solutions to

$$x^2 + (-1)^n 4 = y^2$$

  are $(\pm F_n, \pm L_n)$. The first 2005 team proved an analogous theorem for generalized Fibonacci and Lucas numbers, as well as Fibonacci and Lucas polynomials.

- In order to prove their results, the first team required knowledge of the units in certain quadratic extensions of polynomial and other rings. Some of this was provided by the second team.

- A third 2005 team found the method used to encrypt several sixteenth-century letters to the king of Spain from some of his ambassadors in Italy. This result is of some historical interest, because our collection is only a small part of a larger one in the State Archive at Simancas, Spain. The task was complicated by the fact that, in the words of the students, "the documents are over four hundred years old, the ink has leaked through, and the handwriting finds its closest modern parallels on prescriptions."

- A 2006 team generalized a fundamental theorem of Bandt [3], concerning the construction self-similar periodic tilings, from Euclidean space to what according to Strichartz [10] should be its natural setting: nilpotent Lie groups having certain rationality properties. One reason for interest in such tilings is their connection with wavelets.

- Riedl has been able to boil down one of his own research problems, concerning the classification of monolithic subgroups of wreath product p-groups, into a combinatorial linear algebra problem. Seven students (spanning 2005 and 2006) collectively solved this problem. Since it's part of a much bigger problem, there is plenty left to do in 2007.

Rough assessment. All of the problems that we posed have been solved, sometimes in greater generality than we had expected. Three 2005 teams made well-received presentations at the 2006 AMS-MAA Joint Meetings, and I expect at least one 2006 team to present in January, 2007. One paper [1] on the 2005 results has been accepted pending revision, another is in preparation, and we expect to prepare a paper on the 2006 results in the spring.

Should all of these papers appear, this will make our program look good. However, there is always a potential conflict between the goals of producing the best research possible and producing the best educational experience possible. The latter is more important, but its degree of attainment is harder to measure.

Rigorous assessment. This must involve long-term tracking of our students. However, after only two summers, it is too early for that. So far, we must rely on anonymous student evaluations. But these have been positive, and have included specific suggestions for adjustments. Many students reported a reinforced desire to attend graduate school in mathematics. None reported a decline in interest.

Two students reported having burned out on mathematics between their acceptance of our offer and their arrival in Akron. One is still in mathematics. In time, we will know the extent of our role in that.

Where are they now? The 2006 participates are back in school, of course, and the majority of them are applying to graduate school in mathematics. One intends to teach middle school. Of the 2005 participants who are still in college, one worked for the NSA last summer, and one attended another REU (and gave us lots of feedback!). Both intend to pursue graduate studies in mathematics. Of those who have graduated, two are in Ph.D. programs in mathematics, one works for the NSA, one is an actuary, one is applying to math graduate school after a year abroad, and one is preparing for law school.

4. Future

In the immediate future, our program will expand. First, several of our colleagues in applied math will get involved using their own grant money, thus increasing the number of both students and faculty involved. Second, we will bring in outside faculty mentors (one a professor at a nearby college, the other a former student of ours who will have just received her Ph.D.).

We will seek to work even more closely with UA's polymer science REU. For example, it makes sense for us to offer a joint GRE preparation workshop.

In the longer term, the nature of our program will depend on our ability to obtain continuing external financial support, and our ability to find replacements for faculty members who have retired recently. Among these are Palagallo and Price, although they will be involved for at least one more summer.

References

[1] J. Adler, R. Fuoss, M. Levin, and A. Youell, *Reading encrypted diplomatic correspondence: an undergraduate research project* (submitted).

[2] I. Averill and J. Gregoire, *Tilings of low-genus surfaces by quadrilaterals*, Technical Report 02-13, Rose-Hulman Mathematical Sciences Technical Report Series, 2002.

[3] C. Bandt, *Self-similar sets 5. Integer matrices and fractal tilings of* $\mathbb{R}^n$, Proc. Amer. Math. Soc. **112** (1991), 549–562.

[4] M. Breen and J. Palagallo, *Determining the area of fractal tilings*, Pi Mu Epsilon Journal (Spring, 2003).

[5] I. Gessel, *Problem H-187*, The Fibonacci Quarterly **10** (October 1972), 417–419.

[6] S. Hagey and J. Palagallo, *Complex bases and fractal tilings*, Math. Gazette **85** (2001), 194–201.

[7] B. Marko and J. Riedl, *Igusa local zeta function of the polynomial* $f(x) = x_1^m + \cdots + x_n^m$, Yokohama Math. J. **51** (2005), 117–133.

[8] L. McDonnell, B. Polovick, and T. Price, *Evaluation formulas for a family of integrals* (submitted).

[9] J. Palagallo and M. Palmer, *Analysis of an irregular Sierpinski triangle*, Fractals **8** (2004), 137–144.

[10] R. S. Strichartz, *Self-similarity on nilpotent Lie groups* (1992), 123–157.

DEPARTMENT OF THEORETICAL AND APPLIED MATHEMATICS, THE UNIVERSITY OF AKRON, AKRON, OH 44325-4002

E-mail address: adler@uakron.edu

The Duluth Undergraduate Research Program 1977-2006

Joseph A. Gallian

Introduction

In this article I describe various aspects of my program such as funding, problem selection, recruitment, structure, follow through, and results. It is an update of the one that appeared in the Proceedings of the Conference on Summer Undergraduate Mathematics Research Programs
(see www.ams.org/employment/REUproceedings.html).

Funding

Over the years primary support for my programs has come from the National Science Foundation and the National Security Agency. My department and college have also contributed substantially. In 2006, participants received a stipend of $2250, a travel allowance, housing and a subsistence allowance of $1250.

Problems

Obviously, the selection of appropriate problems is of fundamental importance to a successful research program. I search for problems that meet the following criteria: not much background reading is required; partial results are probable; recently posed; new results will likely be publishable.

Graph theory, combinatorics and number theory provide the source of most of my problems. I find problems by perusing recently published journals, math arXiv, attending conferences, and writing people. As a rule of thumb, I begin the summer with twice the number of problems as I have students. I occasionally have a student continue work begun in a previous program by someone else. Although each student has his or her own problem, I encourage students to discuss their problems with fellow program participants. Matching students with problems is a critical task. The skill with which this is done is a major factor in the success of a program. Undergraduate students, even the most talented ones, have a tendency to become frustrated and want to give up too soon. Here I serve as a counselor

Received by the editor October 3, 2006.

and cheerleader, offering an idea, a reference or a pep talk. Of course, it sometimes happens - about half the time in fact - that a problem is inappropriate. Sometimes a problem is too easy or too hard; sometimes we discover it is already solved by someone else. In these cases I simply assign a new one.

Although I am the only faculty person involved in the REU, two graduate students who were former participants in the program return to act as research advisors to new participants. I choose advisors who have the personality and talent to interact well with others. Another important source of support is the fifteen or so former participants who visit from one to three weeks.

Recruitment

Recruiting well-qualified students has not been difficult. Applicants are generated by a mailing of announcements to a large number of mathematics departments nationwide and a mailing of an announcement and descriptive letter to many students who place in the top 100 of the Putnam Competition. Word-of-mouth advertising by former participants has resulted many outstanding applicants. Detailed information and an application form is available from my web site: www.d.umn.edu/~jgallian. Over the past two years the program web site has been visited about 8,500 time per year. In recent years the program has had 8-10 participants selected from approximately 80 applicants.

Selection of participants is based on letters of recommendation, response to questions on the application form, performance in high school mathematical competitions and the Putnam competitions, previous research experience, reputation of the home school, and course work. I give extra weigh to those applicants who have a particular interest in combinatorics. This can be demonstrated by the courses they have taken, participation in the Budapest Semesters in Mathematics Program (which emphasizes combinatorics), past research projects, or independent study the applicants have done. Students who have participated in mathematics summer programs such as the Hampshire program, the Ross program or the Boston University Promys program are usually well prepared in combinatorics.

It takes more than problem solving skills to do original research. Social skills are also needed to get along well with others in the program. Good "group chemistry" contributes much to the success of the program. Desire to succeed, enthusiasm, outstanding work ethic, and ability to work well with others are as important as raw talent. I try to select people who would be fun to spend the summer with.

Structure

My programs are loosely structured. Each student is given his or her own problem together with an article or two as resource material. Each week the participants give talks on their progress during the previous week to the group. This gives me, the research advisers and visitors an opportunity assess progress, raise questions, make suggestions and identify difficulties. Preparing their talks helps the students organize their work. It also serves as good preparation for presenting their results at a conference after the program is over. Occasionally we have a visitor present a colloquium. Typically, we have lunch as a group three times a week. I meet with students individually upon request and when I feel such a meeting might be beneficial. I have weekly meetings with the advisors to discuss the progress of each participant and plan ahead.

The housing arrangements are a critical part of the program structure. UMD has attractive, furnished three-bedroom apartments with living rooms, kitchens and bathrooms. I reserve six of these that are contiguous (in fact, the students have a wing of the building to themselves). The students move freely among the six apartments so that the housing is like one large apartment complex. Since there are no instructional lectures and few formal meetings, the students spend most of their time in the apartments. The research assistants, returning participants, and visitors live with the new participants or in adjacent apartments. This mixture of old and new people works well. The living arrangements naturally foster interaction and collaboration. I frequently drop by the apartments to see how things are going, to answer questions, and to provide encouragement.

"Field trips" are a component of the program. It is important that the students enjoy their summer. Together, we go alpine sliding, rock climbing, white water rafting, visit beautiful parks in the area, bike, play softball and basketball, and walk along the shore of Lake Superior. On weekends, the participants have access to two university vehicles paid for by the UMD Math department. This makes it convenient for them to see movies, shop and eat out. Occasionally we have lunch at a restaurant. The field trips and group lunches foster a sense of camaraderie among the students, advisers, visitors and me.

Ideally, by the seventh week of the program the students are writing up their work. Papers are written in a style suitable for submission to a research journal. All manuscripts are read by me, the two research advisers, and one or more visitors. The readers make suggestions and comments and eventually I come to an agreement with the students on versions to be submitted for publication. Typically this process is completed by the end of the summer program or shortly thereafter, although there have been instances where it has taken several years to get a paper in publishable form. (Once people leave Duluth there is a natural tendency to concentrate on other matters.)

Follow Through

Except for finding a sufficient number of appropriate research problems, the follow through on manuscript preparation is my most difficult job. When the students leave Duluth I often have, at best, a first draft of their work. It typically takes many letters and phone conversations before the manuscripts are ready to be submitted to journals. Then, many months later, there are the inevitable referees' reports recommending revisions, necessitating another round of letters, phone calls and rewriting. By the time the referees' reports come, the students are busy with other things and are not always eager to follow through.

I view publication of the work done in my REU as a beginning rather than an ending in itself. Indeed, many participants from my program have continued to publish as an undergraduate or graduate student.

I strongly encourage participants to present their work at the annual joint meetings of the American Mathematical Society and the Mathematical Association of America. Rather than participate in the student poster session or the special session for research by undergraduates, I prefer that the Duluth students present their work in the topic-specific contributed paper sessions. In sessions devoted to combinatorics or number theory the audience is interested in the topic and consequently the student will likely meet and have the opportunity to network with others who work in the field. Moreover, I feel that at a conference the undergraduates should be treated like professionals and not be segregated according to experience. Presenting their work at the joint meetings has proved to be a valuable experience for the students. They attend talks, meet people and have people ask about their work. Funding for this typically is provided by the home institution of the participants.

Follow through also includes writing letters of recommendation for fellowships and admission to graduate school and nominating participants for awards. In some cases I have written letters for former participants seeking employment after finishing the Ph. D. degree. I have even served as an external reviewer for tenure and promotion decisions for several participants from my program.

Project Evaluation

The effectiveness of the program is evaluated by a variety of means such as publications in well regarded professional level journals, talks given at conferences, number of students who enter graduate school, quality of graduate schools attended by the participants, number of students who receive Hertz, NSF, or DoD Fellowships, the extend to which participants return to visit future programs, and the long term relationships established among the participants themselves and with me. I track all program participants (see http:www.d.umn.edu/~jgallian) and maintain regular contact with many of them throughout their years of graduate school and after.

Broader Impacts

The development of human resources is the explicit purpose of the program. Taking classes and participating in math competitions provide students with little or no experience with the research process. It is rare for even the best senior thesis to measure up to the standards required for publication by mainstream research journals. In contrast, the Duluth program has an extraordinary record of professional-level research done by undergraduates. Besides learning first-hand the nature of mathematical research and becoming part of the mathematics community, outcomes of participation in the program include: increased self confidence and self esteem; motivation to pursue a Ph.D. degree; enhanced chances of being admitted to a first-rate graduate school and receiving a fellowship; and the development of a network of people who will likely be important members of the mathematics community. The most important contribution the Duluth program makes to the development of human resources is the training of future generations of mathematicians who will foster undergraduate research when they become professionals. Indeed, four participants from the program have been directors of their own REU programs, a fifth is part of a four-school collaboration to involve undergraduates in research, a sixth served as a faculty adviser in the Williams College REU, and two more have been advisers in MIT's Summer Program in Undergraduate Research.

Results

Although the program participants are undergraduates, the level of the research they have done is unequivocally professional. Indeed, more than 100 papers written in the program have been published in professional-level refereed journals. Among them are 44 in the journal *Discrete Mathematics*, 10 in the *Journal of Combinatorial Theory* and 8 in the *Electronic Journal of Combinatorial Theory*. See www.d.umn.edu/~jgallian for a complete list.

Through 2006, 143 students have participated in the program. Of the 129 participants in the program who have received their Bachelor's degrees as of 2006, 116 have gone to graduate school. Of these, 83 have gone to MIT, Harvard, Berkeley, Chicago, Princeton, or Stanford and 87 have won graduate fellowships (Hertz, DoD, NSF). Sixty-four participants now have the Ph.D. degree. Two participants have received a Clay Mathematics Institute Long-term Prize Fellowship (Bhargava and Biss) and two have won the American Institute of Mathematics Five Year Fellowship (Ng and Develin).

Many of the women in my program have had extraordinary success. In the seventeen year history of the Association for Women in Mathematics (AWM) Schafer Prize ten women from the Duluth program have won the award (Ana Caraiana, Alexandra Ovetsky, Melody Chan, Melanie Wood, Ioana Dumitriu, Ruth Britto-Pacumio, Catherine O'Neil, Dana Pascovici, Zvezdelina Stankova and Elizabeth Wilmer) and eight have been named runner up (Margaret Doig, Elena Fuchs, Wei Ho, Karola Meszaros, Beth Robinson, Jessica Wachter, Susan Goldstine and Zvezdelina Stankova. With two exceptions, every winner of the MAA Elizabeth Putnam Prize (for outstanding performance by a woman in the Putnam Competition) has been in the Duluth program.

In the twelve year existence of the AMS/MAA/SIAM Morgan prize for research by an undergraduate the Duluth program has had seven winners (Manjul Bhargava, Daniel Biss, Joshua Green, Melanie Wood, Reid Barton, Jacob Fox and Daniel Kane) and four runner-ups (Kiran Kedlaya, Lenny Ng, Aaron Archer and Samit Dasgupta). With one exception, the research done at the Duluth REU was a major factor in the decision.

Conclusion

Although the term "paradigm shift" is frequently overused, it is appropriate to describe the transformation that research by undergraduates has undergone in the thirty years since I had my first summer research program. What was once considered as "not realistic" or even an oxymoron, is now ubiquitous. I am proud to have had the opportunity to contribute to this change in the mathematics culture.

UNIVERSITY OF MINNESOTA DULUTH, DULUTH, MN 55812
E-mail address: jgallian@d.umn.edu

Proceedings of the Conference on
Promoting Undergraduate Research in Mathematics

Promoting Undergraduate Research in Mathematics at the University of Nebraska – Lincoln

Judy L. Walker, Glenn Ledder, Richard Rebarber and Gordon Woodward

The Department of Mathematics at the University of Nebraska – Lincoln (UNL) has several programs which promote undergraduate research in a variety of ways. Two of these are summer programs which draw from a national applicant pool: The Nebraska REU in Applied Mathematics (Section 1) is a traditional NSF-funded REU site, and Nebraska IMMERSE (Section 2) offers a summer "bridge" program (with a research bent) for students about to start graduate school in mathematics. IMMERSE is a relatively new program, started in 2004 as part of the department's Mentoring through Critical Transition Points (MCTP) grant from NSF. The MCTP grant also is now the primary source of funding for two conferences involving undergraduate research which the department launched in 1999: The Nebraska Conference for Undergraduate Women in Mathematics (NCUWM) (Section 3) and the Regional Workshop in the Mathematical Sciences (Section 4). The bulk of the program at NCUWM consists of talks by undergraduates on their own research, and, while the original goal of the Regional Workshop was to forge and maintain ties between faculty at smaller college and universities, it has recently been expanded to provide a forum for undergraduates to present their research. Finally, we offer several opportunities for our own undergraduates to do research: the MCTP Undergraduate Scholars program (Section 5), the Research for Undergraduates in Theoretical Ecology (RUTE) program (Section 6), the Undergraduate Creative Activities and Research Experiences (UCARE) program (Section 7), and two upper-level undergraduate courses which aim to give students a taste of mathematics research (Section 8). This article provides a brief overview of each of these programs; more details can be found online at `http://www.math.unl.edu`.

1. Nebraska REU in Applied Mathematics

The department has offered the Nebraska Research Experience for Undergraduates in Applied Mathematics as an eight-week NSF-funded REU Site each summer since 2002. We have offered an average of three projects each summer. A typical project group includes three or four undergraduates, one or more UNL faculty members, and a UNL graduate student. Our goal is to give the students as full a research experience as possible, including how to define a good problem, how to investigate

Received by the editor December 1, 2006.

the problem, how to come up with solutions and/or models for the problem, how to write mathematics, and how to give a talk. All of the projects are in applied mathematics, most have an interdisciplinary component (with an emphasis on biology), and many emphasize the role of computer exploration in gaining insight into mathematical problems. A few past project titles have been: "The Spread of Information and Social Interactions," "Exponential Stability of Dynamic Equations on Time Scales," "Chaos Theory in Food Chains," "Dynamics of Fish Populations Dependent on Cannibalism," "Game Theory and Population Dynamics," "Dynamics of Tumor Growth," and "Control Theory Techniques Applied to Population Problems."

Detailed project descriptions, including prerequisites, are posted on the web by the end of the preceding December. The students choose those projects they wish to apply for and rank them according to preference in their applications. Students are selected for individual projects before the Site starts, allowing for preliminary communication between the mentors and the students. When choosing students, we put a premium on how much value our Site can add to their education, rather than making admission decisions based solely on the quality of their previous work. We have had an approximately even gender balance among the student participants. A bit more than half have been from non-Ph.D. granting institutions, almost half have been from the midwest, and about 7% (fewer than one each year) have been from UNL.

There are certain features common to all of the projects. We want the students to have an experience which is very different from taking a course. The students are expected to work together closely. We want them to dive in head-first, learning material as needed. For the first few weeks, the students meet with their faculty and graduate student mentors daily for at least an hour, with the mentors giving informal lectures, exercises, and possibly reading assignments. We do not assume much expertise (if any) in the research topic, and so the material given in these first few weeks provides the students with the background necessary to do their research. The projects are quite demanding, often involving advanced material, and we've learned that most mentors find it desirable to continue to meet most weekdays throughout the eight weeks, initially to help the students stay on track and ask the right questions, and then to guide them in preparing their reports and presentations. During the last week of the Site, we hold a mini-conference, where the students give an oral presentation of their work. In most cases a paper is written, with work being finished after the Site is officially over. If the paper is submitted to an undergraduate journal, most of the writing is done by the students. If the paper is submitted to a professional journal, the faculty mentor takes a larger role in the finished product. Most of the students present their results at conferences during the year after the Site, with their travel being funded by the department.

2. Nebraska IMMERSE

As part of the department's MCTP grant, we run a summer program called Nebraska Intensive Mathematics: a Mentoring, Education and Research Summer Experience (IMMERSE). In a sense, Nebraska IMMERSE is really two programs: one that develops the teaching, research, and mentoring skills of graduate students and early-career faculty, and one that strengthens the preparation of students as they begin the transition from being undergraduate students to being graduate students.

We focus here on the "pre-grad" program; for information on the "early-career faculty" portion of IMMERSE, see the departmental website.

Sixteen "pre-grads" — students who will be starting graduate school, either at UNL or elsewhere, in the fall — comprise the participant list for IMMERSE. The pre-grads emerge from this program with a strong foundation on which their first-year graduate courses can build. They gain experience in working in groups on problems. They develop a solid understanding, with examples in hand, of several of the topics they will encounter in their graduate courses. They gain exposure to mathematical research and start to develop the skill of learning through mathematical research literature. They are also exposed to some of the issues which they will face as teachers, and they develop a network of peers and mentors on whom they can continue to rely as they begin graduate school.

The program lasts six weeks, and the main component consists of two intensive courses: one in algebra and one in analysis. The courses are at the advanced undergraduate/beginning graduate level, with, for example, Herstein's "Topics in Algebra" and Rudin's "Principles of Mathematical Analysis" as resources. However, rather than working through a textbook, the courses are structured around the reading of research papers. The instructors (pre-tenure faculty at four-year colleges who are participating in the early-career faculty portion of IMMERSE) select papers that use as tools some of the topics that typically appear in first-year graduate algebra and analysis courses, and then structure the IMMERSE courses around the material of the papers. The papers are typically relatively recent and reasonably short, and have a single easily-understood result.

For example, the algebra course from IMMERSE 2006 was structured around the 1995 *Houston Journal of Mathematics* paper "Parametric decomposition of monomial ideals (I)" by Heinzer, Ratliff, and Shah [1]. This paper is concerned with existence and uniqueness of decompositions of monomial ideals, that is, ideals generated by monomials in a fixed regular sequence $x_1, \ldots, x_n$ in a commutative ring R with identity. The decompositions under investigation are finite intersections of parameter ideals, which are particularly simple monomial ideals — those of the form $(x_1^{a_1}, \ldots, x_n^{a_n})R$. It is straightforward to show that in order for a monomial ideal J to admit such a decomposition, it is necessary that J contain a power of each element x_i. The authors show that this condition is also sufficient and prove that irredundant parametric decompositions are unique. (Think of unique factorization of integers or polynomials in one variable.) One feature of the article is a geometric interpretation of the parameter ideals occurring in each such decomposition, based on the "staircase diagram" associated to the ideal J. After a bit of time spent on a review of undergraduate-level material, there was an introductory lecture which introduced commutative algebra and explained why decomposing monomial ideals is interesting. The students then read through the paper, probably not understanding much of it beyond the statement of the main result. The intensive algebra course then began, covering the basics of commutative rings, ideals (including the operations of intersection, sum, product, radical, and colon), and the Noetherian property (including the Hilbert Basis Theorem). Notice that while these topics were inspired by the paper chosen by the instructors, most of them are still standard material in a first-year graduate course in algebra. Every so often, the students re-read the paper, understanding more and more each time. By the end of the course, the students had learned quite a bit of algebra, had some good examples in mind related to the topics covered, and completely understood a research paper.

Each course is team-taught by a pair of early-career faculty participants, with a mixed group of first-year and advanced graduate students serving as teaching assistants. Each course has a daily one-hour lecture plus an afternoon problem session run by the graduate student assistants. Students are encouraged to work together on problems and take turns presenting solutions on the board to each other. They also using LaTeX to formally write up their solutions to additional problems. There are several special presentations, colloquia and workshops throughout IMMERSE, each of which provides an additional forum for exploring issues the pre-grads are likely to face as they begin graduate school.

We solicit applications nationwide for thirteen of the sixteen pre-grad positions; the remaining three slots are reserved for incoming UNL graduate students who will be supported the following academic year as First-Year MCTP Graduate Trainees. Selection is based on transcripts, letters of recommendation, and a personal essay on what the student hopes to gain from the program. All acceptances are provisional, with the understanding that the student must actually be accepted to, and commit to attending, a graduate program in mathematics starting that fall.

3. NCUWM

The annual Nebraska Conference for Undergraduate Women in Mathematics (NCUWM) has been held in early February each year, starting in 1999. The conference brings together women undergraduate math majors from all over the United States. The main part of the program is a series of talks by the undergraduate women about their own research. Two prominent women mathematicians give plenary research talks and there are panel discussions relating to graduate school and mathematics careers. When possible, we also have NSF and NSA representatives give presentations on opportunities available through their agencies for undergraduates and graduate students.

Close to 200 students from across the country participated in the most recent (2006) conference. Roughly 50 of the students presented their research, either via talks or posters. Plenary speakers for the first eight conferences have included four members of the National Academy of Sciences, at least three former Vice Presidents of the American Mathematical Society, at least five former Presidents of the Association for Women in Mathematics, at least one former President of the Society for Industrial and Applied Mathematics, and one winner of an individual Presidential Award for Science, Mathematics and Engineering Mentoring.

The conference is now funded by the NSF (via our MCTP grant) and the NSA. Previous conferences have been funded by a mixture of conference grants from both the NSF and the NSA, internal UNL funds, and an NSF grant associated with the 1998 Presidential Award for Excellence in Science, Mathematics, and Engineering Mentoring. In fact, the conference was started in celebration of this award, which the department received in recognition of its success with female graduate students.

4. Regional Workshop in the Mathematical Sciences

UNL hosted its first Regional Workshop in the Mathematical Sciences in the spring of 1999 and we have continued to host a workshop in each fall semester since 1999. The primary goals of the conference are two-fold: to foster research contacts among faculty and students at colleges and universities in the region, and to expose undergraduate students to research-level mathematics across the discipline, while providing them with information about graduate education in the mathematical sciences. The workshop features six 45-minute plenary talks on Friday afternoon on a broad range of mathematical topics, followed by a banquet, a panel discussion on a topic related to graduate education, a social gathering on Friday evening, and a series of parallel sessions of talks on Saturday. Beginning with the 2004 conference, at least one of these sessions has been filled with talks by undergraduates on their own research. All talks at the workshop are aimed at a level appropriate for advanced undergraduates. Undergraduates at the Workshop attend talks by faculty, graduate students, and other undergraduates, and leave with valuable information about how to choose a graduate program.

5. Undergraduate MCTP Scholars

The department's MCTP grant supports five UNL undergraduate students each year as Undergraduate MCTP Scholars. Selection is based on students' interest in attending graduate school in mathematics as well as on their transcripts and letters of recommendation. During the academic year, these students have a research experience (similar to an REU funded as a supplement to an individual investigator NSF grant) under the direction of a UNL faculty member, and they are specifically encouraged to apply to REUs or other programs around the country for the summer. Undergraduate MCTP Scholars are paid up to $2400 for each academic year of participation (typically we expect a two-year commitment) and funds are also available to allow each Undergraduate MCTP Scholar to attend at least one conference to present the results of his or her research.

Undergraduate MCTP Scholars who are seniors also participate in a Graduate School Application Workshop, the Mathematical Landscapes Seminar (required of first-year graduate students), and the Introduction to Teaching Seminar (required of graduate students teaching a non-recitation for the first time).

6. RUTE

The Research for Undergraduates in Theoretical Ecology (RUTE) program is funded by an NSF grant to the Mathematics and Biological Sciences programs at UNL. The purpose is to significantly increase the number of students pursuing advanced degrees in areas that combine mathematics and biology. The primary component of the RUTE program is a structured research experience undertaken by a team consisting of two undergraduates in each of mathematics and biology, at least one faculty member in each of mathematics and biology, and a graduate student in biology. The students begin their project in the spring semester of their sophomore or junior year with a reading course designed by their mentors to familiarize them with the ideas and methods of the project research area. The students do intensive field or laboratory work during the summer under the direct supervision of the graduate student mentor and oversight of their biology faculty mentor. The Cedar Point Biological Station in western Nebraska, managed by the School of Biological Sciences, is an ideal location for such research, but projects can also be conducted elsewhere. The students earn academic credit and a $3500 stipend for their summer work. In the following academic year, the students do mathematical and statistical analyses of their biological data and also work on mathematics problems motivated by their biological work. Some students may be able to obtain academic-year support from the MCTP (see Section 5) or UCARE (see Section 7) programs for their RUTE work. Each project is expected to result in at least one professional paper, along with presentations at conferences. The RUTE program also includes a transitional component designed to introduce students to the possibilities of interdisciplinary research in math and biology while they are still in the early phase of their undergraduate experience. This component takes the form of a 5-week, 3-credit course called "Research Skills in Theoretical Ecology" that is taught in the summer to approximately 12 students who are new high school graduates or have just finished their freshman year of college. We pay for the students' tuition and living expenses, and we also pay a $1500 stipend. The course is team-taught by a biologist and a mathematician. The students conduct a variety of laboratory experiments in population dynamics and learn the mathematical and statistical techniques necessary to develop a mathematical model, fit it to their data, and use it to make testable predictions. To complete the research experience, the students prepare a paper or poster to present their results and write an abstract for a research proposal on a topic of their choice inspired by their summer project. These proposals could lead to an MCTP or UCARE project (see Sections 5 and 7); alternatively, some of these students will participate in the main RUTE program later in their undergraduate career.

7. UCARE

UCARE (Undergraduate Creative Activities and Research Experiences) is a UNL-funded effort to support creative activities and research efforts across all disciplines. This program provides up to $2000 per year for up to two years for juniors and seniors to experience the creative activities of their chosen discipline. Students apply by submitting a proposal approved by a faculty mentor. Students are encouraged to present the results of their work at the annual UNL Research and Creative Activities conference, either orally or as a poster. At least two mathematics students, and sometimes as many as five, participate in UCARE each year.

8. Other local efforts

The department has many students who write honors theses each year; most of these contain some measure of undergraduate research and not all are supported through one of the programs described above. Moreover, we have recently introduced two upper-level undergraduate courses which give students a taste of mathematics research. On the applied side, "Math in the City" gives students the opportunity to develop relevant mathematical models in cooperation with local area businesses. Our "Introduction to Mathematics Research" course focuses on more traditional mathematics research, and aims to help students develop the ability to think creatively and independently on open-ended problems.

References

[1] W. Heinzer, L.J. Ratliff, Jr., and K. Shah. Parametric decomposition of monomial ideals. I. *Houston J. Math.* 21:29–52, 1995.

DEPARTMENT OF MATHEMATICS, UNIVERSITY OF NEBRASKA – LINCOLN, LINCOLN, NE 68588
E-mail address: jwalker@math.unl.edu, gledder@math.unl.edu, rrebarbe@math.unl.edu, gwoodward1@unl.edu

Proceedings of the Conference on
Promoting Undergraduate Research in Mathematics

REU Site: Algorithmic Combinatorics on Words

Francine Blanchet-Sadri

a. Overview

The intellectual focus of this "Research Experiences for Undergraduates (REU)" NSF supported program entitled *Algorithmic Combinatorics on Words* is on interdisciplinary research at the crossroads between Mathematics and Computer Science. *Combinatorics on words* is a rather new field although the first papers were published at the beginning of the 20th century. It grew independently in various areas of mathematics including group theory and number theory, and appears frequently in problems related to automata and formal language theory. In the latest classification of Mathematical Reviews, combinatorics on words constitutes its own section under discrete mathematics related to computer science.

Molecular biology has stimulated considerable interest in the study of *partial words* which are strings that may contain a number of "do not know" symbols or "holes". The motivation behind the notion of a partial word is the comparison of genes. Alignment of two such strings can be viewed as a construction of two partial words that are said to be compatible. While a word can be described by a total function, a partial word can be described by a partial function. More precisely, a partial word of length n over a finite alphabet Σ is a partial function from $\{0, \ldots, n-1\}$ into Σ. Elements of $\{0, \ldots, n-1\}$ without an image are called holes (a word is just a partial word without holes). Research in algorithmic combinatorics on partial words is underway and has the potential for impacts in numerous areas, notably in molecular biology, nano-technology, and DNA computing.

Since the summer of 2005, the University of North Carolina at Greensboro (UNCG) has provided unique opportunities for summer research for eight outstanding and highly motivated students per year for an eigth-week period each year (note that in Summer 2007 there will be support for ten students). Participants work in teams of two under my supervision and in consultation with expert programmers. Admission is competitive and based on motivation, strength of the academic record, and letters of recommendation. As a result of taking part in this program, students become better prepared to enter professional scientific careers. UNCG is classified as a Doctoral/Research-Intensive university and is one of the sixteen branches of the University of North Carolina system.

Received by the editor October 16, 2006.

I have delineated the following five goals or objectives for student activities in this REU site program: A first objective is to introduce undergraduate students to various challenging algorithmic combinatorial problems on partial words related to coding, primitivity testing, and computing periods, through lectures and reading. Two types of research opportunities are emphasized: (1) computer related research, with students writing programs to perform experiments on partial words and to implement algorithms; and (2) combinatorics related research, with students investigating properties on partial words to generate conjectures and to discover algorithms. In addition, students are exposed to the techniques of language theory since this is a natural framework for formalizing and investigating strings and operations on them. While achieving this objective, a second objective of the program is for students to develop superior skills in mathematical writing and oral communication, skills that are essential for conducting research in a variety of scientific disciplines. A third objective of this program is to submit the produced original collaborative research on algorithmic combinatorics on words to leading journals. I am working very closely with the students on various research projects supported by this program. High quality publications involving student co-authors are resulting from this work. Also, students gain experience in communicating mathematics verbally through presentations at national professional meetings and national/international conferences. A fourth objective is for students to gain experience in the use of computers and their interaction in mathematical research. As a result, students establish World Wide Web server interfaces for automated use of the programs related to our combinatorial algorithms. This objective involves extensive computer programming and requires some experience using a programming language such as Java. Although students are selected based on merit after a nationwide search from a broad range of colleges and universities, a fifth objective of the program is to strongly encourage underrepresented groups including minorities, women, and students with disabilities to participate.

b. Student Activities

First, as a research mentor, I provide student participants with theoretical background on words and partial words. I use chapters of Lothaire's three books on combinatorics on words as well as my own book "Algorithmic Combinatorics on Words" (currently under review) as required or suggested reading. Second, students are introduced to the language theory techniques that proved to be useful in my prior investigations on words and that enable them to address open problems on partial words (some examples are discussed below). Most of the procedures are based on similar techniques, as well as techniques related to graph algorithms. Third, participants are introduced to some open problems from algorithmic combinatorics on words which are at the early stages of development and offer many challenging opportunities for future research. I create a booklet on each problem that contains some background material, related papers, etc .... The students are divided into teams of two (some choose to work alone) and allowed to pick one problem on which to focus for the remainder of the REU. Students have access to the computing laboratory where they can experiment on words. This facility (Bryan 330) is reserved from June 1 to July 31 of each summer for the participants in this program. A nearby classroom (Bryan 335) is also reserved and made available to support this program. This arrangement is convenient since my office is near these facilities. Students also have access to the campus library.

One of the goals of the program is to help students attain a higher level of independence in mathematical research. I provide students with some gentle guidance throughout their research projects. This, I believe, gives them independence as researchers and brings them success in their professional lives. The algorithmic combinatorial results get published in major journals, get implemented, and World Wide Web sites get created. Students, who are expert programmers, help me by conducting tutorials on LaTeX, CSS and XHTML, consulting in programming, maintaining the computing laboratory, etc . . . .

An especially important component of this program is afforded by numerous opportunities to meet with and learn from national/international guest lecturers. These speakers not only discuss their own research, but also discuss topics related to their development as mathematicians or computer scientists. The students (and myself) profit from these seminars and fruitful discussions. In the summer of 2005, I invited in particular Paul Duvall who talked about his expertise with the National Security Agency, and in the summer of 2006, I invited the well-known researcher Jeffrey Shallit from the University of Waterloo who talked about the Thue-Morse sequence. Students also have the opportunity to participate in a wide range of activities outside of the computing laboratory and classroom. At the end of each summer, we have a farewell picnic at Hanging Rock, in North Carolina.

c. Research Projects

I am now giving specific examples of major research findings by some of the student participants from Summers 2005 and 2006:

It is well known that some of the most basic properties of words, like the commutativity ($xy = yx$) and the conjugacy ($xz = zy$), can be expressed as solutions of word equations. An important problem is to decide whether or not a given equation on words has a solution. For instance, the equation $x^m y^n = z^p$ has only periodic solutions in a free monoid, that is, if $x^m y^n = z^p$ holds with integers $m, n, p \geq 2$, then there exists a word w such that x, y, z are powers of w. This result, which received a lot of attention, was first proved by Lyndon and Schützenberger for free groups [13]. Dakota Blair and Rebeca Lewis, two REU students from Summer 2005, investigated equations on partial words. When we speak about them, we replace the notion of equality ($=$) with compatibility ($\uparrow$). Among other equations, Dakota, Rebeca and I solved $xy \uparrow yx$, $xz \uparrow zy$, and special cases of $x^m y^n \uparrow z^p$ for integers $m, n, p \geq 2$ [3].

Nathan Wetzler, an REU student from Summer 2005, considered one of the most fundamental results on periodicity of words, namely the critical factorization theorem [9]. More specifically, given a word w and nonempty words u, v satisfying $w = uv$, the *minimal local period* associated to the factorization (u, v) is the length of the shortest square at position $|u| - 1$. The critical factorization theorem shows that for any word, there is always a factorization whose minimal local period is equal to the minimal period (or global period) of the word. Crochemore and Perrin [10] presented a linear time algorithm (in the length of the word) that finds a critical factorization from the computation of the maximal suffixes of the word with respect to two total orderings on words: the lexicographic ordering related to a fixed total ordering on the alphabet, and the lexicographic ordering obtained by reversing the order of letters in the alphabet. By refining Crochemore and Perrin's algorithm, Nathan and I gave a version of the critical factorization theorem for partial words [8]. Our proof provides an efficient algorithm which computes a critical factorization when one exists. Our results extend those of Blanchet-Sadri and Duncan for partial words with one hole [4].

Joshua Gafni and Kevin Wilson, two REU students from Summer 2006, introduced the notions of binary and ternary correlations, which are binary and ternary vectors indicating the periods and weak periods of partial words. Extending a result of Guibas and Odlyzko [12], Joshua, Kevin and I characterized precisely which of these vectors represent the (weak) period sets of partial words and proved that all valid correlations may be taken over the binary alphabet. We showed that the sets of all such vectors of a given length form distributive lattices under inclusion. We also showed that there is a well defined minimal set of generators for any binary correlation of length n and demonstrated that these generating sets are the primitive subsets of $\{1, 2, ..., n - 1\}$. Finally, we investigated the number of correlations of length n and the number of partial words sharing a given correlation [5].

Fine and Wilf's well-known theorem states that any word having periods p and q and length at least $p + q - \gcd(p, q)$ also has $\gcd(p, q)$, the greatest common divisor of p and q, as a period [11]. Moreover, the length $p + q - \gcd(p, q)$ is critical since counterexamples can be provided for shorter words. This result has since been extended to partial words. More precisely, any partial word u with H holes having weak periods p, q and length at least the so-denoted $l_H(p, q)$ also has period $\gcd(p, q)$ provided u is not $(H,(p, q))$-special. This extension was done for one hole by Berstel and Boasson [1] (where the class of $(1,(p, q))$-special partial words is empty), for two or three holes by Blanchet-Sadri and Hegstrom [6], and for an arbitrary number of holes by Blanchet-Sadri [2]. Taktin Oey and Tim Rankin, two REU students from Summer 2006, further extended these results, allowing an arbitrary number of weak periods [7]. In addition to speciality, the concepts of intractable period sets and interference between periods play a role.

d. Student Recruitment

We are attracting outstanding students to participate in this high quality REU program. Eight undergraduate students who have a solid background in mathematical sciences are carefully selected to participate in this eigth-week summer research program in algorithmic combinatorics on words. The students are chosen from a national applicant pool that has exceeded 100 applicants and comes from a broad range of colleges and universities that grant at least the bachelor's degree in Mathematics and/or Computer Science. Institutions targeted include women's colleges and predominantly minority institutions since one objective of this program is to increase participation of women and minorities in science. Outstanding students from universities such as Harvard, Michigan at Ann Arbor, Pennsylvania, Cornell, etc ...have participated so far. I have valuable experience in selecting applicants as I have served on the panel for Mathematical Sciences of the NSF Graduate Research Fellowships Program in 2002, 2003, 2004, 2005, and 2006. A World Wide Web site (continually evolving) has been designed at www.uncg.edu/mat/reu for this REU program that contains the program announcement, my publications with undergraduates, application materials, etc .... I was invited to give a talk on my REU site (as well as a plenary talk on partial words) at the *SCRA 2006-FIM XIII*

13th International Conference of the Forum for Interdisciplinary Mathematics on Interdisciplinary Mathematical and Statistical Techniques that was held in Tomar, Portugal from September 1 to September 4, 2006.

NSF support through this REU program is intended for highly motivated students whose undergraduate study is in Mathematics and/or Computer Science. The ideal candidate for this program will have taken a wide variety of upper-level mathematics and/or computer science courses including some of the following: Discrete Mathematics, Combinatorics, Algorithms, Theoretical Computer Science, and Programming. A distinguished academic record and indication of research interest or potential are essential.

e. Program's Success

Student participants are interviewed regularly in order to determine the effectiveness of the REU approach and to provide input for improvements in the program. For example, our program's success with Objectives 2, 3 and 4 are accomplished through the following:

Objective 2: Develop superior skills in mathematical writing and oral communication. These are important skills that are useful for conducting scientific research and that students should gain through some reading of journal papers, etc .... We evaluate Objective 2 in several ways: I provide students the opportunity to critique their peers' work. Also, in the middle and at the end of their REU summer's research, students formally present their results in a paper (typed in LaTeX) and in an oral powerpoint presentation.

Objective 3: Submit the results of our investigations to appropriate peer-reviewed mathematics/computer science journals. The projects I offer are sufficiently sophisticated to permit publication in respected journals, and this has been a typical outcome of prior research projects with my undergraduate students. The NSF supported research work with my REU students is leading to numerous papers submitted to leading journals such as *Journal of Combinatorial Theory, Series A, Discrete Applied Mathematics* and *Theoretical Computer Science.*

We extend the research experience beyond the eight-week at UNCG by motivating the students to participate in a national professional meeting or a national or international conference. In the falls of 2005 and 2006, several students attended (or will attend) the *Annual Symposium on Foundations of Computer Science* in Pittsburgh, PA and Berkeley, CA respectively. I also encourage the students to present their work at international conferences. The following students have the distinction of having presented their research at such conferences:

- Dakota Blair, Equations on partial words, *MFCS 2006 31st International Conference on Mathematical Foundations of Computer Science*, Stará Lesná, Slovakia, August 28–September 1, 2006.
- Joel Dodge, Counting unbordered partial words, *SCRA 2006-FIM XIII 13th International Conference of the Forum for Interdisciplinary Mathematics on Interdisciplinary Mathematical and Statistical Techniques*, Tomar, Portugal, September 1–4, 2006 (Session on Undergraduate Research in Interdisciplinary Mathematics).
- Nathan Wetzler, Partial words and the critical factorization theorem revisited, *SCRA 2006-FIM XIII* (Session on Semigroups and Languages).

We keep records of participant co-authored publications and presentations arising from this program as a means of evaluating our success in meeting our third
objective.

Objective 4: Establish World Wide Web server interfaces for automated use of
the programs. Specific research products that include software (or netware) have
resulted such as the World Wide Web sites at

www.uncg.edu/mat/border
www.uncg.edu/mat/research/cft2
www.uncg.edu/mat/research/correlations
www.uncg.edu/mat/research/equations
www.uncg.edu/mat/research/finewilf
www.uncg.edu/mat/research/finewilf2
www.uncg.edu/mat/research/finewilf3
www.uncg.edu/mat/research/unavoidablesets

that relate to algorithmic combinatorics on words from work with my participants
from Summers 2005 and 2006.

Our REU participants from Summers 2005 and 2006 are already being accepted
into prestigious Ph.D. programs in Mathematics.

References

[1] J. Berstel and L. Boasson, Partial words and a theorem of Fine and Wilf, *Theoretical Computer Science* **218** (1999) 135–141.
[2] F. Blanchet-Sadri, Periodicity on partial words, *Computers and Mathematics with Applications* **47** (2004) 71–82.
[3] F. Blanchet-Sadri, D. Dakota Blair and Rebeca V. Lewis, Equations on partial words, in R. Královic and P. Urzyczyn (Eds.), *MFCS 2006, 31st International Symposium on Mathematical Foundations of Computer Science, August 28–September 1, 2006, Stará Lesná, Slovakia,* Lecture Notes in Computer Science, Vol. 4162 (Springer-Verlag, Berlin, Heidelberg, 2006) 167–178 (www.uncg.edu/mat/research/equations).
[4] F. Blanchet-Sadri and S. Duncan, Partial words and the Critical Factorization Theorem, *Journal of Combinatorial Theory, Series A* **109** (2005) 221–245 (www.uncg.edu/mat/cft).
[5] F. Blanchet-Sadri, Joshua Gafni and Kevin Wilson, Correlations of partial words, (www.uncg.edu/mat/research/correlations).
[6] F. Blanchet-Sadri and Robert A. Hegstrom, Partial words and a theorem of Fine and Wilf revisited, *Theoretical Computer Science* **270** (2002) 401–419.
[7] F. Blanchet-Sadri, Taktin Oey and Tim Rankin, A generalization of Fine and Wilf's theorem for an arbitrary number of weak periods, (www.uncg.edu/mat/research/finewilf2).
[8] F. Blanchet-Sadri and Nathan D. Wetzler, Partial words and the critical factorization theorem revisited, (www.uncg.edu/mat/research/cft2).
[9] Y. Césari and M. Vincent, Une caractérisation des mots périodiques, *C.R. Acad. Sci. Paris* **268** (1978) 1175–1177.
[10] M. Crochemore and D. Perrin, Two-way string matching, *Journal of the ACM* **38** (1991) 651–675.
[11] N.J. Fine and H.S. Wilf, Uniqueness theorems for periodic functions, *Proceedings of the American Mathematical Society* **16** (1965) 109–114.
[12] L.J. Guibas and A.M. Odlyzko, Periods in strings, *Journal of Combinatorial Theory, Series A* **30** (1981) 19–42.
[13] R.C. Lyndon and M.P. Schützenberger, The equation $a^m = b^n c^p$ in a free group, *Michigan Math. J.* **9** (1962) 289–298.

UNIVERSITY OF NORTH CAROLINA, P.O. BOX 26170, GREENSBORO, NC 27402-6170
E-mail address: blanchet@uncg.edu

Promoting Undergraduate Research

Tuncay Aktosun

1. Introduction

I have been actively involved in promoting undergraduate research in the last ten years. This has been in various forms, such as directing the research of some undergraduate students individually as well as directing some REU programs where groups of students were involved in supervised research. All of these students have been supported by some funds either awarded to me as a PI or directly to the students. Although funding is not necessary to involve undergraduate students in research, the funded research may have some advantages such as providing a structured research atmosphere for the mentor and the mentees, giving a higher priority to the research conducted, requiring the accountability to the funding agency and to the scientific community, and perhaps some incentive and prestige for the student researchers. The funded undergraduate research may also help the mentor to be more selective and careful in choosing the student researchers and in determining their dedication level.

As for directing the research of individual students from one's own institution, I have come to the conclusion that it is the best to identify good candidates as early as possible, start involving them in research at an early stage, and supervise them throughout their undergraduate years. I think a long-term mentor-mentee relationship has a longer lasting impact on the student researcher. Among my undergraduate researchers it was a pleasure for me to direct the research of a bright student, Karolina Sarnowska, starting with her freshman year and almost throughout her undergraduate years. Karolina had an excellent preparation in high school and she was at the junior level when she started her freshman year at the Mississippi State University, majoring in mathematics and computer science. She is now a doctoral student in computer science at the University of Virginia.

I was the PI and director of a ten-week NSF-REU site program at the Mississippi State University during the summers of 2003, 2004, and 2005. I also directed the research of half the participants, and the rest of the participants performed their research under the supervision of my co-PI, Prof. R. Shivaji. Twenty-five undergraduate students were involved in active research; twenty-two participants came from other institutions and three (one each summer) were local students. One of the participants was supported by my own non-REU NSF funds and the rest by

Received by the editor October 4, 2006.

the NSF-REU site grant. The concentration was in applied mathematics empha-
sizing topics in inverse problems, wave propagation in nonhomogeneous media, and
population dynamics. The goals were to provide the participants with meaningful
research experience in applied mathematics, to show them the enjoyment of doing
research, to encourage them to pursue advanced degrees in mathematical sciences,
and to increase participation by women and underrepresented groups. Each year
eight participants were selected among about seventy applicants based on the credentials they submitted. I tried to use the following question as a guide to select
the participants: Which applicants can benefit the most from the program? We
had a diverse group of participants; twelve women and thirteen men, one African
American female and one American Indian male, a seventeen-year old participant
who was among our best, a few from poor families, some from top schools and some
from schools with no opportunities. The research program was supplemented with
some social activities, weekly colloquia, trips, dinners, and other events. The participants were treated like advanced graduate students in a typical research university
working closely with their research supervisors and enjoyed our program.

2. The 2006 Summer MAA-NREUP

I moved to the University of Texas at Arlington in August 2005. With my
colleague Prof. Minerva Cordero-Epperson I ran the 2006 MAA-NREUP (National
Research Experience for Undergraduates Program), a six-week summer research
program for minority undergraduate students in mathematics, sponsored by the
Mathematical Association of America with funds provided by the National Science
Foundation, the National Security Agency, and the Moody's Foundation. Since the
duration of the program was relatively short, we had to arrange all the logistics
ahead of time. My previous experience in directing the NSF-REU site program
was very helpful. We knew exactly what we wanted to do and planned everything
carefully so that the participants would fully concentrate on research.

We chose the participants carefully to have a balance. All the four participants
were Hispanic American, two men and two women, two from our own institution
and two from other institutions in the region. They were all considered to be good
students, two with no previous research experience at all, two with some prior
exposure to research, only one of the four had ever given a public presentation.
I communicated with them frequently before the program so that they had a full
understanding that they would not undertake any other responsibilities during the
program and would fully concentrate on research. I think this is an important
matter that needs to be clearly indicated to the participants before the start of
the program; some participants might assume that they could, for example, have
a part-time job for extra income during the weekends or some evenings. In my
opinion, this would defeat the purpose of the program and the program would
simply be viewed by the participants as nothing but a source of income without
any long-term benefits. We were able to expect and demand that the participants
would not hold any other responsibilities thanks to their generous stipends in our
budget.

The MAA-NREUP research topic was related to the direct and inverse problems
in human speech, one of my current research interests. Every weekday morning I
met with the participants in one of the seminar rooms exclusively reserved for our
program. Each meeting started promptly at 9:00 a.m. and usually lasted till 10:30

a.m. I explained the research problem and identified the goals. Since the topic was interdisciplinary, techniques from various areas of mathematics were used. The participants were pleased for various reasons: They were able to put their knowledge from calculus, linear algebra, and differential equations in use; they were exposed to partial differential equations, complex variables, integral equations, and numerical analysis; they were also exposed to acoustics and linguistics.

Every weekday afternoon during 3:30-4:30 p.m. the participants attended our computer laboratory, staffed by three graduate student mentors, all of whom had some prior experience in teaching and mentoring undergraduate students. The mentors were also familiar with the research topic because they took my special topics course on inverse problems in the spring semester of 2006. We avoided any micromanagement on purpose and let the graduate students interact with the participants to accomplish the clear goals we set: The participants would learn and be fairly proficient in LaTeX, Beamer, Mathematica, and Matlab, they would prepare a joint report in LaTeX based on their research (it would be acceptable if the report could not be put in a final form by the end of the program, the polishing of the report and its preparation for publication could be done after the program), and they would prepare a public presentation in Beamer at the end of the program. We reserved eight of the computers in our computer laboratory to use exclusively in our program. We installed MikTeX, Beamer, and WinEdt on those computers before the program started; the computers were already loaded with Mathematica, Matlab, Excel, and various other software. After the first half of the program, it turned out that the participants were spending more time in the computer laboratory, especially towards the end of the program, when they were doing symbolic and numerical computations and also preparing their presentation.

In the summer of 2006 I offered a seminar class on Monday and Wednesday evenings during 5:00-7:00 p.m. for thirteen of our graduate students. Since this was very relevant to our MAA-NREUP participants, we asked them to attend this class. The seminar class was intended to improve research and presentation skills of the students and to explore some research tools available for mathematical scientists. The topics covered included:

1) mathematical organizations (AMS, MAA, etc.), institutes (MSRI, IMA, etc.)
2) Mathematical Reviews, MathSciNet, Zentralblatt, CMP, MSC, PACS
3) mathematical typesetting, TeX, LaTeX, WinEdt, Beamer
4) mathematical conferences, meetings, attendance, funding, organizers
5) journals, writing papers, submission, refereeing, editorship
6) joint research, interdisciplinary collaborations, communicating with peers
7) teaching, record keeping, interacting with students, grading, mentoring
8) grants, fellowships, funding, proposals, submission, budget, overhead
9) dissemination of research findings, talks, posters, web pages
10) graduate school, application, PhD, advisor, qualifying exams, thesis
11) academic and industrial positions, tenure, promotion, faculty rank
12) software, Matlab, Mathematica, Maple, SAS
13) jobs, how to apply, EIMS, resume, cover letter, cover sheet, interview
14) library resources, information resources, arXiv, etc.

There were various additional activities in our 2006 MAA-NREUP. For example, one of our science librarians gave a special presentation to our participants in using library and online resources. We treated our participants to some lunches and dinners, which provided excellent mentoring opportunities in a friendly environment. Some additional information on our MAA-NREUP is available at the url http://omega.uta.edu/ aktosun/nreup. Overall, this MAA-NREUP was really a rewarding experience for us all. We all worked very hard and provided our participants with an excellent environment to attract them to mathematics at the graduate level and perhaps also beyond that. They were treated like advanced graduate students in our department and they also enjoyed all the perks our faculty and graduate students do (access to printing, photocopying, secretarial help, computer accounts, phone, departmental laptop computers, supplies, etc.) We arranged all these on purpose so that the participants would get a taste of being a graduate student or perhaps a faculty member and see for themselves whether they would like it or not.

3. Evaluation of our 2006 MAA-NREUP

Our program has been evaluated externally by an independent group at the Oregon State University through various questionnaires. Independently from that evaluation, a week after our program ended we requested input from our participants and asked them to send their comments to our secretary, who would remove the identity of the evaluators and forward the comments verbatim to us.

The e-mail I sent to our MAA-NREUP participants requesting their evaluations was as follows:

"Thank you for participating in the 2006 NREUP during June 11-July 22, 2006. Would each one of you please provide us with an evaluation of our program in a few paragraphs? At the end of your evaluation please assign a rating for the program by using the scale 5, 4, 3, 2, 1 (5 excellent, 4 good, 3 average, 2 poor, 1 very poor). In your evaluations you may comment on various aspects of the program such as:

1) The quality of research and supervision in the program.

2) What were your expectations before the program? Have those been met? How may have you benefited from your participation in the program?

3) Logistics (stipend, office space, computer lab, working environment, dormitory, etc.)

4) Has the program had any influence on your plans regarding graduate study, careers, future research, etc.?

5) Have you been able to go from a dependent status to an independent one as far as research is concerned? Has the program been helpful to you to go in that direction in future research endeavors?

6) Please comment on new knowledge or skills you may have learned or gained from this program (research, technical, computational, library, written, oral, critical thinking, etc.)

7) Would you recommend the program to other undergraduate students? Please comment if you have any suggestions for improvements."

All the four participants responded and sent their evaluations to our secretary by e-mail. I quote the four evaluations verbatim, which indicate what and how much the participants have gotten out of our program.

Evaluation by Participant 1:

"The first word that comes to mind when writing this evaluation is outstanding. This six week experience was unbelievable due to the amount of support and dedication from the people from the MAA and the University of Texas at Arlington. Before the program began, I did not set particularly high expectations. Being that this was my first REU program, I did not prepare myself for the invaluable experience and knowledge I would gain. After reading one of the first e-mails from Dr. Aktosun, the program did not sound all that mathematically extensive. He really only stressed that previous experience in linear algebra would be needed; therefore, my first impression was that the research would be focused on applications of lower level college mathematics. As we dipped into upper and graduate level mathematics, I quickly noticed that my expectations were incorrect, but I'm glad to see that they were. This experience has broaden my view of mathematics tremendously and sparked a brighter flame in my enthusiasm for the field. While I was quite seriously contemplating attending graduate school, I was not really firm on that decision and I did not have a clear understanding of the opportunities out there after graduation. After the experience in this program, I have gained strong belief that I belong in graduate school and have something to offer to the mathematical community either professionally or academically. I also must commend all the logistical effort that the people at UTA put into running this successful program. The housing and offices were great, and everybody on campus, students and faculty, gave a really good impression of the university. When compared to the other university environments that I have been around, I think that UTA is a good campus to host this program. In fact, I am quite interested in applying to UTA for graduate school. One of the main goals of this program is to give participants a taste of research. I think that this program has done so, but I am not completely confident in my abilities as an independent researcher. While I really enjoyed the group aspects of the program and would list it as an essential asset to the program, the research independence goes down when working as a group. I think that by working as a group we limited how deep we each connected with the research individually. The overall experience of the program gets a strong 5 from me. My mathematical knowledge and experiences have grown greatly as a result of this program. I have learned LaTeX, Mathematica, MATLAB, Beamer, elementary differential equations, and inverse problem theory. I have gained experience by giving my first mathematical presentation, having a working relationship with four of my peers, writing a paper detailing our research, and being submerged in the lifestyle of research mathematicians.

I would definitely recommend this program and similar programs to all undergraduate math students who have determination to be successful in the math field, and I would tell them to take full advantage of the opportunity."

Evaluation by Participant 2:

"My experience at the NREUP was wonderful because everyone in the department was very encouraging. The directors and students I worked with were helpful in my understanding of the research problem so that I could better contribute to the group in the goals we set for the summer. The supervision given by the directors and mentors was critical for me since I have a hard time concentrating for long periods of time. I didn't expect this program to be easy; I knew it would take work and dedication to complete the desired goal. I benefited from this program by learning what a graduate student does, how math faculty interacts, and what I could achieve in such a short time. I had wanted to go to graduate school before I started this program but didn't know the type of life I would be living. This program gave me that opportunity and experience from talking to the graduate mentors, the office, stipend and spending countless hours in the computer lab. I still don't believe I am capable of doing independent research but this has improved my self-esteem in the math I can do. The program set up was very helpful in that I had a wonderful place to stay, an office to take care of personal business, and a stipend to eat and pay tuition for the fall semester. The mathematical software we learned were very beneficial since I had heard what they were but had never really worked with them. I know I will use them in the future with research and graduate courses. This program is very beneficial to the undergraduate student and I would recommend it to any Hispanic striving to be a mathematician. Overall the program, in my opinion, was a success; I would rate it a 5 (excellent)."

Evaluation by Participant 3:

"My overall rating of the program: 5

Before the beginning of the program I was expecting to have some hands-on experience and to improve my team work skills. Through out the program, my expectation were not only met but surpassed. I really had "a feel" of how it is to work in the field and to work with a team truly willing to contribute to the effort. Also, the software training we received is priceless; I'm already using it to make simulations.

Furthermore, the working environment was very friendly, the housing provided was very nice, and the stipend very generous.

I would certainly recommend this program to other undergraduates."

Evaluation by Participant 4:

"The research topic selected for the program was excellent because it applied mathematics classes that we had taken in the past. Dr. Aktosun's expertise in this topic, his patience and great teaching skills made this experience painless, memorable, and rewarding.

I expected the program to be passive learning, similar to that of class room environment. The program exceeded my expectations by providing an active learning environment where no question was too basic. The main objective was to learn. The program gave me a small insight as to what a Professor and a Graduate student do when they are not teaching. I'm seriously considering pursuing a Masters and then ....

I learned new software programs (Matlab, Mathematica, Latex, Beamer), reviewed and/or re-learned math taken in the past (Differential Equations, Partial Differential Equations, Linear Algebra, Calculus). Our final exam was to give an oral presentation of the mathematical problem/solution using the tools we learned in the program in front of peers, graduate students and professors.

The icing on the cake was that we were paid to attend the program! Students that are considering a higher degree would pay to attend a similar program because the insight would help them decide whether this is what they are interested in.

Overall, the program was excellent! I would definitely recommend the program to other undergraduates. Thank you for the opportunity to participate!"

DEPT. OF MATHEMATICS, UNIV. OF TEXAS AT ARLINGTON, ARLINGTON, TX 76019-0408
E-mail address: aktosun@uta.edu

Proceedings of the Conference on
Promoting Undergraduate Research in Mathematics

Research Experiences for Undergraduates
Inverse Problems For Electrical Networks

James A. Morrow

1. Overview

Our undergraduate research program at the University of Washington typically includes thirteen undergraduates two or three TAs and one full time faculty member. It lasts for eight weeks and begins in late June to coincide with our academic schedule. Most of the research is conducted on inverse problems concerning discrete electrical networks and continuous electrical bodies. In addition to electrical network problems students work on related problems involving such things as the discrete Helmholtz (Schrodinger) network. This REU site has focused on such problems since 1988 and has had great success in developing an extensive archive of results in this area (see `http://www.math.washington.edu/~reu/`).

We have had a wide variety of students in our program. In the past three years 21 of the 24 students supported by NSF have come from outside the state of Washington. Students come from large research universities, universities of moderate size, and colleges with limited resources. All groups of students have benefited from the program.

The objective of the program is to quickly involve students in a research project in an active area of mathematical research. This is done by intense interaction among faculty, students, and TAs. Students are given reading material before the program begins. During the first week, the students are given a crash course on electrical inverse problems and related matters. After this introduction, we help them formulate and work on open problems. Many of the problems we suggest are natural continuations of problems that students have worked on over the years. Students then proceed to investigate these problems, meeting with a faculty or TA adviser every day. Students write up the results of their investigations as progress and final reports. Progress reports are given in weeks three and four and final reports in week eight. All reports are written, distributed to all students in the program, and presented orally. Students are given wide freedom in their choices; they often work on and solve several problems. An active social program is arranged so that students interact intensely with each other.

Received by the editor February 7, 2007.

The University of Washington supports the REU program in a variety of ways: computer facilities, dedicated classroom space, supplies, mailing, photocopying, student travel, and the usual administrative help for such a program. A VIGRE grant contributes support for TAs and some additional students. The Department has awarded a special fellowship, the Bob and Elaine Phelps Fellowship to support a student in the program. The University of Washington Summer Program pays partial salary for a faculty member. We are applying for a grant from Boeing to stabilize the support of TAs and UW students.

2. Student Activities

The main objective of this eight-week program is to involve students quickly and intensely in mathematical research. Except for the first week, students spend full time doing research on unsolved problems. We encourage students to formulate and work on problems as soon as possible. We have a brainstorming session at the end of the first week to come up with a list of problems. This year (2006) students came up with twenty-two research topics. When we first began to involve undergraduates in research, we expected that it would be a long time before they could begin their work. We have have been pleased to find that they actually need only a few days. In 2006, everyone was working on a problem at the end of the first week.

To accomplish this we send them reading material in advance. We send the students a list of references late in April and expect the students to do some reading before the program begins. We also send them a copy of the book [1] as well as selected student papers. A complete collection of student papers is archived on the website http://www.math.washington.edu/~reu/ . The student papers are good indicators of current directions an usually stimulate a lot of interest.

In the first week we give five hours of lecture per day to bring the students up to speed. They are also given instruction by our TAs on the use of our computing facilities and document preparation. They start to learn LaTeX during this period since we require that all papers be written in LaTeX. In the readings and lectures they learn about the conductivity equation, the Dirichlet problem, and related problems for electrical networks. They learn about circular planar graphs, critical graphs, and medial graphs, response matrices, and the $\bigstar - \mathcal{K}$ transformation. They also learn about inverse problems in these areas. They learn about a variety of network functions such as edge and vertex conductivity functions, directed network conductivities, scattering problems for networks, and Schrodinger networks, and Gröbner bases. After this introduction, we help them formulate open problems. Many of the problems we suggest are natural continuations of problems that students have worked on over the years.

They then proceed to investigate these problems, meeting with a faculty adviser or TA every day. Students write up the results of their investigations and give several oral reports. Students often work on and solve several problems. Students can choose to work on their own or in teams. Most of them said they were very happy to be given the freedom to choose their own problems and work on their own time schedules. Several said that this was the most intense learning experience that they had ever had.

They are not given fixed assignments nor are they all expected to study the same material. However, we do suggest problems that we think they will find interesting, and we suggest promising directions. Some students choose to extend work done by former students in the program. In many cases our former students did not completely solve the problems they worked on. Thus we have had students work on parts of the same problem over a period of several years. For example substantial work on non-planar networks has continued during all of the years of the program. We encourage them to be bold with conjectures and not to worry about success or failure. In our recent REU programs this approach has led to a wide variety of ideas for projects. Some worked out quickly to produce good results. Others seemed promising, but could not be completed in the short time of program. Succeeding students often continue previous work during the following year and produce more results. Some of the students have written joint papers with the principal investigator and the results of others has appeared in the monograph [1] co-authored by the REU Director. This monograph appeared in the year 2000 and is the definitive work on discrete inverse problems. It could not have appeared without the contributions of the students in the REU program. Recent progress has been rapid with students building on previous work and answering questions that have been around for many years.

The involve original work in an active area of mathematical research. The AMS meetings at Arcata in 1989 and Seattle in 1995 on inverse problems and the various special meetings of SIAM and IEEE on inverse problems attest to the current interest in inverse problems. There was a semester long program in inverse problems for Autumn, 2001 at MSRI at which there was a week-long PASI on inverse problems aimed at post-docs and graduate students. PIMS has an active collaborative group on Inverse Problems that includes many University of Washington faculty. The director of the program is active in this area, is in constant contact with other workers, and knows the current state of research. The virtue of discrete problems in inverse conductivity is that they can be approached by undergraduate students with a good understanding of calculus, differential equations, linear algebra and elementary physics.

There are guest lectures from local faculty and these faculty are available to consult on questions about graph theory, inverse problems, combinatorial algebra, partial differential equations, numerical analysis and optimization. At the end of the term students submit papers describing their work. These papers are posted on the website http://www.math.washington.edu/~reu/ Some continue to work on their papers after the formal termination of the program. This is encouraged and the proposer will continue to work with these students, correspond with them, and help them further their careers.

The students are told early on that they will be expected to write an exposition of their work and it is suggested that this writing should begin as soon as possible as this is frequently a lengthy and difficult process. We have encouraged the students to show us written work as soon as the second or third week. We start to help them right away in the formulation and of their ideas and solution of their problems. We are continuing to discover what students are capable of doing. The students seem to be more sophisticated and original each year.

We have learned a lot from the student responses to the evaluation forms that we distributed. The students have found that a solid understanding of linear algebra is essential. It is easy for us to overlook the fact that such an elementary subject is so important. We have become more aware of this requirement and have tried to assist the students in learning all the linear algebra that they need. The students said that the paper writing is one of the most difficult parts of the program. We agree with that assessment and we assist them in learning how to write by carefully reviewing their work with them. We ask them to be very critical of their own work and to strive to make it as clear and unambiguous as possible.

Finally we tell them that this program is intended to give them the freedom to experiment. We want them to find a problem, learn all they can about it, attempt to solve it and write up the results. We stress the importance of determination and that they must keep trying even in the face of repeated failures. We tell anecdotes about former students and their work. We point out that often what seems to be a crazy idea turns out in fact to be quite fruitful. We also tell them that sometimes the discoverer never sees the real importance of his or her ideas. What they should be getting from this program is an idea of what it is like to do mathematical research and whether it suits them or not. In that sense the program should be a success for everyone.

A room is reserved for the morning meetings. This room is available to the students all day for unscheduled discussions. The MSCC lab and the Department's computer rooms are open from 7 a.m. to 11 p.m. on weekdays and from 8 a.m. to 10 p.m. on weekends and holidays. There are terminals in the dorms connected to the campus network. Introductory computing classes will be scheduled early in the session to familiarize the students with the computing facilities. Space in a nearby student dormitory will be reserved for the REU students. There is a convenient and comfortable student union which is near the computer lab and the classroom, where our students can gather to socialize. We also invite the students to a local beach club to swim and relax and we have several picnics, regular Frisbee and softball games, hikes, sight-seeing activities, and other parties. There are social events nearly every day. This year's social schedule is archived at http://www.math.washington.edu/~reu/schedule/current. Our students have worked well together and have maintained contact with us and each other after the formal termination of the program.

3. Student Recruitment and Selection

We bring in students from a variety of colleges and university, maintaining the mix of small, medium, and large universities.

The quality of the students in the program has been high. Two of the students have been given honorable mention in the competition for the Alice T. Schafer award. There are seven recent winners of the Goldwater Scholarship. In 2006, one student was awarded both a Rhodes and a Marshall. Most of the students have gone on to graduate school in mathematics or a related discipline and many of them have been awarded fellowships. A number of these students have winners in the International Mathematical Contest in Modeling. The University of Washington has had four winning teams in the last three years (two problems are offered each year). They have won all of the prizes (MAA, SIAM, and INFORMS). The three-person teams were made up entirely from alumni of our REU program. The REU problem taught them how to formulate, solve, and write up solutions to problems. In addition, three winning teams from other universities have included REU alumni.

Several alumni of the REU program are now faculty members at highly regarded universities. Many of these students have said they would not have considered graduate school if it had not been for the REU program and some have switched from another discipline to mathematics because of their experience in the program.

We have a website with information about the program and selected papers for interested parties to read. This year's web address is http://www.math.washington.edu/~morrow/reu06/reu.html. We have made a selected list of universities and colleges in the United States and either by mail or by personal contacts we make the local undergraduate advisers aware of the program. This list includes institutions at which research opportunities are limited and which are located throughout the United States. We distribute flyers and application forms to faculty and students and follow up to attempt to get the best applicants. Since our program has been in operation a number of years we have benefited from the positive publicity given it by the participants of the program. An article about our program appeared in $Math$ Horizons in 2004.

In addition to the usual applications from students on the West Coast, we have many requests from Eastern, Southern and Midwestern students, who probably learned about our program from a faculty member, web site, mailing, or former student. We have had students from more than sixty universities and colleges.In the most recent three years twelve women have participated in the program and we have had one woman TA.

Our program is suitable for students with the following background:

- Differential equations at the level of *Boyce and DiPrima* (typically a sophomore course).
- Linear algebra at the sophomore or junior level including some discussion of numerically solving linear equations
- Advanced calculus, especially Green's theorem
- First year physics (mechanics, electricity and magnetism)
- C, Mathematica, Maple, or Matlab

We ask prospective students to describe their mathematical education, list any special awards and write a short essay describing their interest in the program. We ask for two letters of recommendation. Students will be selected in early April.

4. Project Evaluation

The REU program at the University of Washington has been very successful at introducing students to mathematical research and instilling in them an enthusiasm for research as a career. The faculty and TAs encourage the students to continue the research that they have begun by maintaining contact with the students and helping them write senior theses on the topics that they are interested in. We encourage them to apply to graduate school and counsel them on appropriate schools and write letters of recommendation. As mentioned earlier, graduates of the program include NSF fellows and tenure track faculty at major institutions. We also count among our graduates, medical doctors, Wall Street financial mathematicians, and software researchers. We have encouraged our students to apply for teaching assistantships as undergraduates and many have been accepted as TAs. Some become TAs for our REU program. We are gratified to find that our students are frequent recipients of national and international awards, such as NSF Fellowships, Rhodes Scholarships, Sloan Fellowships, and Mathematical Contest in Modeling awards.

Each year students fill out a questionnaire with information that allows us to keep track of them. They also make comments and suggestions for ways to improve the REU. On the last day of the program we ask the students to meet as a group, select a "scribe" and make a report giving us their advice and commenting on their experiences. We have established a website that allows them to keep up with their fellow students and to see progress on problems. We also ask faculty from other institutions to visit our program and write an evaluation.

BIBLIOGRAPHY

1. E.B. Curtis and J.A. Morrow, *Inverse Problems for Electrical Networks*, $xi + 184$ pp., World Scientific (2000).

DEAPARTMENT OF MATHEMATICS, UNIVERSITY OF WASHINGTON, SEATTLE, WA 98105
E-mail address: morrow@math.washington.edu

Valparaiso Experience in Research for Undergraduates in Mathematics

Richard Gillman and Zsuzsanna Szaniszlo

Introduction

Valparaiso University is a comprehensive university with nationally recognized academic programs. The approximately 3,500 undergraduate students choose to come to Valparaiso based on its reputation for student centered education with small class sizes. While not required for all students, undergraduate research is part of campus life for many. Each spring the university celebrates undergraduate scholarship with a one-day conference at which over 50 undergraduate projects are represented with talks and poster presentations. In addition to this conference, the engineering senior projects, the senior honors projects, and projects involving the performing arts have their own showcases.

Faculty of the Department of Mathematics and Computer Science have been mentoring undergraduate research for over 15 years. For the past ten years, the department has had a formal undergraduate research program that involves students starting at the freshman year. Each year several groups of students work with faculty mentors during the academic year. These projects finish with a written report, many of which result in presentations at conferences and some in professional papers. In addition to the research groups, faculty regularly work with students in a more informal setting on different projects that usually grow out of classroom assignments.

Building on our experience with undergraduate research groups working during the academic year, we established an REU site two years ago. Consistent with the department's philosophy that mathematics is widely accessible and applicable, our goal is to recruit students for whom this experience will be most beneficial.

Recruiting

At the time of this report our REU program has been in existence for only two years. We received funding so late in the funding cycle that we asked for and were granted a year of delay in starting our program after funding. This gave us ample time to prepare for the recruitment process. We distributed flyers at the Joint Mathematics Meetings and sent mailings to all our target institutions. These

Received by the editor October 20, 2006.

included small Midwestern colleges, Lutheran colleges nationwide and minority serving institutions. We had approximately 60 completed applications for the first summer. In the second year we did very little advertising, and received about the same amount of applications. Both years we had many very high quality applicants, and we do not wish to dramatically increase the size of the applicant pool. We look for students who are early in their studies and have not yet decided on attending graduate school yet. We also look for students without previous research experience and whose home institution is unlikely to offer a comparable experience.

The only mathematical requirement for participation is a completed Linear Algebra course to ensure that the students have had an introduction to writing proofs.

Participants

In the first two years of the program we had fourteen rising juniors and four rising seniors, sixteen of whom were from schools without PhD programs. We feel very successful in reaching most of our recruiting goals; it is especially noteworthy that 50% of the participants in the first two years were women. We are trying to identify ways of attracting more minority applicants. We had only two qualified African-American applicants in the second year. One of these applicants was one of only two students who turned our offer down and the other student had a successful research experience at our REU.

Staff

Each summer, three faculty direct the projects. In the first two years we had six different project directors. In addition to the project directors, the PI or the Co-PI of the grant serves as the site director overseeing general aspects of the program. The University of Nebraska-Lincoln supports a graduate student who acts as an assistant for our program. We find the presence of the graduate student extremely useful.

Program Overview

Each year we have nine students working in three groups on three very different research projects. The six projects so far have been in voting theory, algebraic topology, network flow optimization, matrix theory, statistics and graph labelings. We feel that it is a strength of our program that each year the students are exposed to three different areas of mathematics. The program runs for nine weeks, and it is divided into three parts.

During the first week the project leaders give lectures to the whole group introducing the topics. This ensures that every student understand all three projects.

During weeks two through seven the groups work under the direction of their faculty advisor. Each group sets its own schedule, but generally, the advisors meet the students 3-4 times a week. Every Thursday the research groups present their progress to the whole group. The faculty advisors give feedback to their own group and these presentations improve dramatically on a weekly basis. On Thursdays we also schedule special lectures given by faculty advisors, the graduate advisor, or visitors. This lecture is followed by a common lunch with opportunity for conversations and discussions of the cultural and recreational opportunities of the following week.

We also schedule 2-3 visits to nearby graduate schools to raise students' awareness of the requirements for graduate programs. Each year a few of our participants come to the program with the conviction that they will pursue graduate studies in mathematics. These students lead the discussions with questions, while the others generally just observe the conversations. We found that it is important to be very specific in communicating our expectations with the graduate schools regarding our visits in order to provide a meaningful experience for the students. It is also essential for our students to have an opportunity to talk to graduate students without faculty hanging over their shoulders.

The last two weeks of the program are spent on preparing the final document, a poster presentation and an oral conference presentation. Each project must prepare a final report. If the results are deemed publishable, the advisor of the group works with the research group during the rest of the summer and in the following academic year to prepare a paper for submission. Although publishing is not necessarily the goal of any project, four of our six projects have resulted in or will result in submissions to refereed journals. During the last week of our program Valparaiso University organizes a poster presentation where all the undergraduate students conducting summer research in the sciences showcase their work. This mini-conference was very well attended by both faculty and students in the past. We also participate at an undergraduate research conference hosted by Indiana University where all the mathematics REU students working in Indiana get together and present their work.

We encourage all our participants to participate at the Joint Meetings after their summer experience. All three groups presented posters at the Joint Meetings in 2006, and we expect the same for our second cohort in 2007. Most of the first cohort did much more than that by giving presentations at regional and national meetings, including AMS regional meetings, the Nebraska Women Conference, and other undergraduate conferences. Students seem to have no problem locating funding at their home institutions to travel to these conferences.

Facilities

The students and the graduate assistant live in the same residence hall together with other students conducting research on campus. As there is no regular dining hall services during the summer, both years the students organized cooking groups and had at least some dinners together.

Each group has a key to a study room in the library. The study rooms are small rooms with movable white board easels. Most groups chose to work in these rooms, a few preferred the common areas of the residence hall.

The students have access to the university's computer network, including MathSciNet, and inter-library loan services.

The Thursday presentations take place in a small "smart" classroom in the building that is the home to the Department of Mathematics and Computer Science.

Social Activities

During the first two weeks of the program we organize trips to the dunes of Lake Michigan, and to a museum in Chicago. We host a picnic for all undergraduate research students on campus and we invite our participants to the home of one of the advisors for a cook out. These activities help the students become familiar with each other, with other students, and with some of the most attractive tourist destinations. Thereafter, after the common lunch on Thursdays, the graduate assistant presents all the cultural and recreational options available during the weekend and the following week. The students seem to find ample number of interesting activities. Both years they went to baseball games, visited the Taste of Chicago, went to the local drive-in theatre and participated in the departmental Fourth of July picnic.

Student opinion

One of our participants from the first cohort organized a luncheon for undergraduate research in mathematics at his own institution to promote student interest. Together with faculty from his institution and a participant from another REU they discussed the different programs. This is an excerpt from his e-mail, sent to us in March after his REU experience:

"Some key points that stood out about Valpo's REU:
- *Groups of three instead of two.*
- *We were very well prepared for speaking about our work.*
- *Our group interacted on so many more levels beyond just math, (I left with a strong sense of community that people of other programs did not express.)*
- *We visited a variety of graduate schools.*
- *Melissa was an amazing help and having a graduate student around was great.*
- *Our reunion in San Antonio was also another unique thing.*
- *The guest speakers we had was another item that stood out at the meeting.*
- *In general, there was a lot more active emphasis on what it is like to think and work like a mathematician."*

DEPARTMENT OF MATHEMATICS AND COMPUTER SCIENCE, VALPARAISO UNIVERSITY
E-mail address: Rick.Gillman@valpo.edu, Zsuzsanna.Szaniszlo@valpo.edu

Wabash Summer Institute in Algebra (WSIA)

Mike Axtell, J.D. Phillips, and William Turner

1. Introduction

Since 2005, Wabash College has hosted the eight-week Wabash Summer Institute in Algebra (WSIA) in Crawfordsville, Indiana, through a grant from the National Science Foundation. WSIA provides an eight-week learning and research experience for twelve undergraduate students from across the country. The institute focuses on providing a common, cooperative experience for its participants, primarily in algebra, but also in ethics.

WSIA has several goals:

- to provide a meaningful mathematical research opportunity for undergraduates who are less likely to have a research-oriented experience than other student populations;
- to encourage participants to attend graduate school in the mathematical sciences and to help them develop some of the tools and confidence necessary to succeed there;
- to provide the support and framework for the participants to share their results and experiences with the larger mathematical community;
- to begin to provide the tools and knowledge necessary to become an independent, contributing member of the mathematical community;
- to create a highly diverse and supportive environment where the participants learn to work and live with a wide variety of individuals;
- to give students a significant experience studying the ethical dimensions of science and mathematics; and
- to attempt to accurately measure the success or failure of this program through a wide variety of evaluative processes with the assistance of an experienced agency.

As part of the institute, we hope to include up to three Wabash students each year. Because the funding for WSIA comes from an NSF grant, these students must be U.S. citizens or permanent residents. If no eligible Wabash student applies, and if international Wabash students show interest in the program, we may seek other funding from the college to allow an international Wabash student to participate.

Received by the editor October 19, 2006.

2. Research Topics

WSIA focuses on three areas of research in algebra: commutative ring theory, automated theorem proving and applications to loop theory, and computational linear algebra. The three research leaders cull research projects from topics related to their personal research.

Mike Axtell's research focuses on communitative ring theory and zero divisor graphs. He finds the research particularly suited to undergraduate study with comparatively little background needed outside a standard first semester abstract algebra course. It does not even require much experience in or knowledge of graph theory since the area currently uses very few non-basic graph theory ideas. After a brief introduction to graph theory and commutative ring theory, he guides students to investigate properties of the zero divisor graphs of various commutative rings, such as their diameter.

J.D. Phillips uses automated theorem provers such as Otter and Prover9 together with the related finite model builder Mace4 to investigate questions in loop theory and in other areas of algebra. This area, too, requires little knowledge past a standard first semester abstract algebra course. Under his direction, students have worked on the axiomizations of tri-medial quasigroups and digroups and have initiated the structural investigation of LC-loops.

William Turner's research is in symbolic computation, and in particular black box linear algebra, which takes an external view of a matrix, viewing it as a linear operator on a vector space. This view is particularly useful for highly structured matrices, and algorithms in the field use basic concepts such as recurrence relations and generating polynomials. Students who have only taken linear algebra can understand the topic, although some experience in abstract algebra, number theory, or programming can also prove useful. His research students have investigated using standard black box linear algebra finite field algorithms such as those using Wiedemann and Lanczos methods to compute the determinant of a matrix of rational numbers.

3. Program Outline

The eight-week program strives to guide participants from being students of mathematics toward becoming contributing members of the mathematical community. It does this through three relatively distinct phases.

3.1. Introductory Period. Participants find the first few weeks of the program fairly familiar. We dedicate the early weeks of the institute to lectures and problem-solving sessions conducted by the three research leaders and attended by all participants. In this respect, WSIA functions more like a graduate school or institute than a typical REU during the first few weeks. This serves to create a common body of knowledge among participants and foster a sense of shared experience and communal work ethic. Participants use these opening weeks to decide which area they wish to spend the rest of the summer pursuing.

During this phase, the three research leaders take turns working with all twelve participants to introduce them to their particular research projects. All three of us use styles similar to our normal classrooms; however, these styles vary from being lecture-oriented to discussion and problem solving sessions. Some of us focus heavily on students working on problems in groups to encourage the students to forge a strong group identity. This rotation among the three of us means we expose the participants to all three of our styles—and their associated benefits—without any of us having to change our style into something we do not find comfortable.

This phase focuses on providing the participants with the necessary background knowledge and technical tools to participate in the research projects. This serves two purposes. First, it allows the participants to make an informed decision when we ask them which projects interest them. Perhaps more importantly, though, because all the participants have seen all of the research projects, we expect the entire group to be able to discuss all of the research projects to some extent. This gives the program some added unity rather than just comprising three distinct research projects.

At the end of this introductory period, we distribute the participants among the three projects. We ask the participants to rank their interest in the three projects. Using this information, we assign each participant to one of the three projects. In 2005, we assigned each participant to their first choice of projects, but this caused an imbalance in the number of participants in each project. In 2006, to avoid the problems resulting from this imbalance, we asked applicants for their preliminary research interests and used this information to help select participants. We used this preliminary interest statement only to select participants initially. After the introductory period, we asked the participants to rank the projects again, and we balanced the research projects by assigning participants so four or five participants worked on each project. The participants' choices helped inform our decision rather than dictating it. Most participants still received their first choice, and none received their last choice.

3.2. Ethics Component.

3.2. Ethics Component. In the week following the introductory period, David Neidorf conducts a workshop on ethics and research. Neidorf is the vice-president and dean of the college at Deep Springs College in California and the director of educational programs at Bioethics-In-Action, Inc. Previously, he was the director of the integrated studies program at Middlebury College.

The ethics workshop has three goals:

(1) to provide participants a significant and formative exposure to the interface between work in mathematics and ethics;

(2) to ensure that participants are thoughtfully aware of their responsibilities as both professionals and citizens, and are alert to the tensions and conflicts between these two roles; and

(3) to empower the skillful discussion and resolution of concrete ethical problems through the examination of ethical case studies according to varying schema of conceptual evaluation.

Because the range of concrete ethical issues concerning the professional practice of mathematics is not easy to predict, he pursues these goals by using a shared model of ethical issues that involves the work of technical specialists: current ethical controversies in the biotechnological manipulation of human abilities and characteristics.

During the five instructional days of the workshop, participants spend the afternoons starting on their research projects and the mornings devoted to the ethics component. These five mornings are divided roughly into three parts: The analysis of case studies and associated background reading, student presentation of their independent analysis of new case studies, and group discussion of the student presentations.

Participants examine organizational case studies that raise ethical problems in three issue areas in biotechnology: Human Life-Span Extension, Enhancement of Human Performance, and Medical Modification Behavior. As they examine these issues, they identify ethical theories (e.g. utilitarianism, deontology, virtue ethics) and evaluative principles (e.g. autonomy, informed consent, distributive justice, freedom of belief, academic freedom).

We anticipate the ethics component of the REU will start participants down a path of a heightened sense of personal and professional ethical responsibility, and of increased ethical decision-making skills within their organizations. Since many participants will find themselves in leadership positions, this experience can play a significant role in their organizations as well as their personal lives.

3.3. Research Projects. Participants spend the last five or six weeks of the program working on their research projects. In the first week, participants divide their time between research and the ethics component, but they devote all of the remaining four or five weeks to their research project. During this time, participants work in small groups, supervised by a research leader. While each of us has his own style, these groups typically work very independently. We may only meet with the participants a few times each week, unless they run into difficulties and need to meet more frequently. Sometimes we meet with them multiple times in a day.

Research teams present weekly progress updates to the entire group. Since all twelve participants have seen the introductory materials for all of the projects, all participants should find these progress updates accessible.

Participants also prepare a paper and a talk. They present their talks at the Indiana University Undergraduate Research Conference held in Bloomington, Indiana, in late July, and we encourage them to submit their papers to undergraduate research journals or perhaps even professional research journals. We also encourage them to present their findings at the AMS/MAA Joint Mathematics Meetings Undergraduate Research Poster Session in January or at regional opportunities such as an MAA sectional meeting. Our NSF grant provides funding for half of the students to attend the Joint Meetings to present posters, and we encourage all participants to solicit funds from their home institutions to attend the meetings. So far, all participants who have wanted to attend the meetings have received enough funding through WSIA and their home institutions to attend.

4. Outside Speakers

Throughout the program, we invite mathematicians from neighboring institutions to come give a talk to the participants. Typically, the speaker will give a research talk, to which we invite the entire campus, not just the WSIA participants. Afterward, the speaker talks to the WSIA participants about graduate schools; an informal question-and-answer session follows this.

We have had a variety of speakers participate. Graduate programs and department chairs have talked about their particular graduate programs. Current graduate students, including a Wabash College graduate, have talked about their experiences in graduate school. A former WSIA participant talked about her experience applying to graduate schools. Some professors have had no direct involvement in administering graduate programs. We have even had the president of the MAA give a talk about mathematics and biology.

5. Participant Recruitment and Selection

We recruit participants through a variety of methods. In addition to the abstract on the NSF's REU website, we distribute a flyer at the Joint Mathematics Meetings in January, send mass mailings to mathematics departments across the country, and maintain a local website[1]. We recruit participants most heavily from smaller institutions whose students may not have as many mathematical research opportunities.

The typical participant will be a rising senior or junior who has completed at least one semester of abstract algebra, although we may accept an otherwise exceptional applicant who has taken only linear algebra but who has other strengths, such as a good programming background. When reviewing applications, we look closely at the two letters of recommendation for evidence of the applicant's ability to work well in groups. After the imbalanced research groups in the first year, we have also looked at the applicant's non-binding research topic preference to help balance the research groups. We use this preference only in selecting participants initially and not in assigning participants to research groups.

In selecting participants, we also favor women and members of other groups that are underrepresented in mathematics. Besides all the philosophical reasons for doing so, we have discovered a very practical reason to give women preferential treatment. Because Wabash College is a liberal arts college for men, the college is not organized to handle many women students on campus. In fact, the women in WSIA are the only women students on campus during the summer. The college housing makes it impractical to have only one or two women living in a house. In order to house all participants most comfortably, and to not require a third house, which might be difficult to acquire, we must ensure nearly equal groups of men and women participants.

6. Social Activities

The social aspect of WSIA begins with the living arrangements. Wabash College houses the participants in two houses on the edge of the college's small campus. This communal living arrangement allows the participants to bond as a group as well as provides them with common spaces to work and socialize at any time.

We also have several organized social activities. Each of the three research leaders hosts a gathering at his home. Typically, all three of us and our families participate in all of these, helping provide food and entertainment. We host the first of these on the Sunday evening the participants arrive on campus, the day before the institute begins in earnest, to allow the participants to meet each other and us in a relaxed environment. Another professor hosts a gathering during the middle of the program, and the third hosts one about a week before the participants leave. In addition to meals at our homes, we also take the participants on a number of organized outings such as nature hikes, canoe trips, and attending minor league baseball games in nearby Indianapolis.

---

[1]http://www.wabash.edu/academics/math/wsia

The participants have also organized their own social activities. We have heard from the participants that they often prepare meals together. Sometimes this takes the form of one house (either the men's or the women's) preparing a meal for the other house, and then the other house reciprocates a few days later. Participants have traveled to Lafayette, Indianapolis, and Chicago for outings. Participants even organized a trip to the University of Illinois to visit the graduate program there.

7. Assessment

With the aide of Wabash College's Center of Inquiry in the Liberal Arts, we administer several evaluations to the participants throughout the program to gauge the change in the students' attitude toward mathematical research and in their self confidence in their ability to tackle new mathematical problems. These include a base-line evaluation at the beginning of the program, informal sessions throughout the program, and a longer written assessment at the end of the program.

We plan to remain in contact with the participants for five years after their participation in the program to attempt to measure the long-term impact of WSIA. In particular, we are interested in the rate at which WSIA participants enroll and succeed in graduate programs in the mathematical and other sciences. We hope to determine whether the participants continued to mature mathematically through graduate studies in the sciences and if WSIA had any impact on this maturation.

DEPARTMENT OF MATHEMATICS AND COMPUTER SCIENCE, WABASH COLLEGE, P.O. BOX 352, CRAWFORDSVILLE, IN 47933

E-mail address: {axtellm,phillipj,turnerw}@wabash.edu

Proceedings of the Conference on
Promoting Undergraduate Research in Mathematics

The SMALL Program at Williams College

Cesar E. Silva and Frank Morgan

1. Introduction

The SMALL program seeks to introduce undergraduates to the excitement and challenge of doing original research in mathematics. Too often, we have seen talented students lose interest in mathematics because they perceive it as a lifeless subject within which one simply studies the work of long-dead mathematicians. In SMALL, they get to work on current projects under the direction of faculty who are themselves actively engaged in cutting edge research. They see the vitality of mathematics and their own potential for making a contribution. They understand the ultimate goal of learning, namely to utilize that learning in order to make their own advances and further the pursuit of knowledge. Although SMALL began as a program for Williams students, outside students have come to play an ever larger and more important role. Since its founding in 1988, the program has guided over 300 students in mathematics research. SMALL alumni have gone on to earn Ph.D.s in mathematics and related fields, and to other activities such as teaching mathematics in high school. (SMALL is an acronym from the names of the founding faculty, Silva, Morgan, Adams, Lenhart, Levine.) Over 60 research articles coming out of SMALL have been published or been accepted for publication in mathematics journals. More than 25 SMALL alumni have earned Ph.D.s in mathematics or mathematics related fields, and many are on the mathematics regular faculty at colleges and universities, while many others are in postdoctoral or visiting positions. Still other alumni are on university faculties other than mathematics. Of the 80 SMALL participants in the five years 1999 - 2003 more than 50 are enrolled or have completed Ph.D. programs in mathematics or related fields. SMALL has also contributed to the increasing interest in mathematics among our students at Williams, so that currently we have 50 senior mathematics majors, about 10% of our senior class.

The faculty who have participated in the program include: Colin Adams, Duane Bailey, Deborah Bergstrand, Gerald Bope, Carsten Botts, Edward Burger, Elizabeth Camp, Charles Chace, Satyan Devadoss, Richard De Veaux, Thomas Garrity, Gary Lawlor, David Levine, William Lenhart, Susan Loepp, Robert Mizner, Frank Morgan, Allison Pacelli, Cesar Silva, Kris Tapp, Alice Underwood, David Witte, and Janine Wittwer. The first director of the program was Frank Morgan; the

Received by the editor February 7, 2007.

other directors have been Colin Adams, Deborah Bergstrand, Thomas Garrity and Cesar Silva.

The research topics have included knot theory, hyperbolic manifolds, minimal surfaces, symmetry groups, combinatorics, graph theory, computational geometry, algebraic geometry, dynamics and ergodic theory, parallel processing, topology of robotics, CR structures, Riemannian geometry, neural networks, bayesian statistics, and commutative algebra.

The funding for the program has come mainly from the National Science Foundation and Williams College. This article is based in part on previous NSF grant applications.

2. Structure of the Program

The SMALL project lasts for nine weeks in June through August. The students are assigned into groups of about four, each group working with an individual faculty member. The students learn to work as a group, assigning sub-projects to individuals, and helping each other. Each group, along with the faculty advisor, decides on the daily routine. Some meet once or even twice a day at a scheduled time with the faculty member, while others prefer a more open door policy where the students come to see the faculty member whenever they have questions. When not talking to the faculty advisor, the groups are working together or individually in the Mathematics and Statistics Library (which is surrounded by the faculty offices), down in the Math Computer lab or in one of several classrooms reserved for their use. See the pictures at http://www.williams.edu/Mathematics/small_CES.html for a sense of the facilities.

A variety of weekly activities are held for all the students. Every Tuesday at 10:00 am convocation is held. Announcements are made, sometimes followed by short progress talks by the student groups. Tuesday at noon the students, together with at least another 150 students working on research in the sciences over the summer, attend research talks by faculty from the sciences. Lunch is provided by the College. On Wednesdays at 1:00, there is a mathematics colloquium talk by a faculty member from inside or outside Williams. These talks are specifically directed at the students, although the topic is usually current research. At 4:00 on Fridays, there is a tea, giving students a chance to talk to each other and to faculty about their progress over the week. In addition, there are a variety of social events organized by students and faculty that increase the opportunities for interaction. In past summers, the enthusiasm has been so great that students have often worked more than the standard forty-hour work week, including working late into the night and over the weekends.

Over the summer, students present their work to the rest of SMALL. Our students have participated in a poster presentation by all summer science students. In addition, we have had students present their work in numerous other venues, including MAA and AMS national and regional math conferences, as well as the Hudson River Undergraduate Math Conference. Near the end of the summer we have often travelled to MathFest, where the students report on their results in the MAA and PME student talk sessions. Over the years many students have won awards, including the top CUR MAA and PME awards.

Sometimes some or all of the groups participate in other schools or conferences. For example, in 2002 the students spoke at the beginning of the summer about their research project at an MAA meeting at Williams. Many of the students have presented posters at the AMS annual meetings. Some groups have travelled to other meetings; for example, the Geometry Group participated in the Clay Institution summer graduate school at MSRI in 2001, and in a 10-day graduate school in Paris in summer 2004. All the students are housed at a the college. Starting with summer 1991, Williams has provided us with a college house for the exclusive use of our program. Having all of the students living together in a single building allows for the mathematical interaction to expand beyond the boundaries of the workday. The college has also provided us with computing facilities, classroom space, and also often additional funds for supporting students, including supplies, tea, and additional student stipends. See "Is an REU for You?,"by D. Haunsperger and S. Kennedy, *Math Horizons,* February 1998, for a description of the SMALL program from the point of view of one of the participants. A more recent account may be found in "REU Spotlight: Williams College–The SMALL Program," by Jennifer Novak and Eric Schoenfeld, *Math Horizons,* November 2005.

3. Student Selection and Project Evaluation

We aim to have a national representation of students. We mail about 200 posters to various colleges and universities in the region and the nation. We maintain a current web page with information on the program and all the information that is need to apply. We will continue to make strong efforts to recruit minorities and women. About 33% of the 80 students in SMALL during the years 1999 - 2003 were women. We hope to continue to attract talented women and to attract more underrepresented minorities than we have been able to so far. Every summer, we assign two students the duty of formulating an evaluation of SMALL. They create and disseminate a survey filled out by all participants and then write up an evaluation. We have found that information very useful in fine-tuning the program, and we plan to continue to use this system of evaluation. We are tentatively planning to invite an outside faculty member to visit SMALL and write an evaluation from the faculty perspective.

4. Plans of SMALL alumni

SMALL students come from a diverse group of colleges and universities including Carleton, Emory, Stanford, Pomona, Cornell, MIT, Berkeley, Texas A&M, Harvard, Colby, U. Virginia, Missouri Baptist, New College, Wesleyan, U. Chicago, U. Texas, U. Notre Dame, Brandeis, Augustana, College of New Jersey, Swarthmore, U. Richmond, Johns Hopkins, Bucknell, Rose-Hulman, Princeton, and Williams. Out of the 80 students who participated in SMALL during 1999 - 2003, about 50 of them are pursuing or have completed Ph.D.s in mathematics or related fields (such as applied mathematics, physics, computer science, engineering, economics). The graduate schools where SMALL students are currently enrolled or have graduated from include Berkeley, Brown, Caltech, Chicago, Cornell, Columbia, Duke, Georgia Tech, Harvard, Maryland, Michigan, MIT, Northwestern, Penn, Princeton, Rutgers, Stanford, Texas, UCLA, UCSD, Washington and Wisconsin. At least 12 SMALL students have earned NSF Fellowships in mathematics, and another two students have earned NSF Fellowships in related fields. Four have earned honorable mention in the Alice T. Schafer Prize for Excellence in Mathematics by an Undergraduate Woman, one was a runner-up in 2001 and one won the Schafer Prize in

2003. Of the 12 NSF Mathematics Fellowship winners, four are women. We expect these numbers to increase as more students graduate.

5. Research Publications

Every student research group has produced an internal paper, and most of the groups have submitted a paper externally. Many of the papers have already appeared in refereed mathematics journals and many others have been accepted for publication, while others still are in the submittal or preparation stage. For example, of the 80 students who participated in SMALL in the years 1999 - 2003, 64 students are co-authors of one or more published research papers. The result of research that was either generated by the SMALL Project or followed up on work begun in SMALL has appeared or is to appear in journals such as Acta Arithmetica, Acta Crystallographica, American Mathematical Monthly, Annals of Mathematics, Canadian Mathematical Bulletin, Colloquium Mathematicum, Communications in Algebra, Computational Geometry, Ergodic Theory and Dynamical Systems, The Fibonacci Quaterly, Houston Journal of Mathematics, Illinois Journal of Mathematics, Integers: Electronic Journal of Combinatorial Number Theory, Journal of the Australian Mathematical Society, Journal of Combinatorial Theory, Journal of Geometric Analysis, Journal of Knot Theory and its Ramifications, Journal of Discrete and Computational Geometry, Journal of Number Theory, Michigan Journal of Mathematics, New York Journal of Mathematics, Networks, Pacific Journal of Mathematics, Proceedings of the American Mathematical Society, Real Analysis Exchange, Technometrics, Topology and its Applications, Rocky Mountain Journal of Mathematics, Transactions of the American Mathematical Society. All of the papers were either authored or co-authored by students. For a full list of publications the reader may refer to http://www.williams.edu/Mathematics/small_CES.html.

DEPARTMENT OF MATHEMATICS AND STATISTICS, WILLIAMS COLLEGE, WILLIAMSTOWN, MA 01267

E-mail address: `csilva@williams.edu`
E-mail address: `Frank.Morgan@williams.edu`

Industrial Mathematics and Statistics Research for Undergraduates at WPI

Arthur C. Heinricher and Suzanne L. Weekes

For the past nine years, the National Science Foundation has funded the REU program in Industrial Mathematics and Statistics at Worcester Polytechnic Institute (WPI) in Worcester, Massachusetts. During the summers of 1998–2006, the program has hosted 96 students from 28 states and Puerto Rico who have worked on 27 different industrial projects sponsored by 13 different companies. Half of the 96 students were women and almost half of the students came from schools without a PhD program in mathematics.

The goal of the WPI REU program is to provide a unique educational experience by introducing students to the ways that advanced mathematics and statistics are used in the *real world* to analyze and solve complex problems. The students work in teams on problems provided by local business and industry. They work with a company representative to define the problem and to develop solutions of immediate importance to the company; they work closely with a faculty advisor to maintain a clear focus on the mathematics and statistics at the core of the project. When students work with a company on an industrial problem, the problem is real and the company needs a solution. This is usually the first time that students are placed in a situation where someone is going to make a decision, perhaps an expensive decision, based on their mathematical work.

The WPI REU program provides a glimpse of the many career possibilities which are open to students with a strong mathematical background. The hope is that by the end of the summer, the students will have better answers to the questions: (i) *What is the role of a mathematician in business and industry?* (ii) *What is it like to work with technical experts on a problem that requires significant mathematics but also must satisfy real-world constraints?* (iii) *What kind of mathematical and statistical tools are used to solve problems in business and industry?*

The program provides challenges not faced in standard undergraduate programs and strengthens skills not always developed in traditional educational programs. The *SIAM Report on Mathematics in Industry* [1] provides a comprehensive study of these special skills, and they include: (a) *communication at several levels, including reading, writing, speaking, and listening;* (b) *problem formulation as an interactive, evolutionary process;* (c) *the ability to work with a diverse team.*

Received by the editor November 21, 2006.

The REU program at WPI provides an excellent experience for advanced undergraduate students going on to graduate school, whether they choose to specialize in applied mathematics or not. The experience is certainly valuable for students interested in following nonacademic career paths, but it is just as valuable for students who enter "traditional" graduate programs and go into academic careers.

WPI Project-Based Undergraduate Program WPI has a special infrastructure and a long history that supports intense project activities like those in REU's. Project-based learning has been central to the WPI educational program for more than 30 years. In 1971, the **WPI Plan** marked a departure from the conventional approaches to undergraduate education. It introduced as degree requirements three types of projects: the Humanities Sufficiency, the Interactive Qualifying Project (IQP) and the Major Qualifying Project (MQP). The last is a senior-year project completed in the major field of study. It is often the work of a team and spans over $3/4 - 4/4$ of the academic year. The purpose of the MQP is to provide a capstone experience in the student's chosen major that will develop creativity, instill self-confidence and enhance the student's ability to communicate ideas and synthesize fundamental concepts. In completing the MQP, students are expected to:

- formulate a problem, develop a solution and implement it competently and professionally,
- interact with the outside world before starting their careers,
- work in teams and communicate well both orally and in writing.

This project activity has been highly successful at involving WPI undergraduates in significant research with faculty. In 1987, NSF began funding Mathematics REU programs. In 1988, the WPI Mathematical Sciences department hosted its first REU program and was then able to involve non-WPI students in undergraduate research as well.

In order to enhance the industrial project experience for our students and to help make new contacts with business and industry, the Center for Industrial Mathematics and Statistics (CIMS) was established at WPI in 1997. Members of the Center work to establish contacts with industry, businesses and government labs and to develop industrial projects at both the graduate and undergraduate levels for our majors. The industrial mathematics program at WPI has been extremely successful. More than 200 hundred students have completed industrial projects with 30 different companies. We work throughout the year to make new company contacts, maintain existing contacts, and to develop new project opportunities for our WPI and REU students.

REU Recruitment Each summer, we recruit between 10 and 12 undergraduates to take part in our intense, residential REU program which lasts eight and a half weeks. Interested students apply online at `http://www.wpi.edu/+CIMS/REU`. They fill out a standard application form with personal data, education history, courses taken, and are asked to rank a list of preferred project areas. Recommendation letters from faculty are also required. Very importantly, students must write a one-page essay describing his or her interest in participating in the WPI program. After review by the Principal Investigators, promising applications are selected and these students receive a phone call from the program coordinators in a joint information exchanging and interview process. Following NSF guidelines, consideration is given to only U.S. citizens or permanent residents. However, exceptional applicants who do not satisfy these requirements are also considered, provided the availability of additional funding from our industrial partners; we have had one British student from Oxford University and a French student from the Université de Savoie. The final group of students gathered are all interested in applied and industrial mathematics and statistics, and have a diverse range of course backgrounds and interests since the industrial projects usually require a mix of probability, statistics, differential equations, numerical analysis, and optimization.

REU Program Structure The process of meeting a real-world problem, learning to ask good questions and doing the research needed to identify the key mathematical structure, and then refining and redefining the problem, is a crucial part of the industrial mathematics experience. Also, the project is not finished when the problem is "solved." It is important that the students communicate their solution to the company in a form that the company can understand and use. Communication skills, written and oral as well as listening skills, are crucial for a successful industrial mathematician. This is developed via (i) *daily meetings* with faculty advisors, (ii) *weekly presentations* to fellow students and faculty, and (iii) *regular meetings with industrial sponsors*.

The REU students work in teams of 2–6 and each group has at least one faculty advisor plus an industrial advisor. Each team is given an office with 2 or 3 networked computers. (In 2003, while the Deutsche Bank research team was testing their portfolio model, the students had a total of 7 computers working in parallel.) Teamwork is one of the skills required for a mathematician working in industry [1] and one responsibility of the faculty advisor is to observe and guide the team-building process. The students meet with the faculty advisor(s) every day over the course of the 8 and a half week project.

Each team makes periodic progress reports, in the form of written reports and oral presentations, for their fellow students, the faculty and industrial advisors. We also invite faculty from other departments and representatives from local companies to attend these *weekly presentations*. The students receive extensive feedback as the projects evolve through the eight weeks. The students gain valuable practice in presenting their work; the improvement in quality during the summer is quite impressive. At the end of the two months, a *Presentation Day* is organized for the students to formally present their final results for the faculty, invited university administrators, and industrial advisors. This is followed by a special, celebratory lunch.

Each team of participants must complete a *written report* based upon the research they have completed during the summer program. The purpose of this report is to describe the problem considered, the background literature read, the approach(es) taken, the results that have been obtained, and the questions motivated by the research. Participants begin writing parts of this report as early as the first week of the program so that the faculty advisor has an opportunity to assist the students in developing a proper style for writing mathematics.

In addition to the student presentations, we have invited *mathematicians working in industry* to meet the students and discuss their work. For example, Keith Hartt from Bogel Investment described his work as a quantitative analyst; Bruce

Kearnan, Senior Actuarial Associate and General Director at John Hancock Life Insurance described his work as an "actuarial historian"; Derek Kane, mathematician at DEKA Research and Development Corporation speaks often about responsible global scientific research.

There are also at least three special events scheduled during the program. The following are a some of the special events held in the past programs: (i) a visit to the facilities of *DEKA Research and Development Corporation* including fun rides on the iBOT and the Segway; (ii) a tour of *the Mathworks*, with a presentation by technical staff on engineering applications for Matlab as well as career paths for math majors; (iii) participation in the *Math Research Expo* organized by WPI and Boston University as part of the Focus on Mathematics partnership. In 2006, three of the REU students acted as judges in local math research expos in the middle schools. See `http://www.focusonmath.org/FOM/resources/math/mathexpo/`

In order that student-faculty interaction is not limited to the academic dimensions, a group recreational activity is planned for most weeks. Activities have included a lobster dinner in Mystic, Connecticut, Boston Red Sox games (tickets provided by John Hancock), trips to view the Fourth of July celebration in Boston Harbor, as well as barbecues at the faculty advisors' homes.

The industrial projects have become the foundation for several successful outreach programs; this vertical integration of our work will be collected and described in an upcoming article.

REU Projects Completed   Below, we list just a few of the projects completed in our REU program along with the industrial sponsor.
(a) *Modeling Fluid Flow in a Positive Displacement Pump*
   DEKA Research and Development Corporation, Machester, NH
(b) *Mathematical Model for an Electro-Pneumatic Pulsed Actuator*
   Applied Mathematics, Inc., Gales Ferry, CT
(c) *A Continuum Model for the Growth of Brain Tumors*
   IBM Corporation, Boston, MA
(d) *Statistics Procedures for Failure-mode Testing of Diagnostic Equipment*
   Veeder-Root, Simsbury, CT
(e) *Adaptive Risk Score Assignment Model for Underwriting Long-Term Care Insurance*
   John Hancock Life Insurance Company, Boston, MA
(f) *FEMLAB Electromagnetic-Thermal Model of Microwave Thermal Processing*
   Ferrite Corporation, Nashua, NH
(g) *Portfolio Optimization with Non-Smooth Constraints*
   Goldman-Sachs, New York, NY
(h) *Mathematical Model of the Self-Tapping Screw Insertion Process*
   BOSE Corporation, Framingham, MA

Where are they now?   Our WPI REU alumni can be found in academia, as graduate students and faculty, and others have gone on to successful careers in business and industry. From the feedback and correspondence that we get from our students [2], the program has made a positive impact on their lives.

From a WPI REU 2005 alum via email:

Subject: I got a job!

I was offered a job as an "engineering assistant" with the company Applied Research Associates at their lab on Tyndall Air Force Base near Panama City, FL. My position will mostly be mathematical modeling using MATLAB and FEMLAB (and some other random tasks, occasional lab assistancy).

. . .

Clearly, I have the REU to thank. The company actually discovered me through Monster.com because I had FEMLAB in my resume, and that was one of the main reasons they were interested in me. But beyond the "marketable skill" of knowing FEMLAB, the REU is what got me really interested in mathematical modeling–and working in a research environment–in the first place. So to my colleagues and our advisers and everyone else who helped us during those transformative two months last year, thank you!

References

[1] *The SIAM Report on Mathematics in Industry*, Society for Industrial and Applied Mathematics, Philadelphia, Pennsylvania. (Available from the Internet at `http://www.siam.org`.)

[2] Vernescu, B., Heinricher, A., *Research Experiences for Undergraduates in Industrial Mathematics and Statistics at WPI* in Proceedings of the Conference on Summer Undergraduate Mathematics Research Programs, edited by Joseph A. Gallian, 1999, Arlington, VA, pp. 213–219.

DEPARTMENT OF MATHEMATICAL SCIENCES, WORCESTER POLYTECHNIC INSTITUTE, WORCESTER, MA 01609-2280

E-mail address: `heinrich@wpi.edu`

E-mail address: `sweekes@wpi.ed`

Part II

Descriptions of Summer Enrichment Programs

Twelve Years of Summer Program for Women in Mathematics – What Works and Why?

Murli M. Gupta

We have been running the Summer Program for Women in Mathematics (SPWM) at George Washington University for twelve years. In this paper we describe the major components of our program and why it works. We have found, and believe, that the women in general, and our program participants in particular, have strong intrinsic abilities to succeed at high levels in sciences and mathematics; what they sometimes lack is self-confidence. We provide a nurturing environment and role models whose examples and encouragement help these women reach their potential.

1. Brief History

In 1995 we hosted a 4-week pilot program for 10 undergraduate women mathematics majors from around the United States. The program was initiated, in part, to replicate the success of the Mills College Summer Mathematics Institute (SMI) that had at that time been running for almost a decade. A multi-institution proposal was made to the National Science Foundation (NSF) to create similar programs at a number of institutions in the country but that proposal was declined. We were subsequently asked by the National Security Agency (NSA) if we could run a smaller program as a pilot. Through the support of our university administration, we were able to set up a program, invite student applications, find faculty and visitors, and operate a program that, by every measure, was a terrific success. Each of the participants was glowing in support of the program and every one of the directors, instructors, and teaching assistants felt that the program provided a tremendous benefit to each participant. In 1996 and 1997 we hosted a similar 4-week program for 16 undergraduate women. In 1998, the program was expanded to 5 weeks and is continuing in this format for its thirteenth year in 2007 with the continuing support of the National Security Agency.

2. Program Goals

The stated goals of our program are to communicate an enthusiasm for mathematics, to develop research skills, to cultivate mathematical self-confidence and

Received by the editor November 15, 2006.

independence, and to promote success in graduate school. We bring our participants into contact with successful women mathematicians in academia, industry, and government. We provide the students with a broad exposure to mathematical culture, illustrating the beauty and attraction of mathematics, the tools necessary for success in mathematics, applications of mathematics to business and industry, and the career opportunities available to mathematicians. The specific objectives of our program are to provide an immersion program representative of key aspects of graduate school and professional mathematical practice, promote active mathematical thinking, underscore the beauty and enjoyment of mathematics, foster a camaraderie among the participants that emphasizes collaboration and peer support, bring the participants into contact with active mathematical researchers through guest lectures and field trips, provide interaction with a wide variety of successful women in mathematical sciences who serve as role models, illustrate the role of mathematics as the foundation of the sciences and the wide range of mathematical applications in government, business, and industry through first-hand contact with applied mathematicians, and provide students with information about graduate schools and careers in mathematics.

3. Program Statistics

We have, by now, hosted 186 students in the past 12 years (1995-2006). A summary status of the accomplishments of our SPWM Participants from 1995 to 2004 is contained in the following Table. Information on 2005 participants is incomplete and the 2006 participants are still undergraduates.

TABLE 1. Education Data on SPWM Alumnae.

Finished PhD	Finished MS	Working on PhD	Working on MS
21 Participants	26 Participants	52 Participants	8 Participants

Note that there is some overlap between the people who went to MS programs and then continued to PhD programs. Some of our participants are untraceable partly because they may have got married and changed last names. (We use many techniques to track our participants, especially from earlier years. This includes contacting their college professors, the alumni offices, as well as search engines.) The following Table contains data on the employment status of our participants, as known to us.

TABLE 2. Employment Data on SPWM Alumnae.

Professor/ Post-Doc	Industry/Government	HS Teaching	Vet/Law/Others
16 Participants	30 Participants	13 Participants	11 Participants

4. Program Elements: Courses and Teaching Faculty

The main purpose of our summer program is to introduce the participants to graduate level topics not covered in typical undergraduate curricula. The current format consists of two 3 week courses and two 2 week courses. These (noncredit) courses are taught by women mathematicians who design the special courses for our program and use a variety of teaching techniques including individual and group study. The students generally start these courses from scratch – without a prior exposure to the course topics. At the end of these courses, the participants know enough about the topic that they are able to make in-class presentations; some of the presentations are very slick. Each year, we produce proceedings of the summer program containing copies of the student presentations and the students take copies of these proceedings as cherished mementoes of their summer program experience.

We bring teaching faculty from all over the country and sometimes from overseas. In 2006, our teaching faculty consisted of Barbara Csima, Univ. of Waterloo, *Computability Theory*; Lyn Miller, Slippery Rock Univ., *Introduction to Groebner Bases*; Angela Gallegos (SPWM 1998), Occidental College, *Mathematical Modeling in Biology*; and Karma Dajani, Univ. of Utrecht, *An Introduction to Discrete Financial Mathematics: The Binomial Model*. We also invite two graduate students to serve as teaching assistants each year; in 2006, our teaching assistants were Sara Miller, University of Notre Dame and Tiff Troutman, University of California-Riverside; both are former SPWM participants from 2002.

Here is a sampling of our courses from previous years; complete details are posted on the summer program website: $http://www.gwu.edu/\sim spwm$.
Lynne Butler, Haverford College, *Number Theory and Public Key Cryptography*, 2005
Ayse Sahin, DePaul Univ., *An Introduction to dynamical systems*, 2005
Sylvie Hamel, McGill Univ., *Introduction to Automata and Language Theory*, 2003
Janet Talvacchia, Swarthmore College, *Introduction to Symplectic Geometry*, 2003
Joanna Kania-Bartoszynska, Boise State Univ., *Introduction to Knot Theory*, 2002
Leila Schneps, Ecole Normale Superieure, *Fermat's Last Theorem*, 2001.

5. Program Elements: Guest Lectures

The guest lecturer program is an important feature of our summer program. We invite a wide variety of mathematician professionals (mostly women) who come to GWU and spend an afternoon and evening with our participants. The guest lecturers give a specially prepared talk, provide information on their own decisions that took them into the field of mathematics, and entertain discussion on their background, education, and career paths. Listed below are the guest lecturers and the titles of their talks for 2006.

Professor Cathy O'Neil, Barnard College, *Local to global principles*; Professor Linda Smolka, Bucknell University, *Shocks, Waves, Fans and the Method of Characteristics*; Dr. Tad White, National Security Agency, *Algorithmics and Statistics of String Comparison*; Professor Jane Hawkins, University of North Carolina, Chapel Hill, *An introduction to cellular automata*; Professor Allison Pacelli, Williams College, *Algebraic Number Theory: an "Ideal" Subject*; Professor Rebecca Weber, Dartmouth College, *Making randomness rigorous*; and Professor Annalisa Crannell, Franklin & Marshall College, *Math and Art: The Good, the Bad, and the Pretty*.

6. Program Elements: Field Trips and Panel discussions

We have a field trip each week where our participants come into contact with women mathematicians in their own workplace, and are exposed to current issues at the forefront of mathematics, the variety of applications of mathematics, the depth and complexity of the kinds of mathematics involved, and the possibilities for careers related to mathematics. In 2006, the field trips were to U.S. Census Bureau; Dibner Library, Smithsonian Institution; Science Committee Room, U.S. House of Representatives; National Security Agency and National Cryptologic Museum; Northrop Grumman- TASC; and The Aerospace Corporation. Each year we also hold two or three panel discussions, each with several invited guest experts from industry, academia and government. At these panels we consider issues associated with the mathematics community such as careers, the job market, gender issues and graduate schools.

7. Our participants

We bring students from a variety of large universities and smaller liberal arts colleges. We have received applications from large private and public universities including many of the Ivy League institutions, as well as from small four year colleges which may have only 3-4 faculty members in the mathematics department. All of our participants are math majors (some are double majors) and everyone has a strong motivation and interest in graduate school, strong letters of recommendations, and an average GPA from 3.2 to 4.0. Many of the participants have previously attended a Research Experiences for the Undergraduates (REU) program; some have also attended the Budapest semester. Our participants have told us that their reasons to attend SPWM are to:

- *Prepare for the world of graduate school,*
- *Opportunity to explore math,*
- *Increase my knowledge of math,*
- *To meet future colleagues and build lasting academic relationships,*
- *To be exposed to new ideas I may not experience in the regular classroom,*
- *To experience mathematical study outside classroom setting,*
- *Give me the confidence needed to pursue my inherent passion for mathematics,*
- *Curious to see what mathematicians outside of academia do.*

Because of the advance nature of topics covered in our courses, we find it is helpful if our participants have taken courses in abstract/modern algebra and real analysis or advanced calculus. We receive 100+ applications each year; in some years we have received as many as 160 applications from highly qualified and motivated students. We work hard to select the summer program participants, based upon their academic background and their Statement of Interest. We essentially look for a passion for mathematics and their desire to learn and go forward as mathematics professionals. Letters from their professors are often very helpful – in many cases, the professors tell us about their students' achievements and passion for mathematics that the students might be too modest to tell us.

8. Follow-ups: Networking

We attempt to contact our participants at least once a year. We continually update our database to keep track of the participants and have recently started compiling a SPWM Newsletter which is regularly updated and made available to our program participants. The program faculty, teaching assistants, guest lecturers and our hosts at field trip sites provide a valuable resource to the program participants. This includes writing letters of reference for graduate applications, pointers for job applications, help with graduate school search and application process, help with job interviews, and actual jobs. We held a program reunion in January 2000 at the Joint Mathematics Meetings (JMM) which was held in Washington, DC that year. Since 2000, we have sought, and received, extra funding from the program sponsors to invite former participants each year to attend the Joint Mathematics Meetings and have had six annual summer program reunions at JMM. We support a limited number of participants to travel to the meeting site and have found that the most recent participants, who are still undergraduates, get the most benefit from these reunions. However, many of the old timers are happy to get together to share ideas and provide support to one another.

9. Program Evaluation

We carry out program evaluations at mid-point and the end of each program. We also approach our program participants for their impressions about the effectiveness and long term value of our program; many of our participants are happy to give us their positive feedback years after their attendance at the summer program. Here is a sampling of what the students have told us at the program end:

- *I wanted some direction to my future, as well as a better ability to articulate what I love about math and what mathematicians do. This program addressed my main concerns and much, much more. It was amazing.*
- *This has been an excellent experience. I now feel confident about graduate school. I also feel confident about my choice to be in mathematics... I now know I can succeed in math.*
- *It was great reading the research papers. It was very encouraging to realize that I know enough to read other people's work.*
- *I gained a wealth of information and insight. I'm now convinced that I can succeed in grad school.*
- *I think the presentations really helped me gain some confidence in my abilities.*
- *I learned so much math, got a taste of math culture, and grew as a person I think. In fact I grew a lot . . .*
- *I am amazed at how much more aware I am now than I was a month ago.*

Here is what an NSF program manager told us after spending a day visiting our summer program:
I think it's important that we continue to encourage women in the mathematical sciences. Far too often I run into the attitude that the women "problem has been solved" and those who don't think this is necessary aren't always men. I like what GWU does with the courses that it gives the participants. This provides the context for the discussions and talks throughout the rest of the program. Having the participants work on projects is also a very positive aspect of the program. They seem to

enjoy the interaction and they reinforce each other in a very congenial environment. Having them give presentations of their work is a highlight of the program, in my opinion. Presenting the results of their investigations is something that can only benefit them in their future.

Here is a sampling of what our participants told us a few years after their participation:

- *As for the summer program. I found it very effective. It definitely convinced me to attend graduate school. It also gave me confidence in my abilities.*
- *My impressions of the program..... I think it was one of the most important things I ever did. I needed to be with other girls who were math majors. There were a lot of things we had in common that I didn't get to share with people at my own university.*
- *I believe that the summer program helped me tremendously. It gave me an idea of what grad school would be like and it helped encourage me to go on with my studies. I still keep in touch with a few others so I have also made some lasting friendships.*
- *.... the program really helped me a lot to develop more mathematical sophistication and to appreciate the ever-growing world of mathematics. It was also really great to understand math in a more global sense: not only were we taking classes from disciplines we'd never heard of before (Measure Theory), but we were taking classes with people from all over the country, and with teachers from all over the world. So, I really think that this universal aspect of mathematics is something that I only appreciated after the program, and especially now in graduate school.*

10. What have we learned?

We have learned that the women participants are extremely capable of doing graduate studies in mathematics and obtaining advanced degrees at prestigious institutions. Many of our participants come from smaller institutions where they might not have been exposed to a large variety of mathematics; our program shows them that there is a lot more to mathematics than what they may have experienced in their undergraduate career. Many of the women lack self-confidence and often think that graduate school is out of their reach; even after starting the graduate programs they may think that they alone are having the frustrations usually experienced by many graduate students – both male and female. Our networking opportunities are extremely valuable to the participants – the fact that they come to know so many math professionals outside their own institutions and the fact that they can call upon these people for advice and help is priceless. A nurturing and nourishing environment enables these women to remove their self doubts and find the spark needed to go forward and excel as a mathematician. Their visits to the national conferences to see the prominent mathematicians in person leave a lasting impression on the younger participants. The opportunity to get together once a year to compare notes, renew friendships, and offer advice continues to be extremely valuable to our participants.

DEPARTMENT OF MATHEMATICS, THE GEORGE WASHINGTON UNIVERSITY, WASHINGTON, DC 20052

E-mail address: mmg@gwu.edu

Research Experience for Undergraduates in Numerical Analysis and Scientific Computing: An International Program

Graeme Fairweather and Barbara M. Moskal

1. Introduction

In the summer of 2006, five male and five female undergraduates from nine universities across the United States participated in an eight week Research Experiences for Undergraduates (REU) program in Hong Kong that was directed by the Department of Mathematical and Computer Sciences (MCS) at the Colorado School of Mines (CSM). This program was supported by the National Science Foundation (NSF) (DMS-0453600) and is the first of three such opportunities supported through the project, "United States–Hong Kong REU in Numerical Analysis and Scientific Computing"[1]. The participating Hong Kong universities were Hong Kong Baptist University (HKBU), The Chinese University of Hong Kong, and City University of Hong Kong. This REU program is one of forty-seven currently funded NSF mathematics REU site grants nationwide[2]. Only one other site has an international component, Rutgers University, which is collaborating with Charles University in Prague, Czechoslovakia[3]. The mathematical focus of the Rutgers REU is discrete mathematics, computer science, statistics, biomathematics, and biomedical applications. The current paper describes an REU site in Hong Kong, the mathematical focus of which is numerical analysis, scientific computing, and applications in applied science and engineering

The purpose of establishing an REU site in Hong Kong is to provide undergraduate mathematics students with the opportunity to contribute to the exciting research that is being conducted in numerical analysis and scientific computing at an international level. Many students incorrectly assume that researchers in mathematics work in relative isolation. This international REU program is designed to challenge this misconception which may discourage talented students who desire a highly interactive and culturally rich experience from pursuing degrees in mathematics. A growing national concern is that there are few students pursuing graduate studies in science and mathematics [1]. In today's world, mathematics students as

---

Received by the editor November 27, 2006.

[1]http://www.mines.edu/reu-mcs

[2]http://www.nsf.gov/funding/pgm_summ.jsp?pims_id=5517&from=fund

[3]http://www.nsf.gov/awardsearch/showAward.do?AwardNumber=0138973

well as students in other fields must be able to assimilate into different cultures and learn how to function in a global environment [1, 2]. The Hong Kong REU program is designed to encourage more students from diverse backgrounds to pursue higher degrees in the mathematical sciences with an international perspective.

MCS has sponsored prior REU sites on its own campus. The impetus for expanding overseas came from support from NSF's Division of International Programs in the form of a supplement to a previous REU site grant, to investigate the establishment of an REU site in East Asia, an area of strategic importance to the United States.

The purpose of this paper is to describe the first implementation of this international REU program, the specific goals of which are:

(1) to increase undergraduate students' interests in pursuing advanced degrees in mathematical sciences;

(2) to provide participating students with a high quality international research experience in the mathematical sciences.

For a student perspective of the program, see [3].

2. Student Participants

Ten U.S. undergraduates were selected from thirty-three applicants through a competitive application and review process. All participants had completed their third year of college and had a declared major in mathematics. Participants were also required to provide evidence of a strong grounding in ordinary differential equations, linear algebra, and numerical methods, together with knowledge of a programming language such as C, Fortran or Matlab. The students selected were from the following institutions: Colorado School of Mines, Colorado State University at Pueblo, Davidson College, Illinois Institute of Technology, Loyola College in Maryland, New College of Florida, Regis University, Taylor University, and the University of California at Berkeley.

3. Program Design

3.1. Local Logistics. Before departing for Hong Kong, the students were assigned to research teams comprising a faculty member from a participating Hong Kong university and two or three U.S. undergraduates. These individuals began to interact electronically before the trip. In preparation, students were provided background material regarding their project and addressing practical issues, such as cross-cultural understanding, travel arrangements and insurance matters, accommodation and food, and safety and health concerns.

The program was centered at HKBU, where the students stayed at the Ng Tor Tai (NTT) International House, which is conveniently located for shopping, dining and transportation. This on-campus housing provided a safe, pleasant and convenient environment for the REU participants and was within 15 minutes of the business center of Hong Kong via public transportation. A wide variety of food was available, including Western food, at cafeterias on the participating campuses. Since English is an official language in Hong Kong, language barriers were minimal.

The grant covered the students' travel and living expenses, and provided each with a stipend of $2,500. An MCS faculty member traveled to Hong Kong with the students as their U.S. faculty adviser. The host institutions provided each student team with office and computing facilities.

The program ran for eight weeks, beginning in the last week of May 2006. After weeks 3 and 6, the teams met as a group and each team presented an oral progress report, with the U.S. adviser moderating. These reports included a discussion of accomplishments, problems encountered and their solutions, and the remaining goals of the project. Each team was also required to prepare a final written report which was submitted at the end of the program. At that time, each team gave a 30 minute presentation on the completed project to the group and their research mentors as well as other faculty and students from HKBU.

3.2. Research Projects. The four research projects on which the students worked were: 1) the development of wavelet algorithms for high-resolution image reconstruction, 2) subspace clustering methods for high-dimensional categorical data, 3) spectral analysis of differentiation matrices and applications, and 4) numerical challenges for resolving spike dynamics for reaction–diffusion systems. Each project is described on the program's webpage: www.mines.edu/reu-mcs

3.3. Additional Experiences. The students had the opportunity to participate in additional academic activities, such as attending the 2nd International Workshop on Structured Matrices at HKBU. This experience provided them with exposure to international mathematical research that was being conducted beyond their team projects, and a chance to interact with internationally recognized researchers. They also attended seminars presented by visiting speakers and a Ph.D. thesis defense, and had tea with renowned numerical analyst, Dr. Gene Golub of Stanford University.

4. Assessment and Evaluation

Before the program began, the students submitted an essay describing what they hoped would be their future contributions to mathematics and the mathematics community. On their return to the U.S., they submitted an essay on how they would revise their original essay based on their Hong Kong experience. At the end of each of the first seven weeks in Hong Kong, students submitted written responses to the following: "Reflect on the past week's experiences. Explain what you learned both mathematically and culturally." As mentioned earlier, each team was required to submit a final written report and give a presentation which were evaluated by their Hong Kong mentor and the U.S. adviser. This section describes the results of the assessment and evaluation effort. Since the number of participating students was small (ten), only descriptive analyses were completed on the data.

4.1. Pre and Post Program Essay. In their original essay, all students indicated their intention to attend graduate school in a mathematical field (i.e., mathematics, computer science or physics), and half of the students expressed interest in acquiring a Ph.D. in one of these fields. The majority of students also indicated that they had an interest in seeking a career that included college level instruction (seven students) and/or mathematical research (six students).

The majority of the post project essays indicated that the experience had not changed their future aspirations, but rather reinforced them. Six of the eight students felt that the REU experience confirmed that graduate school and a mathematical sciences career was appropriate to them. Two students questioned their mathematical ability as a result of the experience. Neither of these students indicated a change in their desire to pursue a mathematical sciences career, but rather that they needed to acquire a better understanding of the field before making a final decision.

Of the five women who participated in this project, three returned from the REU experience expressing a greater understanding of the impact that they could have on the mathematics community as females. One woman explained, "Their [sic] was only one woman participant in the conference on Structured Matrices, and some of the Chinese professors said that they did not take female Ph.D. Students. The latter fact concerns me the most; how can women feel accepted in the field if they have difficulties finding a professor to take them on as a student based on their gender?"

4.2. Weekly Reflections. At the end of the first week, most students primarily reported cultural observations. For example, "The people here are very nice and friendly", and "People here don't hesitate to cram into the train or a crowded elevator". Very few students addressed what they learned mathematically and those who did, reported that they were still in the start-up process, e.g., "We spent the past week understanding our research project and learning the background material."

In the second week, the students attended the 2nd International Workshop on Structured Matrices at HKBU. The majority of the cultural and mathematical reflections were based on this experience. Students reported that they were surprised that English was the language of the conference and that they felt lucky and somewhat embarrassed to be native speakers. They also reported that interacting with practicing mathematicians allowed them to see the "human" side of the profession.

By the end of the third week, the students' reflections began to focus more on mathematics and on problems they were experiencing with their research, e.g., "We keep trying new things, and sometimes we get a little bit better results, sometimes we get worse results, but never the results we really want." With respect to culture, students began to document differences between the Hong Kong culture and their own, "Asian women wore much more conservative swimwear and Asian men actually wore much less fabric than other men".

In weeks four and five, students began to express satisfaction with the progress of their research, "Our project seems to be moving forward which makes me feel good because there was a while I didn't think it was going anywhere". Another interesting outcome reported in the fourth week was that the students were beginning to judge the culture in which they were participating. Several students expressed concern about the recruitment of Filipino women to Hong Kong to act as poorly paid night club workers or maids.

During the sixth week, a major concern of the students was group work. Several students reported that they were used to being group leaders, and rarely depended on the efforts of others. This was not the case in their research. Instead, they found their own efforts delayed, while waiting for results from other team members.

By the seventh week, the students began expressing satisfaction with the outcomes of their research, e.g., "This past week has really been about tying up loose ends" and "Work-wise, I wish we had another two weeks. I think that we are finally getting some results that make me wish that there were more time...". A number of students were also excited about their meeting with Dr. Golub. Students further expressed excitement and regret at the prospect of returning home.

Excellent	Above Average	Average	Below Average	Unacceptable
This international REU program was well organized as a whole.				
7	3	0	0	0
The international component(s) of the program added value to the underlying scientific experience.				
8	1	1	0	0
This program helped me to improve my research skills.				
8	2	0	0	0
This program helped to increase my general knowledge of mathematics.				
5	5	0	0	0
I will continue my pursuit of a career in mathematics as a result of this international REU experience.				
4	3	3	0	0

TABLE 1. Summary of student responses to a subset of questions from the end of program survey

4.3. Hong Kong Faculty Evaluations. As part of the evaluation process, the research mentors in Hong Kong were asked to rate each of their students with respect to the following: contributions to the research team, final submitted research project, and future research potential. Only two of the mentors completed the feedback form, accounting for six of the ten students. All of these students were evaluated as "Average" or above with respect to each question, the majority being "Above Average. In response to the question concerning research contributions and future research potential, the majority of students received a rating of "Excellent."

5. End of Program Survey

At the conclusion of the REU program, the participating students completed an end of program survey. This survey was based on [**4**] and contained a total of 52 questions. Given the length of this survey, it cannot be summarized here in full. Table 1 provides the results of five key questions that addressed the overall effectiveness of the REU experience. As this table suggests, the majority of students either agreed or strongly agreed with each statement.

6. Discussion

As a result of implementing and evaluating this REU, we are obliged to reconsider goal (1), stated in section 1. The participating students did not uniformly display an increase in their interest in pursuing advanced degrees in the mathematical sciences. In fact, most of the students maintained their original intentions, i.e., to pursue an advanced degree and a career in the mathematical sciences. Students entered the program because they already had interests in this area. Participation in this REU experience did help the students to clarify their career objectives. At the conclusion of the REU, two of the participants questioned their mathematical ability and their future intentions of pursuing a mathematical sciences career. Both of these students indicated that they would seek additional experience and information with respect to mathematics before making their final career decision.

As educators, we must interpret these results as a success. After all, we do not wish to encourage students to enter the mathematical sciences for the simple purpose of increasing the pool. Our goal is, more appropriately, to guide students into a satisfying career that is consistent with their personal goals. With this in mind, we reconsider the first goal of this project and refine it as follows, "To provide students who are interested in pursuing an advanced degree in the mathematical sciences with a better understanding of the nature of mathematical research and mathematical careers". The data presented in this paper supports the assertion that the revised version of the first goal was attained.

The second goal was to provide the participating students with a high quality international research experience in the mathematical sciences. Based on the student feedback as well as the evaluations of the research mentors, this goal appears to have been met. Furthermore, based on the work of [2], the student weekly reflections suggest that the participating students were progressing toward global competency. Downey et al. [2] proposed the following criteria for the evaluation of minimal global competency in engineering students.

- Students will demonstrate substantial knowledge of the similarities and differences among engineers and non-engineers from different countries.
- Students will demonstrate an ability to analyze how people's lives and experiences in other countries may shape or affect what they consider to be at stake in engineering work.
- Students will display a predisposition to treat co-workers from other countries as people who have both knowledge and value, may be likely to hold different perspectives than they do, and may be likely to bring these different perspectives to bear in processes of problem definition and problem solution.

In order to apply these criteria here, we must replace "engineers" with "mathematicians" and "engineering" with "mathematics". The analysis of the weekly reflections suggests that the students in this REU program progressed through the above criteria in almost a linear manner. They began by analyzing similarities and differences, than began to examine the life experiences within the given country and, by the conclusion of the REU program, began to recognize the valuable contributions of all involved in the project.

Acknowledgements. The authors thank the Hong Kong research mentors, Drs. Raymond Chan, Michael Ng, Weiwei Sun and Tao Tang, for their participation in this program.

References

[1] Committee on Prospering in the Global Economy of the 21st Century: An Agenda for American Science and Technology, *Rising Above The Gathering Storm: Energizing and Employing America for a Brighter Economic Future*, National Academy of Sciences, National Academy of Engineering, Institute of Medicine, 2005.

[2] G. L. Downey, J. Lucena, B. Moskal, R. Parkhurst, T. Bigley, C. Hays, B. Jesiek, L. Kelly, J. Miller, and S. Ruff, *The globally competent engineer: working effectively with people who define problems differently*, Journal of Engineering Education, 95 (2006), 107–122.

[3] N. Dovidio and S. Summerson, *An international REU program: a student perspective*, SIAM News, 39 (2006), 20.

[4] *Looking Beyond the Borders: A Project Director's Handbook for Best Practices for International Research Experiences for Undergraduates*, C. A. Loretz, editor, 2002. Available: http://www.nsf.gov/publications/pub_summ.jsp?ods_key=nsf06204

DEPARTMENT OF MATHEMATICAL AND COMPUTER SCIENCES, COLORADO SCHOOL OF MINES, GOLDEN, CO 80401-1887
E-mail address: gfairwea@mines.edu,bmoskal@mines.edu

Part III

Articles

The Long-Term Undergraduate Research (LURE) Model

Sarah Spence Adams, James A. Davis, Nicholas Eugene, Kathy Hoke,
Sivaram Narayan and Ken Smith

The Long-term Undergraduate Research Experience (LURE) model for the mathematical sciences represents a collaboration between the mathematics faculty at four widely different institutions: Central Michigan University (CMU) is a large public comprehensive university, Coppin State University (CSU) is a historically Black university, Olin College (OC) is a new gender-balanced engineering college, and the University of Richmond (UR) is a liberal arts college. We were recently awarded a grant through the National Science Foundation's Mentoring through Critical Transition Points (MCTP) program to pursue the LURE model. This article outlines our rationale for developing the LURE model, including our past experiences and our current thinking on best practices.

The LURE model emphasizes the early recruitment of undergraduates to mathematical research and the cultivation of interest in the mathematical sciences. It builds upon the success of the apprentice model often used in the physical and life sciences, wherein scientists routinely engage first- and second-year undergraduates in laboratory research and then continue to mentor these students until they are prepared to pursue graduate degrees. Although first- and second-year undergraduates generally have little experience or knowledge to contribute to the laboratory research, they perform routine tasks and learn how to use sophisticated equipment. Exposure to the research activities in a particular laboratory prepares them for more substantial contributions in the future. Our scientist colleagues report that their investment in these "apprentices" reaps benefits after the students devote a year or more in their laboratories. This apprentice model is successful in that it recruits early undergraduates to the sciences, retains student interest in the sciences, and prepares students for graduate study in the sciences.

Our study of the science apprenticeship model, as well as our own experiences directing undergraduate research, led us to develop the LURE model for the mathematical sciences. CMU has been a site for an NSF Research Experience for Undergraduates (REU) in mathematics for five years, and UR has enjoyed a variety of industry and government grants supporting undergraduate research in mathematics. Through our involvement at CMU and UR, we have mentored many of the common types of undergraduate research experiences, including summer programs and senior theses. One observation we have made, which appears to be a

---

Received by the editor September 15, 2006.

nationwide trend, is that mathematics students typically wait until their junior year, usually after completion of core courses such as analysis and abstract algebra, before participating in research activities. Thus, research experiences are typically not being used to attract undergraduates to mathematics, although we do believe research experiences assist in retaining those students who have already chosen a major in the mathematical sciences. A second observation is that most undergraduate mathematics research experiences are relatively short, lasting usually for one summer or at most one academic year. Students are often just reaching the point of truly understanding the problem when the experience comes to an end. Our experience indicates that many students and faculty would prefer to have more time to continue their collaboration. Specifically, faculty have noted that the "start-up" cost of involving students in short-term research experiences is high, and that students would be more likely to actually assist them in their research goals if they could keep the students involved for a longer duration.

Through the LURE model, we address the above two observations. First, we change the culture so that faculty begin to recruit early undergraduates to mathematical research experiences, thereby providing opportunities to recruit students to mathematics before losing them to other disciplines. Second, we provide for longer research relationships wherein faculty can train students to fully understand and eventually contribute to the research program, thereby avoiding the frustration that short research experiences can sometimes bring. Specifically, LURE recruits students early in their undergraduate careers and pairs them with faculty who serve as mentors throughout a two-year research experience in the mathematical sciences. Through closely supervised research and independent study activities spanning two summers (10-weeks each) and two academic years, students experience all steps in a research project, from background reading to the professional presentation of results. This allows undergraduates to be involved with mathematics research experiences that are more sophisticated than possible with traditional single-summer research experiences.

We have some evidence that this extension of the science's apprenticeship model will be successful in the mathematical sciences. Over the past four years since the inception of OC, OC faculty have experimented with involving first- and second-year students in mathematical research. Although OC's history of experience is limited, we have seen remarkable results so far. Not only have certain first-year students proven capable of contributing to professionally published papers, but these early experiences have whetted their appetite for research. The result is a culture in which first-year students clamor to be involved in faculty research programs and often continue to work with faculty for several semesters. We have seen the benefits of involving students in long-term research experiences, including facilitating peer-training opportunities, wherein "seasoned" student researchers (for example, juniors who have already spent two years working with a particular faculty member) can help bring first-year students up to speed with the basic ideas in the research program.

The LURE model includes some "best practices" that we have developed through our past experiences at our four respective institutions. Throughout the two-year program, we will focus on developing strong mentoring relationships as well as camaraderie among the students. We will pair students in groups of two or three, to combat feelings of isolation and to build group-working skills. We will also schedule social activities, especially during the summers, to build a sense of community. For example, we plan weekly lunches, augmented by larger activities such as field trips to local parks/beaches, volleyball games, and apple/berry picking. We will improve students' oral and written communication skills through regular oral/written progress reports. The length of the research experience under the LURE model allows us to focus on developing these skills, without sacrificing a significant portion of the research time. A yearly LURE conference will bring together all LURE students and mentors from the four institutions. This conference will serve as a community-building activity and as motivation for students to develop their oral and written presentation skills. Each student will give an oral presentation on his/her work and provide an article for the LURE conference proceedings.

We will proactively recruit women and minorities, with specific aspects of the program having been developed in consultation with literature on cultivating the success of underrepresented groups. Our recruitment strategies include flyers posted around campus and mailed to incoming students, emails to special interest groups (e.g. honors programs, math clubs, science clubs), announcements in introductory mathematics classes, and personal contact. By the end of the NSF funding (four years), eighty students and twenty eight faculty will have participated in the LURE model. Each year, each institution will have two to four teams, with each team consisting of one faculty mentor and two to three students.

The LURE model provides professional development opportunities for faculty, by providing a network of colleagues for trouble-shooting questions and problems, through the optional pairing of inexperienced faculty mentors with seasoned mentors, and through group discussions at the yearly LURE conference. To kick-off the LURE program, we are developing a one-day conference wherein we share best practices and learn best practices from Joseph Gallian and Aparna Higgins, leading mentors of undergraduate research. One of our goals is to increase the number of faculty, with particular focus on women and minorities, who become mentors and role-models in undergraduate research experiences. We hope that the network and training offered by the LURE program will encourage faculty to become involved as undergraduate research mentors. This is particularly important at CSU, a historically Black college, as CSU works to establish a culture encouraging undergraduate research in mathematics.

In our model, we compensate faculty for their time (approximately $7000 per summer). We felt that this was necessary to ensure the continued participation of faculty, since being a mentor under the LURE model requires a significant time commitment. By involving early undergraduates who have less experience than junior or senior mathematics majors, it is necessary for the mentors to meet with their students more often and to do more proactive teaching. However, given the length of the mentoring relationship under the LURE model, the time invested will pay off. The mentors will eventually be able to assign more sophisticated mathematics readings and problems with the understanding that the student can master the techniques over weeks and months rather than days. In time, the students will be more likely to significantly contribute to the faculty member's research program.

By implementing the same basic model at four very different institutions, we will learn what works and what requires modification in the basic model. This will also allow us to more convincingly disseminate our model to a wide variety of institutions. At the yearly LURE conference, participating faculty will discuss and evaluate how the model is working in the different settings. We will also analyze our findings concerning the impact of LURE on participating students through our collected data on a variety of longitudinal measures. For example, we will measure the number of students majoring in the mathematical sciences, the number of mathematics courses elected beyond institutional requirements, and the number of students pursuing a graduate degree in the mathematical sciences; collected data will be compared to previously (pre-LURE) compiled data at each school. We will hire an assessment specialist to assist us in analyzing and presenting the data. We will share our findings broadly and provide guidance on the costs and benefits of initiating a sustainable program. We expect to find concrete evidence of the benefits of involving early undergraduates in mathematical research (in part using numerical measures often considered important by college administrators), and we hope this will help other faculty succeed in securing internal (or external) funding for similar endeavors.

In summary, we have developed the LURE model based on the successful science apprenticeship model and based on our previous experiences mentoring undergraduate mathematics research projects. We will engage undergraduates in mathematical research earlier and aim to keep the students for a longer period of time than is typical in most existing programs. As we gain experience working with the LURE model and collect data on the impact of LURE on participating students, we will report our findings to the mathematics community. We hope that our findings will provide a rationale for other institutions to consider adopting this new model.

FRANKLIN W. OLIN COLLEGE OF ENGINEERING, OLIN CENTER, OLIN WAY, NEEDHAM, MA 02492-1200
    E-mail address: **sarah.adams@olin.edu**

DEPARTMENT OF MATHEMATICS AND COMPUTER SCIENCE, UNIVERSITY OF RICHMOND, RICH-MOND, VA 23173
    E-mail address: **jdavis@richmond.edu**

DEPARTMENT OF MATHEMATICS, COPPIN STATE UNIVERSITY, 2500 WEST NORTH AVENUE, BALTIMORE, MD 21216
    E-mail address: **NEugene@coppin.edu**

DEPARTMENT OF MATHEMATICS AND COMPUTER SCIENCE, UNIVERSITY OF RICHMOND, RICH-MOND, VA 23173
    E-mail address: **khoke@richmond.edu**

DEPARTMENT OF MATHEMATICS, CENTRAL MICHIGAN UNIVERSITY, MT. PLEASANT, MI 48859
    E-mail address: **Ken.W.Smith@cmich.edu**

DEPARTMENT OF MATHEMATICS, CENTRAL MICHIGAN UNIVERSITY, MT. PLEASANT, MI 48859
    E-mail address: **naray1sk@cmich.edu**

Research With Students from Underrepresented Groups

R. Ashley, A. Ayela-Uwangue, F. Cabrera, C. Callesano, D. A. Narayan

Introduction

In 2003 the *MAA National Research Experience Program for Undergraduates* was created for students from underrepresented groups. The program provides grants to institutions that support four or five students and a project mentor. To insure that minorities are chosen, prospective students must be named in each incoming proposal. In 2005 Darren Narayan was awarded one of twelve grants to run this program at the Rochester Institute of Technology (RIT). Dr. Narayan conducted his project for six weeks during the summer with two African-American and two Hispanic student researchers.

Although RIT has a diverse population of students, it was difficult to find four mathematics majors from underrepresented groups. Two outstanding mathematics majors were selected from Dr. Narayan's discrete mathematics and graph theory courses: Rachell Ashley, a 4th year Applied Mathematics Major from Ft. Worth TX and Carol Callesano, a 2nd year Applied Mathematics major from Ganesvoort, NY. Two more were selected from his calculus classes: Aisosa Ayela-Uwangue, a 2nd year Electrical Engineering major from Benin, Nigeria and Frances Cabrera, a 1st year Environmental Management major from Virginia Beach, VA

The four students bonded quickly despite different backgrounds and being in different stages of their academic careers. Each student brought their unique talents to the table and a strong research group developed. Not only did the group nearly solve an open problem in combinatorics, but their work culminated in an award-winning poster session at the National Meetings in San Antonio in 2006. Publication of their work in refereed research journal is also in the works.

Selecting a Research Problem

Since two of the students were not mathematics majors, Dr. Narayan searched for a project that was easy to explain to non-mathematicians. On the first day he told them, "I want you to have a research problem that you can explain to your mother, father, sister, brother, aunt, and uncle. While I am not all that interested in whether your family members understand, I do care if one of your potential employers does. This research experience will be a prominent item on your resume

Received by the editor December 1, 2006.

that could come up during an interview. You might have a 30-second window to explain your research. You have to make it count."

Our project in combinatorics is a rectangular tiling problem using tiles with dimensions 4×6 and 5×7. For a proper tiling the entire rectangle must be covered so that no tile extends beyond the rectangle and no pair of tiles overlap.

We considered the following problem: Determine all natural numbers m and n such that an $m \times n$ rectangle can be tiled with 4×6 and 5×7 rectangles.

It was important that the students become interested and excited about solving this problem. Six weeks is a short period of time for research. Potential applications gave this problem extra appeal. Many companies use two-dimensional load software to obtain optimal packing configurations for shipping. Another application unique to the dimensions 4×6 and 5×7 involves optimal cutting of photo paper. For example it was shown in [2] that a 34×33 rectangle can be tiled and hence a 34×33 sheet of photo paper can be partitioned into popular sizes 4×6 and 5×7 with no waste. Since the methods of solving the problem could be grasped quickly, the students could produce results quickly which greatly added to their enthusiasm. One student, Frances Cabrera, commented "I never imagined I'd get so excited, even passionate, over a seemingly obscure theoretical combinatorics problem, but I did. This made the entire experience much more enriching. It made me realize how much I enjoy mathematics and decide not to let math leave me."

The student's excitement was not confined to the group. They wanted to share their experiences with others. In two cases friends of the student researchers mentioned that they were intrigued by the problem. One non-NREUP student commented. "My friend was excited about the research problem she was working on. She explained it to me and I got excited."

The Research

Tiling problems of this specific nature were introduced on the 1991 William Lowell Putnam Examination. Problem B-3 asked "Does there exist a real number L such that if m and n are integers greater than L, then an $m \times n$ rectangle may be expressed as a union of 4×6 and 5×7 rectangles, any two of which intersect at most along their boundaries?" Clearly some rectangles, such as the 7×9 and 8×10 rectangles, can not be tiled. Allen Schwenk and Darren Narayan showed that any rectangle with both dimensions at least 34 can be tiled [2]. However this does not cover cases involving rectangles, such as 23×91, 17×200, etc. where one dimension is less than 34. For any fixed width, the length can be arbitrarily large so the group had to consider infinitely many rectangles. This was in no doubt an ambitious project.

Allen Schwenk mentioned that he had considered working on this particular problem years ago, but was hesitant with the large number of cases involved. However the students were enthusiastic about tackling a challenging problem.

After four weeks, we were able reduce the number of rectangles under consideration from an infinite number to a mere 54. This was a tremendous accomplishment. A computer search could now be brought in to finish the work. However, we wanted to obtain a solution that was independent of a computer. However many of the cases proved to be quite difficult. We contacted Allen Schwenk, notified him of our progress, and offered to include him in the project. He was delighted and provided tips on further reducing our list of 54 cases.

During the final two weeks we focused our attention on these cases. Our conjecture is that many of these rectangles can not be tiled, but this is difficult to prove. Using various combinatorial arguments, we made steady progress and showed one by one that many cases could not be tiled. This process gave the program an incredibly nice feature. We would write the number of remaining cases up on the board, and watch as the number steadily decreased. We would often be asking each other, "What number are we down to?" Not only did this add excitement, but it also gave us an additional measure of our progress.

Program Activities

In addition to the program's research component, we spent a considerable amount of time talking about graduate school. We discussed various programs, as well as graduate studies in general. Midway through the program we attended a master student's defense. As a result the student researchers were not as intimidated by the thought of producing and defending a thesis. They commented afterwards that a defense should not be scary. By that time the student knows their work better than anyone.

We also benefited from interactions with more than 100 other undergraduates and mentors working on research projects in the College of Science at RIT. Activities included a seminar series featuring student presenters. The NREUP students enjoyed presentations from other scientific fields, and gave their own dynamic presentation during the program's final week to an audience of nearly 100 students and faculty.

The last day of the program was spent on a field trip to Letchworth State Park. This provided an informal setting to talk about graduate school, interviewing for permanent jobs, and future career paths. It also gave us an opportunity to reflect on everything we had accomplished.

Follow Up Activities

The program officially ended in the summer, but we continued working on the problem. There are now only 18 cases to resolve. The four students presented their work at the 19th Midwest Conference on Combinatorics, Cryptography, and Computing, held October 7-9, 2005 at RIT. They presented their work again in the 2006 AMS-MAA-SIAM Special Session on Research in Mathematics by Undergraduates in San Antonio, TX. The group also presented in the Undergraduate Student Poster Session at the same meeting. To cover the students' travel expenses Dr. Narayan secured funds from RIT alumni, the RIT Honors Program, and the students' home departments. In addition, the students received support from the *MAA Diversity Initiative*, a program designed to increase the number of women and minority students attending the Joint Mathematics Meetings.

The national meetings were a tremendous success. Both their oral presentation and poster presentation were well received. The students were named as one of the winners in the Undergraduate Poster Session and received a $100 cash prize. A picture of the group celebrating their win appeared on the cover of March Issue of FOCUS, the newsletter of the MAA. We are currently preparing their results for publication in a refereed journal [1].

References

[1] R. Ashley, A. Ayela-Uwangue, F. Cabrera, C. Callesano, D. A. Narayan, and A. J. Schwenk, Necessary and Sufficient Conditions for Tiling with 4×6 and 5×7 Rectangles, to be submitted.

[2] D. A. Narayan and A. J. Schwenk, Tiling Large Rectangles, Mathematics Magazine, 75, No. 5. (2002), 372-380.

NARAYAN, SCHOOL OF MATHEMATICAL SCIENCES, ROCHESTER INSTUTUTE OF TECHNOLOGY, ROCHESTER, NY 14623-5603

E-mail address: Darren.Narayan@rit.edu

Research Classes at Gettysburg College

Béla Bajnok

Introduction. I have been supervising undergraduate research projects in mathematics for over fifteen years. Three years ago I had the opportunity to formalize this (and give, as well as receive, course credit) by introducing three levels of research courses at Gettysburg:

Math 201: *Introduction to Research in Mathematics*,

Math 301: *Intermediate Research in Mathematics*, and

Math 401: *Advanced Research in Mathematics*.

Introduction to Research in Mathematics has no course prerequisites, and is usually taken by students whose career plans do not involve mathematics, who but like the subject and have a desire to take the course as an elective. *Intermediate Research in Mathematics* has Math 215 (*Abstract Mathematics I*) as a prerequisite; it is taken mostly by students majoring or minoring in mathematics. Finally, *Advanced Research in Mathematics* is taken by math majors who already took Math 301 and intend to base their capstone experience in mathematics (a requirement for all majors) on their research project. Roughly speaking, Math 201 expects students to learn new material and make discoveries; Math 301, in addition, asks them to provide proofs for their conjectures; and Math 401 focuses on writing a professional paper as well.

First we offered these courses once a year; last year, we began to offer each course every semester. However, due to some obvious reasons (e.g. not having sufficient faculty resources), we offer the courses in "one-room-schoolhouse"-style: students are together in the same class. The combined enrollment in the three classes usually ranges from fifteen to twenty.

I have been largely satisfied with these courses and, more importantly, the students seem to be as well. Below I attempt to summarize my experiences and contrast our program with more traditional undergraduate mathematics research programs.

The sums. Each research class is really the sum of three classes, and each class is, in essence, the sum of several independent research courses. I truly believe that what we have is greater than the sum of the parts. However, special care needs to be taken when coming up with research projects.

An ideal research project needs to be

Received by the editor September 15, 2006.

- based on substantial topics – students should be engaged in the study of non-trivial and not-too-esoteric mathematics;
- challenging at a variety of different levels – students with different backgrounds and interests will be able to engage in each project;
- approachable with a variety of different methods – students interested in theoretical, computational, abstract, or concrete work will be able to choose their own approaches;
- incrementally attainable, where at least partial results are within reach yet complete solutions are not easy – no student would want to spend long hours of hard work and not feel productive; and
- new and unsolved – the results attained by the students would have to be, at least in theory, publishable.

During the past thirteen years, I have had an opportunity to work with a very large number of research students (perhaps as many as one hundred students). Coming up with so many research projects is not easy. The topics of the research projects have come mainly from my own research interests in algebraic and geometric combinatorics, finite set theory, and additive and combinatorial number theory – but they also included questions from probability, statistics, computer networks, game theory, and quite a few other topics. I believe that my carefully selected projects yielded meaningful and enjoyable research opportunities for many students.

The differences. Our program structure is rather unusual and thus requires unusual approaches. Some of the main differences are the following.

- Our courses run parallel to other classes that students take. This gives them a somewhat longer time frame (fourteen weeks), but prevents them from truly immersing themselves in their research. (Nevertheless, students report spending a very large number of hours with the class – occasionally in excess of twenty hours a week.)
- The courses are taught together, and therefore the students in the same room have a very diverse set of objectives, interests, backgrounds, and talents. This gives special challenges for the instructor.
- Since the research classes yield course credit in mathematics, an important feature needs to be that students learn a substantial amount of new mathematics. Therefore, especially during the first month or so during the semester, I lecture on topics which are typically beyond regular courses.
- Students receive a grade in the course. Since research results can be unpredictable, most components of the grade need to come from effort, proficiency in learning of new material, frequency and level of class presentations, and willingness and ability to participate in the projects of other students.

I believe that these differences, while necessitating certain limitations on the research, provide unique opportunities for a diverse group of students.

The products. The primary products of the courses are the research results achieved by the students. These results must be communicated both verbally and in writing. Therefore, I ask each student to

- give weekly presentations on his or her progress;
- give a final formal presentation on the research – this presentation is open to the public;

- write a formal research paper – these papers appear in the proceedings of the course entitled *Research Projects in Mathematics*, published locally at the end of the course.

In addition, each student is encouraged to

- give an off-campus presentation on his or her research;
- submit (an improved version) of his or her paper to a refereed journal.

We have had numerous students give presentations at conferences, including the national meetings of the American Mathematical Society, sectional meetings of the Mathematical Association of America, the Nebraska Conference for Undergraduate Women in Mathematics, the Rose-Hulman Undergraduate Mathematics Conference, and the Moravian College Mathematics Conference. Two students also gave talks at a research conference where they were the only undergraduates present. Additionally, two students gave a poster presentation at the annual meeting of the American Mathematical Society; their work won Honorable Mention.

Although I strongly encourage students to revise their paper so that it can be submitted to a refereed journal, no student has been able to do this so far. Even if they don't win awards or get applauded after giving a successful presentation, it is very important to give recognition to every student who performed quality research.

The quotients. I have not carried out a systematic assessment of these classes; to be honest, I am not entirely sure how to perform such an assessment. For now, here are some statistics on the students so far.

- about 1 out of 25 students at Gettysburg takes a research course in mathematics, including about a third of those minoring and a half of those majoring in mathematics;
- about 4 out of 5 students generate new research results that can be considered non-trivial (although, in all honesty, most results can be expected from a talented and motivated student);
- about 1 out of 3 students receives an A in the course, and about 3 out of 4 students receive a B or higher;
- about 3 out of 4 students rate the course as "excellent"; all have rated the course as "very good" or "excellent";
- about 2 out of 3 students who did not originally intend to major or minor in math decide to do so after having taken the course.

I intend to develop appropriate assessment tools and evaluation criteria for these courses.

The powers. Our corridor in front of the Department has a huge poster with a fist and the sign "Math is power", and I believe that few courses empower students as much as a research course. In addition to learning about exciting, current, and lively topics, students have the opportunity to perfect a wide variety of skills from dealing with setbacks to presenting their accomplishments.

Ways in which research activities benefit students with a serious interest in mathematics – especially those intending to go to graduate school – have been described by others. But I feel strongly that every student should engage in an undergraduate research experience. Whether they will go on to graduate school, enroll in professional studies, or take jobs in education, government, or industry, students will benefit from the opportunities for perfecting a variety of skills that a research experience provides.

An invitation. I am looking forward to hearing anyone's comments on this article whether before, during, or after the conference. It would be particularly interesting to hear from colleagues who are engaged in similar mathematical research experiences at other institutions.

GETTYSBURG COLLEGE, GETTYSBURG, PA 17325-1486 USA
E-mail address: bbajnok@gettysburg.edu

Proceedings of the Conference on
Promoting Undergraduate Research in Mathematics

Research in Industrial Projects for Students: A Unique Undergraduate Experience

Stacey Beggs

The Institute for Pure and Applied Mathematics at UCLA, with support from The National Science Foundation (NSF), The National Security Agency (NSA), and its industry sponsors has offered the Research in Industrial Projects for Students (RIPS) program every summer since 2001.

The mission of the Institute for Pure and Applied Mathematics (IPAM) is to promote and facilitate cross-disciplinary connections between a broad spectrum of mathematicians and scientists, to launch new collaborations, to better inform mathematicians and scientists about interdisciplinary problems, and to broaden the range of applications in which mathematics is used. IPAM holds programs throughout the academic year for junior and senior mathematicians and scientists who work in academia, the national laboratories, and other private or public organizations, as well as for students.

The RIPS Program was modeled after The Math Clinic at Harvey Mudd College, part of Claremont Colleges. Unlike conventional mathematics education where students typically solve problems independently, it emphasized a more structured and personalized team format to enhance the research training experience. RIPS features the same concept in a 9-week format.

In RIPS, undergraduates work in teams on a research project formulated by a government or industry sponsor in consultation with IPAM. Each RIPS team is comprised of four students, a faculty mentor, and one or several industry mentors. During the nine week program, the students study the problem and master the latest analytical approaches and techniques to solve it. The students also develop the report-writing and public-speaking skills needed to present their work to a scientific audience. Industry mentors provide regular contact between the team and the sponsor, while monitoring and helping to guide student work. Participation in RIPS provides valuable real-world technical and managerial experience for students and valuable R&D for sponsors.

Recruitment and Selection of Students

IPAM makes extraordinary effort to promote RIPS, especially to women and to underrepresented ethnic groups. An announcement is submitted to national organizations and journals. A poster advertising RIPS is distributed at IPAM events,

Received by the editor December 1, 2006.

national conferences (Society for Advancement of Chicanos and Native Americans, Historically Black Colleges and Universities, etc.), and professional meetings. IPAM board members assist in distributing information. We send emails to mathematics faculty, student advisors, and undergraduate club officers across the country. Undergraduate engineering and computer science advisors are also routinely included in mailings. A poster promoting RIPS to young women is disseminated widely. Finally, all past participants of RIPS are asked to help promote RIPS to undergraduate students at their institutions. Word of mouth is probably our most successful means of recruitment.

RIPS students are academically in the top tier of their class. While an extensive mathematics background is a prerequisite for selection, the RIPS students represent a wide range of backgrounds and academic pursuits. For example, while the majority of the RIPS 2006 participants were mathematics majors, a number of them studied computer science, electrical engineering, physics, and computational biology. The group also represented a variety of institutions, including prestigious technical colleges (Caltech, MIT, RPI), Ivy League schools (Yale, Harvard, Princeton), other liberal arts colleges (Gettysburg College, Carleton College), and public universities (UCLA, San Jose State, North Carolina State). Most of the students were rising seniors or 2006 graduates.

Applications are due mid-February. The IPAM directorate initially screens the applications for general suitability. A subset of the total applications is then presented to each industrial sponsor, which then assigns grades to each applicant to indicate their preferences and needs of the project. The directorate then considers industry sponsor preferences as well as other goals for RIPS participation to make final selections. In the event that a student declines the invitation to participate, alternates are also chosen. The goal is to put together teams whose members have the appropriate combination of skills and experience for the proposed project.

The selection process is highly competitive. Out of 225 applicants for RIPS 2006, only 36 students (16%) were ultimately chosen to participate. The competition for high-achieving minority students is always quite high; nevertheless, we were pleased that four students who are members of underrepresented groups did participate. We were also delighted at our success at recruiting women. See Table 1.

Table 1: RIPS 2006 Participants							
	Total	Female		Male		Members of Underrepresented Groups	
Student Participants	36	14	39%	22	61%	4	11%

Interested students may (and often do) apply again if they are not chosen the first year they apply. Due to demand for RIPS, we do not allow a student to participate in RIPS twice, although many express a desire to do so.

IPAM Directorate

IPAM Director Mark Green is largely responsible for the scientific content and operation of RIPS. Director of Special Projects Stanley Osher and Associate Director Christian Ratsch assist Dr. Green with student selection, recruiting industrial sponsors, reviewing the Sponsors' project descriptions, and identifying and recruiting faculty mentors.

Program Director

Michael Raugh, retired Professor of Mathematics at Harvey Mudd College and cofounder of Interconnect Technologies Corporation, has served as the RIPS Program Director each summer. The Program Director duties include weekly meetings with faculty mentors and project managers to provide support and guidance, reviewing each team's reports and presentations, and active participation in Orientation Day and Projects Day.

Faculty Mentors

One faculty mentor is assigned to each team. Faculty mentors are typically recently graduated post-doctoral scholars in mathematics or other relevant disciplines. They provide methodological guidance, review and critique work, offer encouragement, support, and sometimes career counseling and advice. Some have returned for a second or third year because they enjoy working with RIPS students.

Sponsors and Industry Mentors

RIPS sponsors represent industries in the government and private sectors, such as biotechnology, computer animation, and anti-virus software development. All are engaged in research and development in leading-edge technologies that require advanced mathematics knowledge and skill. Sponsors are typically recruited through IPAM's scientific programs. IPAM invites prospective sponsors to RIPS Projects Day (see RIPS Schedule, below) to demonstrate the impressive outcomes of current projects. See table 2 below for a list of RIPS 2006 sponsors.

Sponsors are intimately involved with the design of projects and selection process. During the program, the team's industry mentor or team of mentors representing the sponsor meet with their team weekly either in person or by telephone conference call if the sponsor is not local. All industry mentors attend Orientation Day and Projects Day. In addition, the sponsor hosts the team for at least one site visit.

Many organizations are serial sponsors. IPAM recruited two new sponsors in 2006 (see table 2), and is in conversation with several private companies and a state agency about sponsoring RIPS projects in 2007 or 2008. IPAM will continue to seek out new industrial sponsors with interesting projects, as well as retain the ones that have participated in the past.

Projects

RIPS projects are carefully designed to challenge students and expose them to a range of mathematical theories and concepts relevant to the project. For example, the Symantec team's project in 2006 involved the use of image compression algorithms to identify image variants found in spam. See Table 2.

Prior to the start of the RIPS Program, each industrial sponsor prepares a Project Description explaining the problem. The statements are written by the sponsor in consultation with the faculty mentor and IPAM directorate.

Table 2: RIPS 2006 Sponsors and Projects

Sponsor	Title of Project	New/ Returning
Areté Associates, Inc.	Video Images Mapping and Compression for Efficient Data Downlink	Returning
Hewlett-Packard, Inc.	Procurement Is Possible	Returning
Jet Propulsion Laboratory	Impulsive Low Energy Transfers Between the Earth and the Moon	Returning
Los Alamos National Laboratory	Robotic Path Planning and Visibility with Limited Sensor Data	Returning
Lawrence Livermore National Laboratory	The Disambiguation Problem	Returning
NASA Goddard Space Flight Center	Methods for Detecting Gravitational Wave Signals in LISA Data	New
Pixar Animation Studios	Simulation of Many Colliding Deformable Solids for "Set Dressing" and Arrangement	Returning
Symantec Corp.	Image Similarity for Detecting Image-Based Spam and Phishing Attacks	New
TimeLogic, Inc.	Determining the Sequence of Peptides from Tandem Masspectra	Returning

RIPS Schedule

RIPS begins with Orientation Day, at which students are introduced to the structure and expectations of the program and have the opportunity to discuss their Project Description with their industry sponsors.

Shortly thereafter, the students prepare a work statement, a contract between the student team and sponsor, which is forwarded to the industrial sponsor for approval or negotiation. Once completed, the team begins its research and conducts regular meetings with its industry mentor and faculty mentor. Each team also selects one student to act as project manager.

During the fourth week, each team presents their project findings to date and submits a brief mid-term report to the Program Director. Faculty mentors and industrial sponsors provide feedback and direction. Throughout the next few weeks, each team visits their industrial sponsor's site, often to present their research to a group working on similar problems.

On Projects Day, held at the end of the eighth week of the program, each team gives a professional-quality slide presentation of their work. Industrial sponsors, faculty mentors, UCLA mathematics faculty and graduate students, prospective industrial sponsors, and some family members of RIPS participants typically attend. The students also begin preparing their final reports.

During the ninth week, final revisions are made to the reports which are then carefully reviewed by faculty mentors and the Program Director. The report and slide presentation are sent to the sponsor upon completion. On the last day of the program, students "graduate" from RIPS and receive a certificate of participation.

Impact and Results

In five years, RIPS has become a highly selective program. It began in 2001 with 43 applications, 12 students and four projects. By 2006, RIPS received 225 applications for 36 openings and 9 projects. Our recruitment of women in 2005 and 2006 was particularly successful, representing approximately 40% of total participation each summer. While recruiting members of underrepresented ethnic groups continues to be a challenge, we have achieved a participation rate of 11% to 12.5% for the past three years.

The number of projects each summer has gradually expanded from four to nine, demonstrating our ability to recruit and retain sponsors as well as the increased demand for the program. We are pleased with the mix of private companies and national labs and with the quality of the prospective new sponsors for 2007 and 2008. Many of the sponsors return to participate multiple times; in fact, four industrial sponsors have participated for four or more consecutive summers through 2006. The projects that they propose are of high quality, challenging the students yet presenting a problem that is appropriate for a nine-week undergraduate program.

The quality of the students' work is impressive. Sponsors have been pleased with the results, and have consistently praised the program in our closing survey. Here are some comments from past industry mentors:

- Peter Eltgroth and Peter Brown, LLNL, RIPS 2002 and 2004: "I was impressed by the quality of the students, not just in terms of their credentials coming into the program, but by what they produced, as undergraduates, over the course of two months ... The final report for this project is a remarkable piece of work that benefits us here at LLNL."
- Terence Kelly, HP Labs, RIPS 2006: "RIPS was a very rewarding experience for us as industrial sponsors. Our team shed new light on known aspects of our problem. More importantly, our team discovered new aspects of the problem and suggested new directions for future research."

Many students tell us that RIPS had a significant influence on their decisions about graduate school and attitude towards math and industrial research. Here are a few quotes from RIPS students that demonstrate the immediate and profound ways the program impacts its students:

- Jennifer Garcia, RIPS 2001: "If it were not for IPAM and a small list of caring professors, I would never have learned of my passion for research."
- Jason Geertz, RIPS 2002: "After attending the IPAM program RIPS 2002, I decided to attend graduate school and continue with research. I believe my experience there taught me a lot about managing and doing scientific research."
- Jacob Macke, RIPS 2004: "[RIPS] has greatly strengthened my resolve to pursue a career in academia, and I will embark on graduate studies in computational neuroscience next year. I have plenty of offers to choose from ... and am convinced that without RIPS at IPAM, I would not have been able to achieve this."
- Lisa McFerrin, RIPS 2004: "My IPAM experience was a huge motivator to my current situation. I was placed in the BioDiscovery group and was introduced to the field of bioinformatics. I loved the ideas that we were working with and the implementation process which coincided exactly with my expertise and interest."

The impact of RIPS on undergraduate student academic and professional careers is broad. For some students, RIPS opens doors to graduate study, for others to new career pathways, and for still others, the possibility of job opportunities with an industrial sponsor. Here is a sample list of achievements of past RIPS students, directly or indirectly resulting from their participation in RIPS:

- Miranda Lee (RIPS 2001), now a graduate student at Stanford, won National Defense Science and Engineering Graduate (NDSEG), National Science Foundation (NSF) and Stanford Graduate Fellowships.
- Dan Shaevitz (RIPS 2002) began working at Areté Associates as a Research Analyst after participating in RIPS.
- Two of the four students on the 2002 LLNL team went on to pursue their graduate studies in Computational Transport at Princeton and the University of Colorado at Boulder.
- Two RIPS students on the LANL team (from different summers) subsequently completed LANL internships.
- Linda Hung (RIPS 2003) participated in the Cryptanalysis and Exploitation Services program at NSA the following summer. She then entered the PhD Program in Applied Math at Princeton.
- Che Smith (RIPS 2004), an African-American woman from Spelman College, was accepted to graduate school in Biostatistics at Harvard, UCLA, and North Carolina State.
- Jeff Aristoff (RIPS 2004) says that his participation in RIPS helped him obtain an NSF Graduate Research Fellowship in applied mathematics.
- Lauren Anderson (RIPS 2004) has since completed an internship with Northrop Grumman, which offered her a position after she graduates.

RIPS also influences the careers of its faculty mentors. For example, four-time RIPS faculty mentor Shawn Cokus, who had a background in combinatorics and knew no biology when he was selected to mentor a RIPS bioinformatics projects, now has a job working in a bioinformatics lab at UCLA.

Finally, IPAM has offered its support and materials to those who wish to replicate RIPS at their own institutions. Several interested faculty attended 2005 and 2006 Projects Day to see first-hand the final products of the student teams.

Future Plans

IPAM is pleased with the success of RIPS and intends to continue running the program for the rest of IPAM's current NSF grant, which continues through 2010. We will maintain the current size of the program: 9 projects and 36 students.

At the same time, we will expand RIPS by offering RIPS-Beijing in 2007. Dr. Harry Shum, Managing Director of Microsoft Research Asia (MSRA), has agreed to host five RIPS projects at their Beijing facility. Ten US students and ten Chinese students will work in teams of four (with two members from each country) on five projects sponsored by MSRA research groups. The idea is to integrate an international research experience with the RIPS concept. IPAM recently applied for an International Research Experiences for Students (IRES) grant through NSF to support the U.S. participants of this program.

UNIVERSITY OF CALIFORNIA, LOS ANGELES
E-mail address: sbeggs@ipam.ucla.edu

Proceedings of the Conference on
Promoting Undergraduate Research in Mathematics

What Students Say About Their REU Experience

Frank Connolly and Joseph A. Gallian

The 2006 AMS Survey of REU Graduates

Do REU's do any good?

At a time when more than 80 mathematics REU sites are running; and at a time when the term *undergraduate research* has become a solemn mantra among university administrators; and at a time when some undergraduates believe a refereed article in a mathematics journal is *necessary* for admittance into a good graduate program, this question might seem more than somewhat impolite.

Yet the very pervasiveness of good feeling toward the REU movement seems to create a perceptible murmur of skepticism, a suspicion in some parts of the mathematical community that this whole REU business is wildly overblown. A second reason for this suspicion is that, for many older mathematicians, REU's constitute an unknown quantity.

Thus it is especially timely that in the early months of 2006, the American Mathematical Society did a survey of graduates of REU's in mathematics. The purpose of the survey was to discover what effect REU's in mathematics were having, for good or for bad.

The survey asked what students thought was useful about their REU and sought to find out what had happened to a large echelon of REU students several years after their REU experience. The questions were designed by a committee of five people, including the second author.

At the request of the organizing committee of the AMS-NSA sponsored 1999 Conference on the Summer Undergraduate Mathematics Research Programs the AMS assembled a list of names and addresses of 560 students who participated in an REU program in mathematics between 1997 and 2001. Despite annual appeals from the AMS to update addresses, by 2006 only 444 addresses remained current. In early 2006 the organzing committee of the 2006 AMS-NSA Conference on Promoting Research in Mathematics by Undergraduates sent a survey consisting of 30 questions

---

Received by the editor February 7, 2007.

about their REU experience to these 444 students. Because the respondents were commenting on an event at least five years past, they had time to gain perspective on their undergraduate research experience, and were unlikely to be swayed in their responses by an emotional afterglow of a summer's camaraderie. Responses were received from 262 students (59%). Of these, 55% were male, 9% were Latino, and 2% were African Americans. Almost without exception, these respondents were reporting on a summer research program, and for a large majority (75%) this program was held at a college or university different from their own.

The first striking result emerges when these students are asked whether they think their REU was valuable to them. They were offered three choices: *Not Valuable, Somewhat Valuable*, or *Valuable*. The answers exhibit a strong consensus. Fully 84% answer that it was *Valuable*; only one respondent says it was *Not Valuable*; the rest answer *Somewhat Valuable*. (Interestingly, 100% of Hispanic students declared their REU experience to be *Valuable*). This is a clear endorsement of REU programs.

The most important question about REU's is whether they really are successful in achieving their goals. Of course goals vary. But the goal of the National Science Foundation and the National Security Agency, (the two principal financial sponsors of the REU programs), is to enhance the research infrastructure of mathematics within the United States. It is therefore impressive that 78% of all of these respondents entered graduate school intending to obtain a Ph.D., and 14% more entered graduate school with the intention of obtaining a Master's degree. Nearly all of those entering graduate school (93%) were seeking degrees in one of the mathematical sciences. These are remarkably high numbers.

Some REU directors believe in the need for a strong instructional element in their program, but apparently they are a minority. Only 31% of the respondents said their REU had a strong instructional component. Moreover, the responses of this sub-population of graduates of REU's with strong instructional elements seemed not significantly different from those whose REU focused exclusively on research. This suggests that both approaches can work well.

It is sometimes said that REU's are only getting the students who have already decided to go to graduate school in mathematics. But the survey results do not support that viewpoint. Out of 262 respondents, a total of 231 entered graduate school (82%), and of these, 78% said that the REU was definitely a factor, or somewhat a factor, in their decision to do so. Only 32% of those who did not go to graduate school say the REU was definitely or somewhat a factor in their decision. This might be the most significant finding of the whole survey. It suggests that REU's are valuable precisely in the way they nurture the *commitment* of a student to pursue a career in mathematics.

The respondents seem quite clear-eyed, however, about what an REU can and cannot accomplish. An overwhelming majority (82%) of those going to graduate school say the REU did not shorten their time there. Another clear majority (53%) of these say the REU did not influence their choice of a thesis area (although 36%

report their REU had at least some influence on their choice). But two out of every three of these say that the REU had some effect in accelerating their development as research mathematicians.

Most respondents believe that graduate admissions committees saw their REU experience as valuable. Three out of every four say that it helped them get into a better graduate program. Moreover two out of every three respondents who had won a fellowship thought that the REU was a factor in the award.

Snippets From Student Comments

But numbers cannot wholly convey the force of the feeling that comes through from those 79 respondents who chose to append a comment to their response. These comments were sometimes quite extensive. To give the reader some sense of the flavor of these, we have chosen one snippet from each, which seems most nearly to capture the overall opinion expressed. With only five exceptions, these are quite favorable to REU's. The less favorable five appear at the end.

...built my self-confidence ....
... a good experience, learned so much.
...learned a huge amount, enjoyed it immensely.
...greatly influenced my desire to do research.
...an excellent experience; fun and challenging.
...REU is good.
...REU's are wonderful experiences.
I'm glad I participated in it.
...pivotal in my attending grad school.
...good to experience the frustration of dead ends and elation of breakthroughs.
...learned at the REU that I enjoyed [research].
...really thankful for the experience....
...a great experience.
...a fantastic experience....
...helped me understand what research IS....
...one of the most valuable aspects of my undergraduate career.
...changed my entire career.
...its value can not be overstated.
...pushed me ...to pursue a Ph.D. Thank you!
REU was very helpful in getting to grad school.
I view my REU as a good experience....
I'd recommend it to anyone....
...helped me ...see myself as an equal among all my peers and not as the minority student.
...got a sense of what mathematical research would be like.
...taught me how to chip away at research ideas....
My REU experience was very positive....
...an invaluable experience....
I would not have attended graduate school had I not participated in an REU....
...REU experience was invaluable.
REU's ...were life-changing experiences.

...a wonderful experience.
...made me excited about doing research....
It was very motivating and encouraging.
It also planted the idea of graduate school....
I enjoyed my REU experience....
...helped me later on in ...graduate school.
...during my REU experience I first realized I could actually do mathematics....
Thanks for the wonderful experience....
I found that I liked [research], so I went to graduate school.
...one of the principal factors behind my decision to attend graduate school....
...extremely fortunate that I had the opportunity....
...REU was a great experience.
...an invaluable research experience.
REU I attended was very helpful.
...greatly influenced my decision to pursue the Ph.D. that I will receive in May.
I enjoyed my REU experience.
...it seems to be almost necessary to have done an REU to get an external fellowship.
...was very instrumental in helping me decide to go to graduate school....
Through my REU, I was greatly encouraged.
...technical skills I picked up during my REU have been very useful....
...a life-altering experience.
...REU was very interactive and collaborative.
Were it not for ...the REU, I likely would not have considered this career path.
...REU was crucial for my grad school experience.
My REU was great fun–but not much work.
...wish my REU had had a more instructional component.
...more should be done in REU's to encourage students to dive into the literature as soon as possible.
Unfortunately, bad professors do exist.
...a rigorous undergraduate education in mathematics is much more useful than an eight-week summer program.

The survey provides overwhelming evidence that REUs are achieving the goals for which they were created: to encourage mathematically talented students to pursue graduate degrees and to provide participants a meaningful research experience at the undergraduate level that accelerates their development as research mathematicians. A side benefit to the mathematical community is that REUs provide an early introduction to the profession.

DEPARTMENT OF MATHEMATICS, UNIVERSITY OF NOTRE DAME, NOTRE DAME IN 46556
E-mail address: `connolly.1@nd.edu`

DEPARTMENT OF MATHEMATICS AND STATISTICS, UNIVERSITY OF MINNESOTA DULUTH, DULUTH, MN 55812
E-mail address: `jgallian@d.umn.edu`

Diversity Issues in Undergraduate Research

R. Cortez, D. Davenport, H. Medina, and D. Narayan

1. Introduction

This article summarizes the presentations given by the authors on the panel "Diversity issues in undergraduate research" held at the conference Promoting Undergraduate Research in Mathematics. The panelists discussed the importance of involving underrepresented communities in undergraduate research in mathematics as a method for increasing their representation in the profession. Data on U.S. mathematics degrees (especially for underrepresented groups), effective recruiting techniques and successful program strategies were discussed.

2. Herbert Medina: Comments on Demographics of U.S. Mathematics Degrees

The percentage of U.S. Ph.D.s in the mathematical sciences awarded to U.S. citizens and permanent residents continues to decline. Indeed, by 2005 the percentage of doctoral degrees in the mathematical sciences awarded to U.S. citizens had dipped to 41%, the lowest that it has been in many years (perhaps ever) [1]. As a comparison, we note that it was close to 70% twenty-five years ago [3]. The percentage of degrees going to U.S. citizens and permanent residents has dropped from 65% in 1995 to 52% 2004 [4].

The declining number of undergraduates majoring in the mathematical sciences suggests that, without intervention, these percentages will continue their dramatic decrease in the next few years. For example, the number of undergraduate mathematics degrees awarded to U.S. citizens and permanent residents in the U.S. decreased from 14,771 in 1989 to 11,673 in 2000 [2]. The data strongly suggest that the country is not producing enough native mathematicians and that the country's dependence on foreign mathematical talent will continue to increase.

In discussions at this and other mathematics conferences, individuals in mathematical leadership positions at the American Mathematical Society (AMS), the National Science Foundation (NSF) and the National Security Agency (NSA) have expressed concern about the country's inability to meet its mathematical needs from within. This challenge leads one to ask the question, *Which groups within the country have not been tapped sufficiently to contribute towards the mathematical sciences?*

Received by the editor December 1, 2006.

There is no one answer, but there are easily identifiable groups within which the cultivation of mathematical talent has been largely absent. These are African Americans, Hispanic/Latinos, and Native Americans. Indeed, from 1993 – 2004, the percentage of U.S. citizen doctorates earned by members of these groups was 5.1% [**3, 4**]. By 2005, these groups made up 28% of the U.S. population, and by 2060 the percentage is projected to be 41% [**5, 6**]. There is a severe underrepresentation of doctoral degree recipients from these groups and the changing demographics of the U.S. population suggests that if the U.S. is to alleviate the shortage of U.S. citizens and permanent residents earning mathematics doctorates, then mathematical talent from these groups is going to have to be better cultivated. Perhaps a one can voice this concern by asking, *Can a country expect to be a scientific leader when 40% of its citizens do not contribute significantly to a field as important as mathematics?*

This observation leads one to the question of whether or not there is mathematical talent within these ethnic groups to cultivate. The answer is affirmative. While the number of total U.S. bachelor degrees in mathematics has been declining, the number of degrees awarded to members of these ethnic groups increased from 1,218 in 1989 to 1,682 in 2000 [**2**] so there is a significant quantity of students from which to find potential doctoral students.

Programs like the Summer Institute in Mathematics for Undergraduates (SIMU), which ran at the University of Puerto Rico - Humacao, have demonstrated that this last statement is more than an assumption. Indeed, SIMU served 115 students, 91% of them were Hispanic/Latino U.S. citizens from the U.S. and Puerto Rico; 59 of these were accepted to Ph.D. programs in the mathematical sciences; 23 were accepted to masters programs in engineering, mathematics or education programs; and 4 students were accepted to "other" graduate programs. Five of these students have already finished Ph.D.s; 21 have finished masters degrees; 45 are still in Ph.D. programs; and 8 are still in masters programs. Other programs, like the Mathematical and Theoretical Biology Institute (MTBI), a program that also targets U.S. ethnic minorities, also have enjoyed success with their students going to and completing graduate programs.

In the long-term, an overarching plan for dealing with the underrepresentation is for the entire U.S. mathematics community to identify early promising Black, Hispanic/Latino and Native American undergraduates and invest time, money and energy in developing and cultivating their mathematical talent. Certainly this practice, although perhaps not a coordinated effort, has resulted in significant progress in solving the underrepresentation of women earning doctorates in the recent past. Indeed, in 1991 21.6% of doctorates in the mathematical sciences awarded to U.S. citizens/permanent residents went to women; that had increased to 30.4% in 2005 [**3, 1**].

One idea on how to deal with the underrepresentation of certain groups in mathematics is to motivate the students and cultivate their talent through participation on undergraduate research. Suggestions of models and strategies to recruit, introduce and sustain minority participation on undergraduate research in mathematics are offered in the next sections.

3. Ricardo Cortez: Strategies for Introducing and Sustaining Minority Participation in Research

The fact that African Americans, Latinos and Native Americans, as a group, are severely underrepresented in higher mathematics leads us to ask the questions: *How are these students different from others?* and *What are the strategies that can increase the number of minority students in mathematics graduate programs?* Though the answers are not simple, a partial answer is that minority students often come from low income families, are first-generation college students who do not have family members or close friends that can advise them on graduate school, and have pressure to join work force as soon as possible. Here we describe a strategy for introducing undergraduate students from underrepresented groups to mathematics research and creating a network of advisors that mentors them over time and sustains their participation in mathematics activities that can ultimately place them in a more competitive position for graduate school. What follows is the collective experience of many people accumulated over more than 15 years through research programs for minorities such as the Summer Math Institute at UC Berkeley (1991-1997) and the Summer Institute in Mathematics for Undergraduates, SIMU, in Puerto Rico (1998-2002).

When to recruit: If the goal is to increase the number of well prepared students in mathematical sciences graduate programs, it is ideal to introduce the students to research during the sophomore or junior year. This allows enough time for the interested students to make appropriate adjustments to their remaining undergraduate classes.

The first research experience: Through an REU or some other research program, the students can have a rewarding first research experience. The topic of the project should be interesting and should not be limited to inconsequential projects specifically made for undergraduates. On the other hand, the projects must include challenges at various levels of difficulty so that there is a high probability that some results will be reached. Faculty can stress the difference between coursework and open-ended problems with no *a priori* answers. When possible, it is a good idea to stress the importance of basic courses to connect the students' classes with the research. The students must learn the importance of reading and communicating mathematics, while at the same time, discover that they meet the challenges posed by the project. Having students with a diverse academic background is helpful but students should not be underprepared for the program.

It is important to create the "right" atmosphere right away. Students have different expectations and often the wrong ideas about a research program. They are not used to doing math all day long. If the research program involves students from several universities, this will be the first time that many will compare themselves to peers from other schools. Consequently, some students will invariable intimidate others. In order to diffuse misconceptions, it usually helps to introduce cooperative learning, give them more work than any single student can do alone, and create groups that mix students with different academic backgrounds.

It is a good idea to start with the structure that undergraduates are used to (having specific assignments, problems with a clear beginning and end, having faculty ask the questions, etc.) and slowly reduce it. The students will have to make the transition to posing the questions themselves, dynamically adjusting the direction of the research, and dealing with open-ended questions. Starting from a structure they understand and letting them know that a transition must occur is a good way to build self confidence.

One can take advantage of the students' inexperience with research to establish a work ethic that demands 15-hour days. It is surprising how the students accept this and make it a choice, especially when the projects are interesting and relevant to them and when they have resources available 24/7. This way, the students can be allowed to follow every lead and to discover which ideas pan out. Of course, the faculty mentor must lead by example. This is very difficult but the faculty must put in a similar amount of effort and dedication, be available at the time the students have breakthroughs and other proud moments.

All students have strengths and must be guided to find their role within the group. For example, some students are strong in analysis and others are strong in scientific computation. It is important that the students learn from each other and, at the same time, develop their own strengths. Getting to know them well helps to align the research challenges with their interests. The students will learn the balance between collaboration (exchanging ideas and developing synergy) and individual contributions based on their particular strengths. One word of caution is to never give students a false sense of ability.

As the faculty assumes the role of mentor to the students, the research program provides a unique opportunity to talk to them about graduate school options and research careers in the mathematical sciences. Many of the students will be hearing this for the first time and may not see themselves as graduate school material. Often students have misconceptions about the type of students who are competitive in graduate school and those who are unlikely to receive funding.

Introduction to a network of mentors: There are several organizations with strong mentoring components for students. Three societies of minority scientists that distinguish themselves in this area are the Society for Advancement of Chicanos and Native Americans in Science (`www.sacnas.org`), the American Indian Science and Engineering Society (`www.aises.org`), and the National Association of Mathematicians (`www.nam-math.org`). Hundreds of undergraduates present research posters at these societies' annual meetings which are also full of mentoring activities. Other opportunities for undergraduates can present their research and become part of a network of scientists are the Joint Math Meetings, the annual meeting of the Society for Industrial and Applied Mathematics (SIAM) and the National Conferences for Undergraduate Research (NCUR). It is a good idea to make it part of the research program for students to present a poster or give a talk at a national conference. During the conference, it is up to the mentor to introduce them to key people who will become part of their network of mentors.

Continuity in subsequent years: In order for students to continue along a successful path, it is necessary to stay in touch with them; to assist them to continue improving their reports to publishable level; to keep the students in mind for other conferences, programs, and fellowships; to continue talking to them about graduate school and offer to write recommendation letters; and to keep them as part of a large network of mentors.

4. Darren Narayan: Models for successful NREUP programs

A central goal of many REU programs is to encourage students to apply to graduate programs in the mathematical sciences. It is in these REU programs that students get their first taste of mathematical research and discover their abilities for leading a successful career in the mathematical sciences. Currently the percentage of university faculty in mathematics that are African American, Hispanic, or Native American, is very small and a far cry from the demographics in the overall population. Inclusion of students from underrepresented groups in all REU programs is a critical step for increasing the number of minority faculty.

In 2003, the Mathematical Association of America launched the MAA National Research Experience for Undergraduates Program, which was designed to provide a research experience for students from underrepresented groups. The central goal of the program is to "increase undergraduate completion rates and encourage more students to pursue graduate study by exposing them to research experiences after they complete their sophomore year" [7]. Participation in an NREUP program can help make the student more competitive for acceptance into REU programs at other universities during ensuing years, and better prepare them for when they start.

Intelligent, well prepared, minority students are welcomed by REU programs, and as a result many of these students receive multiple offers. However less prepared minority students fare no better than less prepared students in general, and often find themselves not accepted into any REU programs. In order to boost the participation of minority students in REU programs, the pool of well prepared, mathematically talented, minority students must be increased. A strategy to do this is to invite minority students to participate in undergraduate research at their home institutions either during the summer or academic year. Not only does this provide an empowering experience for students, it also can serve as a means for discovering untapped potential. Some universities have internal funds available to support such endeavors. Externally funded programs include the MAA NREUP program and National Science Foundation REU Supplement grants. In addition to providing the student with their first research experience, the bond with a faculty member is also likely to be beneficial in terms of retention as well as in success beyond graduation.

It is also wise to also recruit minority students for undergraduate research in mathematics that are not mathematics majors, but are majoring in subjects where mathematics plays a central role, such as engineering, computer science, business, and management. The 2005 MAA NREUP program at RIT is an example of a successful undergraduate research program where two of the four students were not mathematics majors; one majored in electrical engineering, and the other in environmental management [8]. Including such students may encourage them to change majors once they discover their talents for mathematics, and the many career opportunities for mathematics majors.

The presence of minority students in all REU programs can not be underestimated. In order to increase the number of minority faculty in the mathematical sciences, the inclusion of minority students in REU programs is the best place to start.

5. Dennis Davenport: Recruiting Students from Underrepresented Groups

Because of the limited number of minorities who major in the mathematical sciences, one has to be persistent and creative when searching for qualified members from underrepresented groups. Although generating applicants is important, it is also very important that a number of the applicants be qualified. A goal of most REUs is to encourage students to pursue advanced degrees in the mathematical sciences. A student who is unqualified this year may be qualified next year, given a year of mathematical training. There are several strategies one can use to recruit minorities from underrepresented groups.

We strongly suggest that you attend conferences which cater to minority students and faculty members, such as The National Association of Mathematicians (NAM) which is a non-profit professional organization whose main objective is the promotion of the mathematical development of underrepresented American minorities. Started in 1969, NAM aims to address the shortage of underrepresented minorities in the mathematical sciences by operating several educational and research programs, including NAM's MathFest. The goal of NAM's MathFest is to give guidance to minority undergraduate mathematical science majors and to introduce them to mathematical concepts they may not see at their home institution. They are also given the chance to give oral presentations to an audience of mathematicians and fellow mathematics majors. Visibility at MathFest can be enhanced by sending a student from your program to give a presentation and publishing an article about your program in NAM's fall newsletter, which is usually handed out during MathFest. This exposure has been a tremendous asset for some who have tried this strategy. Also the faculty members at minority serving institutions will find out about your program at MathFest. Surveys given by some programs have shown that as high as 78% of the students heard about them from a professor. Hence, we believe it is very important to reach as many faculty members as possible and NAM is an excellent vehicle to do so. There are similar conferences where these strategies have been tried and have shown some success, for example the Society for the Advancement of Chicanos and Native Americans in Science (SACNAS). Although this conference is for all sciences, there is a mathematical science component. There is also an undergraduate student poster session. Another excellent conference to try these strategies is the SIDIM Conference in Puerto Rico. This annual conference is held in late February, an excellent time to leave the frigid north for some sun. While this may pose some problems with the application deadline (most REUs have March 1 or earlier as deadline), this can be overcome by allowing students from the island a week grace period.

To reach faculty members we suggest doing a mass mailing, both through email and snail mail. There is a website (http://www.diverseeducation.com/index.asp) which gives the top fifty colleges for graduating minority students in any discipline. Make sure the institutions that graduate the highest number of members from underrepresented groups in the mathematical sciences are on your mailing list. Also include minority serving institutions and institutions with a large enrollment of minority students. Do not rely on the chairperson at these institutions to relay your information to interested faculty members or students. While the chairperson should be on your contact list, also include other faculty members at the institutions. Try to find faculty members who are interested in diversity issues. This can be done by calling the department or by searching the institutions website.

Also try contacting institutions that are involved in NSF's Louis Stokes Alliance for Minority Participation (LSAMP) program. Alliance members work diligently to increase the number of minorities from underrepresented groups in mathematics, science, engineering and technology. A list of Alliances can be found at the following site: http://www.ehr.nsf.gov/EHR/HRD/amp.asp. Send a message to NSF's lead program director of LSAMP, Dr. A. James Hicks (ahicks@nsf.gov), letting him know about your program and commitment to recruit students from underrepresented groups. He will send a message to all alliances.

Today's students are very technology oriented. If you survey your students you will find that a large number of students are learning about your program through your website. For some programs this number has been increasing from year to year. A good webpage is essential to recruiting; it's inexpensive and easy to access.

Recruiting minority students is difficult, but doable. One must be persistent and creative. You may also want to take a chance on a student with good grades (not very good to excellent like most REU students) who has other positive qualities. Your decision could enhance the student's chances of success and provide the mathematics community with a capable graduate student.

References

[1] *2005 Annual Survey of the Mathematical Sciences in the United States*, Notices of the AMS, **53**(7), 2006, pp. 775 – 786.

[2] *National Science Foundation, Division of Science Resources Statistics, Science and Engineering Degrees: 1966-2000*, NSF 02-327, Author Susan T. Hill (Arlington, VA 2002).

[3] *National Science Foundation, Division of Science Resources Statistics, Science and Engineering Doctorate Awards: 2001*, NSF 03-300, Susan T. Hill, Project Officer (Arlington, VA 2002).

[4] *National Science Foundation, Division of Science Resources Statistics, Science and Engineering Doctorate Awards: 2004*, NSF 06-308, Susan T. Hill, Project Officer (Arlington, VA 2006)

[5] United States Census Bureau,
http://www.census.gov/population/estimates/nation/intfile3-1.txt.

[6] United States Census Bureau,
http://www.census.gov/population/www/projections/natsum-T5.html.

[7] http://www.maa.org/nreup/

[8] R. Ashely, A. Ayela-Uwangue, F. Cabrera, C. Callesano, and D. A. Narayan, Research with Students from Underrepresented Groups, Proceedings of the Conference on Promoting Undergraduate Research in Mathematics.

[9] Alderete, J. F., February 16, 1998. Absence of Minorities from Research Fields Will Result in Grave Consequences in U.S. The Scientist 12[4]:8.

[10] Alexander, J., W. A. Hawkins. 1997. Survey of Minority Graduate Students in U.S. Mathematical Sciences Departments, MAA-NAM Publication.

[11] Davenport, D., Porter, B. Starting and Running an REU for Minorities and Women, Problems, Resources, and Issues in Mathematics Undergraduate Studies, to appear (with B. Porter)

[12] Davenport, D., An REU Designed for Underrepresented Minorities and Women, the Mathematicians and Education Reform Forum Newsletter, Vol. 15, #3 (2003), 6-8.

DEPARTMENT OF MATHEMATICS, TULANE UNIVERSITY
E-mail address: cortez@math.tulane.edu

DEPARTMENT OF MATHEMATICS AND STATISTICS, MIAMI UNIVERSITY, OHIO
E-mail address: davenpde@muohio.edu

DEPARTMENT OF MATHEMATICS, LOYOLA MARYMOUNT UNIVERSITY
E-mail address: hmedina@lmu.edu

SCHOOL OF MATHEMATICAL SCIENCES, ROCHESTER INSTITUTE OF TECHNOLOGY
E-mail address: dansma@rit.edu

Center for Undergraduate Research in Mathematics (CURM) at Brigham Young University

Michael Dorff

1. Introduction

In September 2006, we established a year-round "Center for Mentoring Undergraduate Research in Mathematics" (CURM) at Brigham Young University (BYU). CURM is supported by a $1.3 million grant from NSF, and will promote undergraduate research projects in mathematics throughout the U.S. by: (1) training faculty members as mentors for undergraduate research projects; (2) having these faculty members mentor undergraduate students in research groups working together as a team on one research project during the academic year; (3) advising faculty members at other institutions on how to establish consistent funding to support undergraduate research at their own institution; and (4) preparing undergraduate students to succeed in graduate studies in mathematics. To help achieve these objectives, CURM will administer mini-grants each year to approximately 15 professors at various institutions across the U.S to assist them both financially and organizationally in operating successful undergraduate research groups at their own institutions. There will be an emphasis on targeting women (both in faculty who receive a mini-grant and undergraduates who are mentored by faculty) and participants from undergraduate institutions (i.e., institutions that do not offer PhD's in mathematics).

This is a new direction in undergraduate research by providing support for small research groups during the academic year, and we are very excited about the possibilities and benefits it provides for students and faculty. CURM can help professors who are interested in initiating projects to do undergraduate research with their own students but who need guidance. Also, it can assist mathematics departments that are interested in expanding undergraduate research beyond a model of loosely structured independent projects. In addition, many younger professors at undergraduate institutions want to do undergraduate research projects with students, but there is too little time to devote to the students, because they are already teaching three to four courses each semester. This program can help them by providing funds for professors to buy out some of their teaching, freeing up more time to work with an undergraduate research group. Further, it is important for many young faculty members to be able to say that they have received external support

Received by the editor October 5, 2006.

for their research. Because these professors will receive funding from CURM, this may make it easier for these faculty members to convince a provost or dean to provide additional funds for their research projects. Finally, with an available stipend students will have the opportunity to spend more time and energy in a research project that is meaningful to their education by receiving a stipend, instead of having to juggle school and a part-time job that is not related to mathematics.

2. CURM

CURM will be based on a model that has been successful at BYU for several years. BYU has a strong emphasis on quality undergraduate teaching, and recently there has been institution-wide emphasis on mentoring undergraduates in research projects awarding more than $500,000 annually to over 400 undergraduates to do research projects during the academic year. These awards are up to $1,500 per semester per student, allowing students to work on self-designed research projects that are overseen by a professor. In addition to this, BYU awards over $1.5 million annually to faculty members specifically for projects involving undergraduates. These projects allow faculty researchers to create research groups using undergraduates who are supported by stipends of $1,500 per semester. Additional funds are designated for student travel and group supplies.

During the past four years, the BYU Department of Mathematics has successfully operated four different academic-year undergraduate research groups in the areas of algebraic geometry, geometric analysis, mathematical physics, and applied and interdisciplinary mathematics. These groups typically consist of 3-5 undergraduate students, 1 graduate student, and 1-3 BYU faculty members. All of these groups are supported by either university mini-grants, donations from alumni, or funding from local businesses and industries. These groups are very effective in helping students to make the transition from being an undergraduate to a graduate student, because they involve undergraduate students in mathematical research, challenge them beyond what they see in the classroom, and give them personal attention from faculty mentors. During the past four years, 53 BYU undergraduate students have participated in these mathematics research groups resulting in 44 undergraduate presentations at conferences, 15 joint student-authored refereed research papers, and 37 out of a possible 39 participants going on to graduate school (the other 14 participants are still undergraduates).

2.1. Mini-grants. CURM will serve to help faculty members set up a similar undergraduate research program at their own institution. Each year, CURM will administer about 15 mini-grants to professors who apply to the program. These mini-grants will consist of financial support for undergraduate research groups consisting of 2-4 undergraduate students and 1 faculty member. CURM will provide a $3,000 stipend for each student in the group ($2,000 to be paid initially and $1,000 to be paid at the completion of the research project). The faculty mentor needs to be actively involved with the group. However, many of them will be at institutions with a teaching load of 3-4 courses/semester. Hence, CURM will offer $5,000 for the professor to buy out at least one course from his/her teaching load during the academic year in order to free up time to spend working with these mentored groups. When a faculty member applies to this program for a mini-grant, we will require a letter from his/her department chair, dean, or appropriate administrator explicitly stating what the $5,000 faculty stipend will be used for. At a minimum, this must be a reduction of one teaching course. However, we suggest some additional items, such as a commitment to partially support the faculty member to attend the Joint AMS/MAA Meetings in January at which we will have a "Mid-year Project" meeting or a commitment to send undergraduate students to a regional conference where they can present their research. Also, each research group will receive travel funds (up to $1,000 for the faculty member and up to $400 for each undergraduate) to help cover the cost of attending the faculty training workshop and a spring research conference at BYU. Finally, each research group will receive $250 for supplies.

2.2. Application process. We anticipate that the application material will consist of: (a) an application form (available at the CURM web site www.curm.byu.edu); (b) a letter of support from the department chair, dean, or appropriate administrator explicitly stating what the $5,000 faculty stipend will be used for; (c) a personal statement describing the faculty member's experience and interest in undergraduate research; and (d) a description of the research project. The combined length of the personal statement and the research description should be no more than 1 page. In addition to contact and other general information on the application form, there will be a place to list the specific names of prospective undergraduates who will be involved with the faculty member in the research project. We realize that the specific students may choose to not be involved in the research after the application has been submitted, so a list of alternate students will be encouraged. Applications for the upcoming academic year will be due in March. We anticipate that initial notification of the mini-grant awards will be emailed before the end of March.

2.3. Undergraduate research groups. During the fall semester, the undergraduate research groups will start and continue through the academic year with the students committing 10 hours per week to the research project for two semesters. Each undergraduate research group will meet together at least one hour a week and the students will meet and work together at least three hours a week. The rest of the time each individual student will work on his/her research problem. Students working in groups tend to motivate each other and also they learn to become more independent of the faculty mentor. We feel that this research experience will be a strong motivating factor for participants to choose to attend and succeed in graduate school. Hence, we will require that students must not be graduating at the end of the academic year in which they participate in the program.

2.4. Faculty training. To obtain the goals of CURM, it is necessary to give intensive, preparatory, and ongoing training for the participating faculty members. We plan to accomplish this through an intensive summer pre-workshop, a mid-year meeting, a culminating spring research conference, and an ongoing electronic listserv for the group to discuss undergraduate research throughout the year. The 15 selected professors will attend a two-day summer workshop at BYU to discuss some logistics for the program, present ideas about effectively working with undergraduate research groups, and provide resources to help undergraduates prepare for graduate school. During this workshop there will also be some social activities, such as a banquet, a hike in the local mountains, and a possible excursion to a national park. In January, there will be a mid-year meeting at the Joint AMS/MAA meetings at which the faculty members will report on how their research group is progressing. In addition, we will have discussions concerning topics that the mentors will address in the following semester, such as "How to help undergraduate students present quality talks at a conference," "Where undergraduate students can publish research papers," "How to get support from your institution to continue to do undergraduate research mentoring," etc. In March, CURM will hold a spring research conference at BYU for students and their faculty mentors from these undergraduate research groups. During this conference, there will be a session in which the faculty mentors report on their research groups, evaluate the effectiveness of the program, and discuss issues concerning future undergraduate research.

2.5. Activities to prepare undergraduates to succeed. Since a major objective of CURM is to encourage and prepare undergraduate students to attend and succeed in graduate school in mathematics, we will plan activities to help students achieve this goal. Students will participate in a conference in which they will learn about the advantages of attending graduate school, receive information about what they need to do to prepare for graduate school, and give a presentation on their research projects. Also, students will submit a final research report. In March, CURM will hold a spring research conference at BYU for students and their faculty mentors. The conference will consist of three components: (1) Friday sessions in which the student participants will be motivated and intellectual stimulated to continue to study mathematics and prepare for graduate school; (2) Friday sessions in which the faculty mentors report on their research groups, evaluate the effectiveness of the program, and discuss issues concerning future undergraduate research; and (3) Saturday sessions in which the student participants will present their research with written feedback and guidance from faculty judges. There will be awards for the best research presentations. Finally, we will require all undergraduate research groups to submit to us a final written research paper and will assist the groups in submitting the paper to a journal for publication. These activities will help the undergraduates be better prepared to apply for and succeed in graduate school in mathematics.

2.6. National Advisory Board. In addition, we will organized a 6 member National Advisory Board for our project. These individuals will be experts in the various components of our project. They will act as consultants during the project and have agreed to evaluate the project. So far, the following individuals have committed to be members of the Advisory Board:

Name	Institution
Erika Camacho	Loyola Marymount University
Joe Gallian	University of Minnesota, Duluth
Aparna Higgins	University of Dayton
Darren Narayan	Rochester Institute of Technology
Zsuzsanna Szaniszlo	Valparaiso University
Judy Walker	University of Nebraska, Lincoln

2.7. Recruitment. Participants will be recruited nationally. We will distribute flyers at mathematics meetings such as the Joint Meetings, MathFest, and regional MAA meeting, and advertise in various magazines, such as AMS Notices, the MAA Focus, and the AWM newsletter. Also, CURM has a web site (http://curm.byu.edu) with program announcements, suggestions for background reading, information about the previous programs, photos, participants' comments, and information about applying for the program. Finally, the program will be advertised on the Project NExT listserv. Applications will be accepted in the spring from faculty members who would like to participate in this program.

DEPARTMENT OF MATHEMATICS, BRIGHAM YOUNG UNIVERSITY, PROVO, UTAH 84602
E-mail address: mdorff@math.byu.edu

The MAA-NSF Undergraduate Student Conferences Program

J. Douglas Faires

The Mathematical Association of America was awarded the National Science Foundation grant DMS-0241090 in the Spring of 2003. The purpose of the grant was to provide a simple procedure for awarding small amounts of funding to conferences in mathematics whose aim was encouraging presentations on mathematical subjects by undergraduates. The primary objective of the grant was to give undergraduate students an opportunity to see how exciting our discipline is by providing a nurturing environment in which to present and listen to a broad range of talks on mathematical subjects. The expectation was that this experience would encourage more students to continue their education in mathematics at the graduate level. The awards were expected to be quite modest, in the range of $1000 to $4000, and were open to both new endeavors and to established conferences that wanted to expand their programs but were limited because of funding considerations.

The first call for proposals was announced in March of 2003 with a deadline date of 1 June 2003 for the academic year 2003–2004. A committee consisting of the co-PIs of the grant, Colin Adams, Doug Faires, Joe Gallian, Michael Pearson, and Dan Schaal, reviewed the 11 proposals that were received that year and awarded a total of $21,000 to 10 institutions. In total, the grants were expected to fund conferences in which there would be a total of 260 undergraduate speakers and 965 undergraduate attendees from approximately 160 institutions. The institutions receiving the initial grants were Arizona State University, Colorado State University-Pueblo, University of Dayton, Embry-Riddle Aeronautical University, University of Nebraska-Lincoln, Rose-Hulman Institute, Rowan University (for the MAA New Jersey Section), St. Norbert College, Western Kentucky University, and Youngstown State University.

A major goal of the grant was to eventually support enough conferences so that every interested student in the United States would have the opportunity to talk to their peers formally, and informally, about mathematics within their geographic region.

A second round of proposals was solicited in August of 2003 primarily to encourage people to propose new conferences for the Spring of 2004. This resulted in the awarding of an additional 5 grants to conferences totally $11,900. These grants were awarded to conferences at Mount Holyoke (for the Hudson River Conference),

Received by the editor 15 December 2006.

Morehouse University, Boston University, Furman University, and Dordt College. These five conferences expected to attract 190 undergraduate speakers and 500 undergraduate attendees from approximately 115 institutions.

The total funding for the first year to the grant was $32,900 to support 15 conferences, at which nearly 1500 undergraduates were expected to attend and 450 undergraduates were expected to give presentations. It was expected that about 275 institutions would send students to one or more of the conferences. The results from surveys received from the conferences indicated that these expectations were met in nearly each instance. In total there were approximately 1332 students from approximately 250 institutions attending the conferences, of whom 463 were speakers.

During this second academic year of the three-year grant, the academic year 2004–2005, support was provided to 27 conferences. Eleven of these were conferences that were supported in the initial year, and the remainder were new. By that time we were supporting 14 conferences that were not in existence before the NSF grant was awarded to the MAA. More detailed information regarding the conferences supported during the first granting period can be found in [1].

The NSF renewed the grant last Spring to extend the period to the academic years 2006–07, 2007–08, and 2008–09, so our current approved proposals are for the first year of this extended grant. Through the 2006-07 academic year we will have supported 96 conferences for a total amount exceeding $150,000. You can obtain information about the supported conferences for the Spring of 2007 at

`http://www.maa.org/rumc/upcoming.html`

Contact information for the Conference Directors is listed on this site so that you can contact them if you would like information about their conferences or if you would like advice on starting your own program. There have also been articles written about a number of the conferences and published various issues of the MAA Focus (See, for example, [2], [3], and [4].). A review of these experience should excite you with the possibilities.

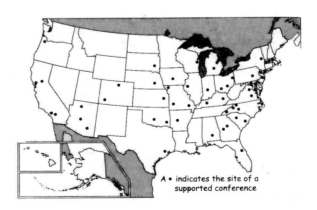

A • indicates the site of a
supported conference

The map gives an indication of the locations in the country that have held conferences and received grants from our program since its inception. As you can see, most regions east of the Mississippi river have had a sponsored conference within reasonable driving distance. However, we have been less successful in promoting our program in the Western part of the country.

We are hoping to expand the number of conferences for the final two years of the extended grant in order to meet our goal of giving every undergraduate an opportunity to attend and present a mathematical talk at a regional conference. If you are in an area that does not have a history of this type of activity, we encourage you to take the initiative and organize a conference. Those who have done so will tell you that it is not only personally rewarding, but makes a very positive impression on their administrators with regard to their interest in undergraduate students and in promoting the institution. This is something administrators can tell trustees and alumni about the mathematics department that can be appreciated by those not in the discipline. Be prepared, however, to have a possible professional life-changing experience. You may find that your students are much better and more interested in mathematics than you thought they were.

The review deadline for conference proposals for the academic year 2007–2008 is 1 May 2007. Proposals will be considered later than this date if funds are available, and a second call for proposals for conferences being held in the Spring of 2008 will go out later in the summer. The proposal process is quite simple; we need to know how you are planning to attract students to give and listen to talks with mathematical content. Let me emphasize that these are not expected to be original research talks. Talks on standard subject matter are not generally appropriate, but many of the most well-received talks by undergraduates often involved relatively elementary mathematics involving some mathematical topic that the student did outside the classroom. The student speakers need not be mathematics majors, but all the talks should be accessible and interesting to an undergraduate mathematics audience.

When preparing a proposal, keep in mind that our objective is to provide an opportunity for undergraduate students to give talks with mathematical content. Our program is not designed to support faculty presentations, although these could be a portion of the conference program. It you have questions about anything regarding our program, please contact me. I can provide sample proposals from conferences that were awarded grants, programs from conferences that have been successful, and contact sources that can give you alternative prospectives about this exciting program.

Bibliography

1. Faires, Doug, *The MAA-NSF Undergraduate Student Conferences Program*, MAA Focus, January 2005, pages 18-19.
2. Jones, Julie C. and Jacqueline A. Jensen, *Starting an Undergraduate Research Program*, MAA Focus, November 2003, pages 6-7.
3. Soto-Johnson, Hortensia and Mike Brilleslyper, *First Annual Pikes Peak Regional Undergraduate Mathematics Research Conference*, MAA Focus, August/September 2004, page 25.
4. Spalsbury, Angela *The Youngstown State University Undergraduate Mathematics Conference*, MAA Focus, January 2005, page 12.

DEPARTMENT OF MATHEMATICS YOUNGSTOWN STATE UNIVERSITY, YOUNGSTOWN, OH 44555
E-mail address: faires@math.ysu.edu

The YSU Center for Undergraduate Research in Mathematics

J. Douglas Faires

History of Undergraduate Mathematics Research at Youngstown State University

In the later 1970's Milt Cox at the Miami University of Ohio and I proposed to the Ohio Section of the MAA that a formal program of undergraduate mathematics presentations be established at the Spring Section meetings. before that time there had been a few student presentations but no formal structure. The program that Milt and I proposed required the Section to make the following commitments.

- There would be separate sessions for student talks, on late Friday afternoon, if possible, which would run parallel to faculty talks. Faculty were welcome to attend.
- A pizza party, or similar, would be held on Friday evening at no cost to the students.
- Lodging would be provided at no cost to the students.
- There would be no registration fee for students.
- Doughnuts, pastries, etc., would be available for breakfast at no cost on Saturday morning.
- Invited talks at the Spring meeting would be accessible to undergraduate students and have mathematical, rather than pedagogical, content.

There was some initial objection to the plan. There were faculty who felt that the Section meetings gave them a chance to, essentially, get away from their students and meet alone with their colleagues. The response to this objection was that they still had this opportunity at the Fall meeting. It was also pointed out that this argument would not likely hold much credence with their administrations who were subsidizing their travel. The second objection concerned the expense of the program to the Section. However, the expense would not be as great as it first appeared since the students could likely be accommodated in dormitories (times have changed somewhat), the breakfast entries were already being provided and generally subsidized by publishers, so the only substantial additional expense would be the pizza party. The point that Milt and I emphasized was that this program could be important for bringing new students into the profession, and that this was

Received by the editor September 29, 2006.

the likely market for future MAA members. The program was adopted, and by the 1980's the number of students at the Spring MAA Ohio Section meetings nearly approached the number of faculty.

Milt, who at the time was President-Elect of Pi Mu Epsilon, the National Mathematics Honorary, was very much involved with undergraduate student research and student presentations at the summer meetings. I was a recent past MAA Ohio Section Chair and interested in the program because I felt that the students at YSU needed to get away from the Youngstown area to see how they compared with other students in mathematics. YSU students generally come from the geographical region of Youngstown, Ohio, and traditionally commute to campus and hold part-time jobs off campus. Since they have also done their precollege training in the region, many of them do not have much contact with any other area of the country or even of the region. I was a product of this educational system and did not gain an appreciation of my abilities until a graduated, worked in the aerospace industry and then went to graduate school. What I found was that I was very well prepared relative to others at my level. My feeling was that if I could get our undergraduate students to see how they compared nationally, it might encourage more of them to do graduate work in mathematics or a related area.

In the Spring of 1982 Milt inducted the Pi Mu Epsilon Chapter at Youngstown State, and we established goals for the organization that included having at least one student speaker at the annual PME meeting, which was held in conjunction with the Joint Summer Meetings (the predecessor to MathFest). In the summer of 1982 I accompanied seven YSU students to Toronto where one of our students gave a talk. Since that time YSU students have given presentations at every annual PME meeting.

In the late 1980's PME was given a grant by the American Mathematical Society to award monetary prizes for Outstanding talks at the National Meetings. By that time YSU was taking at least 6 student speakers to each National Meeting, and quite a few of our students became award winners. The administration of the University was particulary pleased with this since it gave them something to tell alumni, and in consequence, the mathematics department became a model for undergraduate programs on campus.

In order to better prepare our students for their talks at MathFest, in 1998 we began the YSU Regional Undergraduate Mathematics Conference. This conference is held each year on the second Saturday after the COMAP Mathematics Modeling Competition. It is designed to attract students from about a 120 mile radius of Youngstown, which includes both the Cleveland and Pittsburgh areas, as well as Akron in Ohio and Erie in Pennsylvania. In total there area about 40 Colleges and Universities within this region.

In order to permit people to easily arrive and return on the same day, the conference does not begin until 10:30 in the morning and is finished by 3:00 in the afternoon. We provide coffee, juice, and doughnuts in the morning and pizza and sodas for lunch, which is the major expense. This has been defrayed at times by donations from the Giant Eagle grocery stores and Papa Johns pizza, and for the past four years has been supported by the MAA-NSF grant DMS-0241090, which gives support for regional undergraduate mathematics conference at which students give presentations. All communication at the YSU conference is done through e-mail and the web, so the cost of the conference is minimal. The conference is unique in the sense that it is totally students speaking to students, and is organized and run by the students in the YSU PME Chapter. This gives our students experience organizing meetings in a manner that gives maximum effect with minimal labor.

The YSU Regional Conference annually attracts approximately 120 students and 40 faculty from about 15 of the schools in the region. The schools attending vary; about 30 schools have attended in the first 8 years of the conference. About 40 of the students give presentations, YSU students dominating with about 15-20 of the talks. Another dozen of so students are involved in the COMAP session that runs in the morning, which gives teams of students an opportunity to tell how they approached the solution to the problems that were recently given. I understand that these sessions can at times be quite animated, but I have never attended since I am a COMAP judge.

The expectation for our students is that they will give the first presentation of their talk at our conference in February, give a revised and improved version at the MAA Ohio Section Meeting in April, and then present a very polished version at the summer MathFest. By that time they will likely have additional material to discuss and be very comfortable with presenting the material. It doesn't always go according to plan, of course, but in the past decade YSU students have won 31 Outstanding PME awards for their presentations at MathFest of a total of 86 awards given, and an additional 8 awards at the MAA Student paper sessions.

Most importantly, all but one of the 19 YSU award winners in the period 1996 through 2003 have advanced degrees, 4 have received the Ph. D., and an additional 12 are currently at the dissertation level.

The Importance of Recognition of our Efforts

Faculty in mathematics often understate the importance of their accomplishments to the administration of their institutions. In mathematics we are accustom to making statements only we can back them with a logic that is irrefutable, and when all cases are known to be logically correct. We need to recognize that while this is the nature of our profession, it is not what the public, and in particular our administrators, are accustom to hearing. If we, and more importantly our students, are to gain the deserved recognition for our work we need to present our case to the our institutions in terms that are better understood. This does not have to come at the expense of remaining true to our profession and our consciences, although we might at times have to grimace at the way our Public Relations Offices presents what we have told them.

We need to develop better relationships with our Public Relations Office, our Development Officers, the upper-level administration, and the Director of any Foundation that supports programs within the institution. These people need to present the institution to their Boards and to the public in a way that is both complementary and understandable. They often do not have a sophisticated mathematical background so they cannot do this without our help, and we have found that at all levels these people have been extremely grateful for our efforts of communicating the success of our undergraduate students. It has helped our program in numerous ways, including helping us gain additional faculty over other programs for which they do not have the same level of respect.

In the next Section I will describe the latest effort that we are pursuing to bring credit to the program that we feel has made a great impact on our students over the past two decades, and will hopefully ensure that this program and the respect it brings the Mathematics Department will continue.

Establishment of the Center

Because of the success and recognition of our students, it was decided to establish The Center for Undergraduate Research in Mathematics to ensure that this activity would continue beyond the tenure of those who initiated the program. The Center will have a Director with one course per term release time, a dedicated Undergraduate Mathematics Study Room, and a dedicated Undergraduate Research classroom with state of the art technology. An endowment was established at the YSU Foundation with current funding of about $125,000 and additional pledges of about $75,000. It is hoped that the endowment will eventually total $300,000, which will provide $15,000 annually for the operation of the Center.

The primary goal of the Center is to continue to provide opportunities for YSU students to work on undergraduate research in mathematics and to prepare high quality presentations, both oral and text, based on this research. A second goal of the Center is to attract high-quality undergraduate students to the program. Currently most of the YSU undergraduate mathematics majors initially enter the University intended to major in some other area. After they see the excitement of the upper-level mathematics students and the opportunities provided to them, they first add mathematics as a dual major, and then often switch entirely to mathematics. Of the 12 students from YSU who gave presentations at MathFest this past summer, for example, all were double majors except one, and that student was majoring in Mathematics Education. Since only one of these was a senior, a very strong double major in mathematics and physics who is doing graduate work in physics, there is ample time for the remaining students to consider mathematics as their primary endeavor.

An Oversight Board is being established for the Center that consists of alumni of the department, retired faculty who have been involved in undergraduate research activities, faculty from other institutions who have made substantial contributions to the Center, and some prominent members of the Youngstown community. The purpose of the Oversight Board is to ensure that any money directed to the Center is spent in a manner consistent with the goals of the Center, and that money directed to the Center not be used to support activities, such as student research activities and the presentation of this research at professional conferences, that are generally supported from other University sources.

Other mathematical activities that the Center intends to pursue revolve around increasing the written presentations of our students. We currently have a number of our students solving problems and submitting their solutions to various mathematical journals. We have also had students with published articles in the PME Journal, Math Horizons, and Mathematics Magazine, but very infrequently. Our students are generally LaTeX capable early in their first year, and they have creditable performances on the COMAP competition, so they have the writing skills that are required. However, if they are to be successful professional mathematicians, they need to better develop these writing skills so that they prepare written articles without undue strain.

Conclusions

The YSU Mathematics Department has gained a substantial reputation on campus, and I think off campus as well, by the involvement of undergraduates in mathematics research. For most of the faculty in the department the effort has been minimal. Undergraduate projects are required and faculty direct them. We require both oral and written presentations and most faculty require that the written presentations be done using LaTeX. From that point a rather small group of faculty (originally only one, last year three or four, next year about six) take over and work with the students to prepare their talks. There is a rather rigid scheme for this, which involves one person who is concerned with structure and clarity, others involved with technical details, followed by extensive peer review.

This has been a rewarding activity for me in the following sense. In my first 12 years as a YSU faculty member, I did not have a single student who eventually received a Ph. D. in the mathematical sciences, although I must admit that some of these did exceedingly well in other areas. Since 1982, when we established our PME Chapter and began taking students to the national meetings, more than 30 of our PME students have the Ph. D. degree in the mathematical sciences or a mathematics related area like quantitative economics, and an additional 12 are at the dissertation level. If you want to train mathematicians, this seems to be a viable way to start.

DEPARTMENT OF MATHEMATICS & STATISTICS, YOUNGSTOWN STATE UNIVERSITY, YOUNGSTOWN, OH 44555

E-mail address: faires@math.ysu.edu

UBM and REU: Unique Approaches at Tennessee

Hem Raj Joshi, Louis J. Gross, Suzanne Lenhart, and René Salinas

We feature two programs for research by undergraduates at the University of Tennessee, both funded by the National Science Foundation. The Undergraduate Biology and Math (UBM) program brought together math and biology undergraduates and was held from 2003-2006. The Research Experience for Undergraduates (REU) program was a summer program for math majors during 1987 to 2005. Both of these programs recently ended, and we remark on lessons learned and our different approaches. Our experiences aid us in continuing efforts in research with undergraduates and may assist the development of similar programs elsewhere.

1. UBM

As a supplement to an NSF award (DMS 0110920, Spatially-distributed population models with external forcing and spatial control), we developed a 2-year project involved 3 math and 3 biology undergraduates. The research focus was spatial aspects of ecological systems with emphasis on invasive species. This project began in the fall of 2003 and, with an extension, the program was completed in winter 2006. The faculty mentors were Jake Weltzin (a plant ecologist), John Drake (a community ecologist with a laboratory utilizing microcosms), Louis Gross (mathematical ecologist) and Suzanne Lenhart (applied mathematician). René Salinas and Andrew Whittle, post-doctoral fellows on the NSF award, were involved in mentoring and organization of activities. Our goals were:

- to provide interdisciplinary education and training for students at the intersection of math and biology,
- provide every participant with a mixture of hands-on biological field and laboratory experience and mathematical analysis,
- have students themselves develop the research questions and design experiments, with advice and support of their mentors.

Our procedures to reach these goals were to have the mixed group of students carry through the entire process of a research project by developing a research idea, planning the work required to do the research, carrying out the research, and preparing the results for dissemination. Throughout both semesters of the first year we met with the students twice a week. The sessions were a mix of brainstorming, discussing papers and lectures. We read through two books to

---

Received by the editor February 7, 2007.

provide some background: "A Primer on Ecology" (N. J. Gotelli) and "A Primer on Ecological Statistics" (A. M. Ellison and Gotelli). The students were encouraged to get to know about each others interests, to bring up potential research projects, to query faculty about these projects and to decide for themselves projects to pursue. All that we required was that the projects had to include a mixture of field/lab and mathematical components.

The first project idea seriously considered involved analysis and distribution of fish species in streams of the region. Though this project was driven in part by the strong interest of one of the participants in fish ecology, the students eventually dropped the idea due to the difficulty of collecting appropriate field data (this required electro-shocking fish at various locations, which the students decided was not feasible). Concurrently with the fish project consideration, the group had considered possible laboratory analyses of aquatic community assembly spurred by a visit to the lab of one of the faculty mentors. The students began with a pilot study of the growth of protists (*Colpidium* and *Paramecium*) using microcosms to investigate how competitive outcomes are affected by initial densities. The protists were first grown in monocultures to estimate growth rates and carrying capacities and then grown in competition. All six students learned to "count" the protists using a microscope to monitor the progress of the populations. The data collected were fit to a Lotka-Volterra model and several poster presentations were given on this work and talks were given by the students on the results of this project.

During the summer, the group met once a week and began looking for ideas for new projects. In the second year, the students designed and started on two separate plant experiments. Four students continued onto the second year and we recruited two new biology students since two biology students had graduated. Having a core of continuing students was essential. The students divided themselves into two groups, according to their interests. The projects were

- The role of trait diversity on community invasibility
- Quantifying competitive hierarchies for groups of species.

Then in some meetings, the multivariate statistics required to analyze data arising from the planned experiments was discussed. After a successful preliminary experiment with 355 *Arabidopsis* plants, the trait diversity experiment (e.g. differing competitive abilities based upon differences in traits derived from different genotypes of the same species) failed twice due to failures of equipment in the control room. The second group of students (though all students assisted with each of these two experiments) developed a new mathematical method for constructing competitive hierarchies and this was evaluated in a greenhouse experiment with two grasses and two herbaceous dicots. This resulted in a recently submitted paper by two students (Alex Perkins and Bill Holmes) and Weltzin.

The students were paid by the hour on a biweekly payroll which was raised from $10 per hour to $12 in the second year. The faculty members received from two weeks to one month of summer salary depending on their level of participation. The grant included funds for seeds and equipment for the experiments. The students all participated in determining the supplies needed, learned about procedures for equipment purchases, and established the work schedules for the experiments. While the biology students were better prepared to ask questions concerning experimental design, the math students often brought up significant factors on these matters as well.

In summary, while either having pre-defined research projects for the students or having more faculty-guided development of the projects would have increased the the amount of publishable research carried out, this would not have exposed the students to the range of difficulties in the full scientific process. The mixture of students was successful in assisting them to learn a common language from each other. While the biology students certainly developed more math appreciation, the projects were perhaps more beneficial to the math students who had not previously gotten their hands dirty with either experiments or data. The students who have graduated have moved on to graduate programs.

Jacob Kendrick (biology student) remarked on the first experiment: "Though time consuming, cultures were counted daily, the experience proved a valuable lesson in the constraints placed on experimental ecology by the time and effort necessary to complete an experiment and certainly influenced how future projects were planned."

We end this section with a quote from Bill Holmes (math student): "The two year time length and interdisciplinary nature of it have presented unique situations and opportunities. In the past, I have worked in summer research programs, and feel that I have taken more from this program than any other. I have learned a great deal about mathematics, the research process, and in general how to work with others. ... One of the truly unique aspects of this program was its interdisciplinary nature. ... The program gave me the opportunity to work with five people of various backgrounds."

2. REU

Our REU program, funded by the NSF (DMS 0243774) and the UT Science Alliance, has completed 16 years under the directorship of Suzanne Lenhart. Hem Raj Joshi was a past graduate assistant and a faculty advisor for the program in 2003. Our program was based on the theme - "one advisor with one student." Each student in our program worked with an advisor on a research problem. We believe that this "one-on-one" approach simulates most of the research experiences in mathematics and it definitely mimics the experience of working on a PhD dissertation. Although some mathematics is investigated in larger groups of researchers, usually tasks are broken down into subtasks for one or two researchers. This feature made our program distinct from most other REU programs which emphasized working in groups perhaps giving students a choice among several groups. In our case, each student was informed of their advisor and the area of the project at the time that an offer was made.

Our 8-week summer program usually had 10 to 12 students and 10 to 12 advisors. How did the program pay for the large number of research advisors involved? Our funding from the National Science Foundation was supplemented by the University's Science Alliance program, allowing us to pay all advisers at some level. We had a strong group of faculty members who wanted to be involved.

Our application process required an information form, a letter of interest, two letters of recommendation, and a transcript. Students were asked to express preferences among a list of possible research areas. Each year we received well over 100 applications and accepted 10-12 students. The student selection was done by the faculty mentors, with oversight by the director to encourage diversity. We tried to recruit from a variety of types of undergraduate institutions. Over the years, our student participants were about 50% female and 50% male; about 90% of the students went on to attend graduate school.

The main emphasis was working on research projects with each student meeting with his/her advisor at least three times a week. The researchers involved created excellent opportunities for students to work on diverse projects varying from numerical analysis to algebraic structures. The group met together on Monday, Wednesday, and Friday mornings. There were two short courses each summer - one on a "pure" math topic and one on an "applied" math topic. Each short course consisted of ten lectures. These courses gave the students a chance to interact with each other and to learn something new and sometimes unusual. Course topics were varied, including ones not typical in undergraduate math programs, such as "Circle Packing" and "Cryptography." There were no tests or grades in these courses, but the students were encouraged to work together on some problems and present solutions to the group. Each faculty advisor presented a seminar on his/her research field. Three times during the program we discussed the progress on research projects; sometimes students suggested ideas for other projects or the director saw the need for further assistance (including more help with software such as MATLAB). One group meeting was devoted to discussing graduate school opportunities, jobs, the structure of academic departments and professorial ranks, and mathematics organizations. The aim was to educate students about the spectrum of the mathematics community but not to recruit them for a particular department. Our program helped these students make a well-informed choice about their mathematical future.

In the fifth week of the program, the students gave short practice talks, reporting on their progress. The director gave an informational session a few days before this on "how to give talks." The director and the graduate student assistant attended the practice talks and gave advice and constructive criticism. The last two days were reserved for 30 minute finale talks by each student. Rather than focusing on the publication of results, we chose to make the finale talks as the high point of the program. However many publications have resulted from this research program, an estimate being that about 20 percent of the projects resulted in publications. We encouraged students to write up their results and the steps of their research journey. Our program encouraged and supported partially the REU participants to present their work at regional and national math meetings.

Each summer, two student participants worked with scientists at Oak Ridge National Laboratory. ORNL is nearby and those students went to the Laboratory on Tuesdays and Thursdays. Lenhart is a part-time employee of the laboratory and was able to coordinate this situation. This interaction with ORNL gave two students an in-depth national laboratory experience, but the others went as a group to visited ORNL once each summer and met some researchers there. We discussed the differences between working at a government laboratory and working at a college or university. This unique lab connection feature was a draw for our program because the students have the opportunity to work with researchers not directly associated with academia and to use sophisticated computer equipment. Another unique feature was the strong participation of our Math Ecology group.

Our REU students received free lodging in university apartments, a stipend of $3000 and transportation reimbursement up to $500. They also received access to university facilities such as computer labs, the library, gym and health center. Our social events included: hike, picnic, volleyball games, white water rafting, zoo trip, and several lunch outings with faculty members. The program began and ended with parties.

We completed formative evaluations at the beginning of the program and did midway and final evaluations. Below are the remarks from Rachael Miller from our REU program in 2003: "In general I think REU's offer a great opportunity to gain exposure to areas of math that are not as common at many smaller schools with limited faculty or research funds, as well as interact with other faculty. I was fortunate enough to thoroughly enjoy my research project, my interactions with coworkers and other students, and the ability to travel. One of the important things to learn is that in two months, in addition to discovering a great deal of new mathematics, it might be possible to discover areas of mathematics that are not for you, rather than waiting until graduate school or much farther down the line to come up with this conclusion. ... Of the many memorable seminars, trips, and other events, one we were quite proud of was the painting of the 'Fraternity Rock.' To say thanks to the university and our advisors, we painted "Good-bye MATH CAMP 2003" on the rock along with some key concepts of our projects. We were pretty sure that was the first time that anybody painting the rock actually knew what the Greek alphabet meant."

3. CONCLUSION

Working with researchers at the frontiers of exciting mathematics and biology, our participants had experiences which had a positive impact on their education, career decisions and aspirations. Participation in our programs helped these students make well-informed choices about their scientific future.

Joshi, Xavier University, Department of Math and CS, Cincinnati, OH 45207-4441

Gross, University of Tennessee, Dept. of Ecology and Evolutionary Biology, Knoxville, TN 37999-1610

Lenhart, University of Tennessee, Department of Mathematics, Knoxville, TN 37996-1300
E-mail address: lenhart@math.utk.edu

Salinas, Appalachian State University, Mathematical Sciences Department, Boone, NC 28608

Highlights in Undergraduate Research in Mathematics 1987-2006

Joseph A. Gallian

The year 2006 marks the 20th anniversary of the NSF Research Experience for Undergraduates (REU) program. During this period there has been an enormous change in the quantity and quality of research done by undergraduates in the United States. The tenor of the times when REUs began is well illustrated by the following statement made to the National Science Board by Lynn Steen, President of the Mathematical Association of America, in November 1985 [3]:

> Research in mathematics is not like research in the laboratory sciences. Whereas undergraduate research can thrive in most chemistry, biology, or physics research laboratories, research in mathematics is so far removed from the undergraduate curriculum that little if any immediate benefit to the undergraduate program ever trickles down from standard NSF research grants. Publication patterns provide vivid proof: hardly every does one see papers in mathematics jointly authored with students, either graduate or undergraduate. There are a few exceptions–in applied mathematics, in statistics, and in new areas of combinatorial mathematics. But as a general rule, undergraduates can neither participate in nor even understand the research activity of their mathematics professors.

Another statement that perfectly reflects the attitude of the math community had about involving undergraduates in research twenty years ago is the announcement in 1987 in the Notices of the American Mathematical Society of the first REU program sponsored by the NSF [1].

> To clarify the range of activities eligible for support under this program, the DMS has formulated the following examples.
> - Direct involvement of a student in a research project operating in an experimental mode, e.g., generating data or working out examples in order to develop conjectures.
> - Independent study activities where the student is expected to carry out literature searches that indicate the development

Received by the editor December 1, 2006

This article is an expanded version of a joint opening address I and my friend and colleague Aparna Higgins gave at the conference.

> *over time of the area under study, possibly working through*
> *the details in seminal papers. Depth and difficulty of the*
> *material could be adjusted to meet the student's background.*

The striking thing about this announcement is how modest the expectations are. Today most people who run REUs would consider these activities as the starting point, not the end product of an REU.

To document the growth in interest in fostering research in mathematics by undergraduates in the past 20 years I provide the following list of some of the important events along the way.

1987

- The NSF initiates the REU program. Site awards were granted to Harvey Mudd College, the University of Colorado, Oklahoma State University, the University of Minnesota Duluth, Oregon State University, the University of Tennessee, Rice University and the University of Utah. In addition to REU sites the REU program permits individuals with standard NSF research grants to request a supplement to support a few undergraduate students.[1][2]
- The MAA sponsors an "experimental" contributed paper session for undergraduates at the summer meeting that attracts four speakers. Based on the reaction the session becomes an annual event. (See the article in these proceedings by Betty Mayfield for an historical discussion of the MAA contributed paper session at the summer meetings.)
- Pi Mu Epsilon sponsors a contributed paper session for undergraduates at the summer meeting that attracts 30 speakers. (See the article in these proceedings by Terry Jo Leiterman and Rick Poss for an historical discussion of the PME contributed paper session at the summer meetings.)

1988

- Under the leadership of John Greever the MAA forms a subcommittee on Research by Undergraduates. This committee is responsible for many of the things that appear on this list.
- NSF funds 14 REUs. The University of Washington and Worcester Polytechnic Institute have their first.

1989

- Williams College, Mount Holyoke College and the Rose-Hulman Institute of Technology have their first REU.

1990

- The NSA Director's Summer Program is established. It provides a 12 week REU-like program for top level undergraduates.
- The Council of Undergraduate Research (CUR) establishes a division for mathematics and computer science.
- The AWM Alice Schafer Prize is established. (The prize was not created to recognize outstanding research but over the years research has become the decisive factor in the selection of the winner.)

---

[1]Prior to REUs, which are funded from the NSF Research Division, the NSF Education Division funded a program called "Undergraduate Research Participation" (URP) as far back as the late 1960s. Duluth had one in 1977 and Indiana University and Clemson University had them even earlier.

- Aparna Higgins has her first REU at the University of Dayton.

1991

- The first MAA poster session is held at the joint math meetings in San Francisco. There were 12 posters. The judges pitched in for three cash prizes.
- The MAA committee on Research by Undergraduates sponsors a panel on "Models for Undergraduate Research" at the joint meetings.
- The AMS and the MAA jointly sponsor a three part special session for research papers by undergraduates that features 22 talks representing the work of 54 students.

1993

- The MAA and CUR sponsor a panel discussion on research by undergraduates at the joint meetings in San Antonio.

1994

- The MAA sponsors the first minicourse on undergraduate research at the joint meetings at Cincinnati.
- The MAA and CUR sponsor the second poster session on undergrad research with 19 students participating. The poster session now becomes an annual event.
- The AMS sponsors a four part special session for research by undergraduates at joint meetings with 38 talks representing the work of 68 students.[2]

1995

- The AMS sponsors a three part special session for research by undergraduates at joint meetings in San Francisco with 39 talks representing the work of 54 students.
- Aparna Higgins gives the first Project NExT course on undergraduate research. This course becomes an annual event and draws 25 or so participants each year.
- The MAA, AMS and SIAM jointly establish the Morgan Prize for Outstanding Research by an Undergraduate.
- The poster session at the joint meetings has 13 entries.

1996

- The Notices of the American Mathematical Society begins identifying talks by undergraduates at the joint meetings in the program announcement. There was not a special session for research by undergraduates but six undergraduates spoke at contributed paper sessions at the joint meetings in Orlando.

1997

- The MAA sponsors a minicourse on undergraduate research at joint meetings at San Diego.
- The AMS sponsored three special sessions on undergraduate research with 22 presentations representing the work of 29 students.
- Fifteen papers are presented in a contributed paper session on establishing and maintaining undergraduate research programs in mathematics.

---

[2]Two of the student speakers, Dan Isaksen and Darren Narayan, are faculty participants in this conference.

1998

- 171 undergraduate students attend the joint meetings at Baltimore.
- Twenty people contribute papers in an MAA session on establishing and maintaining an undergraduate research program.

1999

- The number of students involved in the poster session is a record 68.
- The National Security Agency and the AMS sponsor a conference on summer research programs in mathematics for undergraduates. The proceedings are published by the AMS (see http://www.ams.org/employment/REUproceedings.pdf).
- The AMS and MAA sponsor a special session for undergraduates at the joint meetings in San Antonio.
- The MAA minicourse on getting undergraduates involved in research at joint meetings becomes annual event.

2000

- 140 students are involved in the poster session at Washington D. C. In subsequent years the number is limited by the size of the room available as demand exceeds the space available. The poster session annually becomes one of the best attended events at the joint meetings.
- The AMS sponsors a four part special session on research by undergraduate students with 38 talks representing 70 students at the joint meetings.
- The MAA sponsors a special session for professors on how to establish and maintain a research program for undergraduates.

2001

- The MAA includes in its Mission Statement: "We support research, scholarship, and its exposition at all appropriate levels and venues, including research by undergraduates."
- The poster session has 77 posters representing the work of 148 students with fifteen $100 prizes.

2002

- At the joint meetings in San Diego SIAM sponsors a minisymposium on undergraduate programs and research projects in applied and computational mathematics.
- Project NExT organizes a swap session on involving undergraduates in research at the San Diego meetings.
- The following events become standard at all successive annual joint mathematics meetings:
  - A special session for research by undergraduates;
  - A contributed paper session for professors on involving undergraduates in research;
  - A poster session for research by undergraduates.

2003

- The poster session at the joint meetings has 200 participants.

- With funding from the NSF, the MAA initiates a minigrants program for conferences that focus on talks by undergraduates.
- With initial funding from the NSA to support 8 students, the MAA establishes a national REU with emphasis on providing opportunities for underrepresented groups. With subsequent support from the Moody's Foundation and the NSF, the program grows to 52 students in 2006.
- For the first time SIAM joins the AMS and MAA as cosponsor of a special session for research by undergraduates.

2006

- 377 undergraduates attend annual joint meetings at San Antonio with 44 of them giving talks.
- The AMS, MAA, AWM, CUR and the EAF (Educational Advancement Foundation) combined for 37 $100 prizes for the poster session.
- Michael Dorff from BYU receives $1.28 million from the NSF to provide 12-15 minigrants to fund faculty and 2-4 undergraduate students during the academic year to work together on research.
- The MAA sponsors it 20th annual contributed paper session at Mathfest in Knoxville. Sixty students give talks in six sessions. Twelve prizes are awarded by CUR, SIAM and the MAA.
- Pi Mu Epsilon sponsors 43 students talks at Mathfest.

Although there are many reasons for the dramatic rise in the number of undergraduates doing research in mathematics, I wish to conclude by identifying the ones I feel are the most significant.

- **NSF and NSA funding**
  By a far margin, the generous support from the NSF and the NSA for summer REUs and REU-like programs has been the impetus for nearly all other efforts.

- **The Council on Undergraduate Research**
  The lobbying efforts by the Council on Undergraduate Research (CUR) were largely responsible for the NSF creating the REU programs. CUR has provided some professional development opportunities for mathematicians who wish to involve undergraduates in research.

- **MAA minicourses**
  Over the past dozen years several hundred people have taken the MAA minicourse on how to involve undergraduates in research. Most of these were offered jointly by Aparna Higgins and the author.

- **MAA poster session**
  The MAA poster session at the joint meetings has become a showcase event for undergraduate students to exhibit their work.

- **Project NExT**
  For a dozen years Aparna Higgins has offered a very popular Project NExT course on getting undergraduates involved in research. About 300 new faculty have taken the course.

- **MAA undergraduate conferences**
  Since 2003 the MAA has provided funding through an NSF grant for regional conferences that focus on research by undergraduate students.

- **Pi Mu Epsilon**
  For many years Pi Mu Epsilon has sponsored regional conferences and paper sessions at Mathfest dedicated to undergraduate research.

- **MAA paper session at Mathfest**
  The MAA has sponsored a student paper session at Mathfest for 20 years.

- **NSA Director's Summer Program**
  The NSA Director's Summer Program, now in its 17th year, is one of the largest research programs for undergraduates in existence. Moreover, it is one of the few that provides a research opportunity for students in the summer prior to their entering graduate school.

- **VIGRE programs**
  Several major Ph. D. granting institutions have run REU-like programs as part of their VIGRE program.

- **Pipeline effect**
  REUs have been around long enough that REU alumni are now providing research opportunities for their own students.

- **Deans are demanding it**
  More and more Deans are demanding that faculty in all disciplines provide research opportunities to undergraduates. This is even the case at many schools where faculty have very high teaching loads. It was commonplace for math departments to be exempt from this pressure but this is becoming less so.

With Lynn Steen's statement made in 1986 given at the beginning of this article in mind, I wish to close this essay with my own made at the 2006 AMS-NSA Conference on Promoting Undergraduate Research in Mathematics.

"Under the right circumstances, undergraduates CAN participate in mathematics research."

References

1. New program for undergraduates, Notices Amer. Math. Society 34 (1987) 297-298.
2. Undergraduate students participate in research, Notices Amer. Math. Society 34 (1987) 911-912.
3. Lynn A. Steen, Restoring scholarship to collegiate mathematics, FOCUS, 6:1 (1986) 1-2, 7.

DEPARTMENT OF MATHEMATICS AND STATISTICS, UNIVERSITY OF MINNESOTA DULUTH, DULUTH, MN 55812
E-mail address: `jgallian@d.umn.edu`

Helping Students Present Their Research

Joseph A. Gallian and Aparna W. Higgins

INTRODUCTION

With the guidance of an experienced research mathematician, there are many undergraduate students who are capable of professional level work in mathematics. The intent of this article is to assist mathematicians in helping their talented students become fully involved in the research experience. Such an experience will benefit the student, the adviser and the mathematics community.

An essential part of a good research experience is the discipline of writing up the results of the research and presenting them. Presenting one's research provides a good capstone event for a research experience. It helps the student to organize and focus the research results. Fortunately, the number of ways of doing this has increased in recent years, and each student researcher can find one that appeals to her/him personally. Apart from the traditional communication of publication in a printed journal, there are electronic journals, poster sessions and paper sessions at meetings. You can help your undergraduate research students choose the most appropriate methods of communicating their results depending on the depth of the results obtained, the personality of the undergraduate, and the logistical difficulties involved. As an item on a student's resume, communicating the results of a research experience, whether in a journal, or at a mathematics meeting, is valued by graduate schools and potential employers alike.

WRITING UP RESULTS

A well written paper should be a goal of every research endeavor. Such a paper tells the mathematical "story" that the student has created – the origins, the context, the results, the methods, the applications, and possible future investigations. At the same time the student learns to use mathematical type-setting software, such as LaTeX ([5] is an excellent introduction). We strongly recommend that research students write up a paper documenting their results and their proofs and any interesting motivation or applications. It is a good exercise for students to be

Received by the editor December 28, 2006

This article is an updated version of the one with the same title published in [4]. We include it for the convenience of the reader.

able to write their results so that they can be read by others – this changes the research from a collection of isolated computations or proofs into a coherent whole. Moreover, it is often the case that while in the process of formally writing one's results better proofs are found and new questions arise. As with any well written document, attention to good grammar, proper punctuation and correct spelling provide a rewarding pay-off in the final product. Whether or not the paper will be submitted for publication, it should be written as though it will be submitted. In particular, it should include a title, co-authors, an abstract, an introduction, acknowledgments and references.

Such a document is useful in several ways – it provides a starting point for a future publication or poster presentation; it can be included as evidence of accomplishment in reports to administrators who supported the research effort; and it will be a convenient reminder of the student's results when you are called upon to write letters of recommendation for employers, graduate schools and fellowships a year or two later. For those students whose research efforts occur at an institution different from their own schools, the paper is a professional way of showing the department at their home institution exactly what they did during their research experience.

PUBLICATION

Advisers can assist students in preparing papers in several ways. First, a model of a well written paper should be given to the student as a guide. Emphasize to the student that the introduction should tell the reader what the author has done and how it relates to the existing knowledge on the subject. The introduction should persuade the reader that the entire article is worth reading. Advice should be given about the extent of details needed in arguments and the number and kinds of examples to include. Two excellent sources for advice on preparing manuscripts for publication that should be made available to students are [6] and [7]. Since students are not familiar with the literature, the adviser must play a critical role in the decision as to whether and where the article will be submitted. Our advice is to make a realistic assessment of the chances of the paper being accepted by a few particular journals but to be a bit on the conservative side in the final choice. A rejection is a psychological blow to a student. In many cases a student who has a paper rejected by one journal will not submit it to another journal even when an adviser assures the student that paper is worthy of publication. It is not clear whether or not the author of a paper should be identified to the editor of a journal as an undergraduate student. In some cases editors and referees will take pains to help a student get a paper published or, at least, will write a tactful rejection letter, while in other cases they will take the paper less seriously than they do papers written by professionals. Our feeling is that as it becomes more common for undergraduates to publish professional level papers the former scenario will become the norm.

A widely held misconception is that papers written by students are natural candidates for the MAA journals. On the contrary, the MAA journals desire articles that are of broad interest and are exceptionally well written. Like most papers written by professionals, most papers written by students will not meet these criteria. In fact, the rejection rates for the MAA journals are over 80%.

On the other hand, there are some journals such as the Pi Mu Epsilon Journal, the Pentagon (the official journal of Kappa Mu Epsilon) and the Missouri Journal of Mathematical Sciences that have a policy of welcoming undergraduate work. Several newer journals have been started to showcase student work. Among these are the Rose-Hulman Undergraduate Mathematics Journal, Journal of Undergraduate Sciences, Journal of Young Investigators, and the Furman University Electronic Journal of Undergraduate Mathematics. Information about these are provided at the end of this article.

A good resource article on possible publication outlets is Paul Campbell's "Where Else to Publish" [1].

TALKS

There are increasingly many opportunities for students to present their results at meetings. If attending a meeting is not a viable option for a student, an adviser can arrange for the student to give a talk to his or her department or even at a nearby school.

Careful preparation is required for a successful talk. The best way of ensuring that an oral presentation is good is to practice giving it, and then to practice some more. Advice on how to give a good talk can be found in the articles, "How to Give a Good Talk" [2] and "Advice on Giving a Good Powerpoint Presentation" [3]. Common mistakes made by students and professionals alike are the "too's": writing too small, assuming too much, talking too fast and trying to do too much.

The first transparency for a talk should include the title of the talk and the speaker's name and affiliation. Students should bring several copies of a one-page description of their talk and have it ready to give to anyone in the audience who expresses an interest in knowing more about the results. The information on the sheet should include the title, an extended abstract and all ways of getting in touch with the presenter. A web site where a preprint of the article is available is desirable.

A talk should tell a carefully thought out story. Ten or fifteen-minute talks should be used to disseminate new results, to give a context for the results, to provide an outline of the method of proof, and to suggest future lines of inquiry. Computations should be avoided.

Lack of departmental funds should not deter you from seeking other funds to allow for student travel. Many Deans, Provosts and institutional programs have funds that can be used for academic opportunities such as undergraduate participation in professional meetings. Some Student Governments have funds that they award to clubs. Many math clubs and MAA Student Chapters have fund-raising activities to finance student travel to meetings. Some MAA Sections and conferences that focus on undergraduate presentations have funds from mini-grants that provide for student travel to their meetings. Most meetings have drastically reduced registration fees for undergraduates and many MAA Sections waive registration fees for undergraduates.

There are many opportunities for students to present a paper. The summer meeting of the MAA (MathFest) includes Pi Mu Epsilon and MAA Student Paper Sessions. A member of either organization's Student Chapters can present a paper. Some travel money is available for speakers, and each organization gives five to ten cash prizes for the best presentations (made possible by the generosity of the

AMS, MAA and NSA). Papers can be presented at the AMS or SIAM meetings. An MAA Section meeting that accepts contributed papers is another opportunity for presentation. There are many regional meetings of groups like PME and Kappa Mu Epsilon. There are also many conferences whose main purpose is to promote undergraduate research. Some, such as the Hudson River Undergraduate Mathematics Conference allow paper presentations by students and faculty, while others, such as the Michigan Undergraduate Mathematics Conference, accept contributed papers only from students. Many schools host annual undergraduate conferences. Details are available at

http://www.pme-math.org/conferences/regionalconferences.html
and http://www.maa.org/rumc/upcoming.html.

The National Conference on Undergraduate Research sponsored by CUR (Council on Undergraduate Research) welcomes student papers, as does the annual Argonne Symposium for Undergraduates in Science, Engineering and Mathematics, although these conferences are not limited to mathematics.

In general, we believe that it is preferable for undergraduate research to be presented in a session that is topic-specific, rather than presenter-specific. That is, it is preferable to present a paper on differential equations in a session devoted to differential equations, rather than to present it in a session devoted to undergraduate research. The main reason we feel this way is that in the former kind of session the audience is interested in the topic and consequently the student will likely meet and have the opportunity to network with others who work in the field. Moreover, we feel that at a conference the undergraduates should be treated like professionals and not be segregated according to experience.

We also believe that it is beneficial for students to present at a meeting where there is far more going on than merely contributed papers by other undergraduates. Attending a national meeting of the AMS, the MAA or SIAM provides students the chance to hear some well-known mathematicians give talks, allows them to witness their professors as learners, and gives them the opportunity to meet and network with a variety of mathematicians in different areas and varied types of institutions.

POSTER PRESENTATIONS

Increasingly, mathematics conferences are including poster presentations as a method of communicating research results. As a matter of fact, many students are more comfortable talking informally about their work to one or two people at a time, as is the case in a poster session, than they are giving a formal talk in front of an audience of experts. Good advice on poster presentations can be found at the web site http://www.maa.org/students/undergrad/meetings.html.

A poster must tell its story by itself. It is usually best to have several sheets of 8-1/2 x 11 inch paper with the results, motivation, applications and indications of methods of proof, mounted on contrasting colored construction paper. Much of what was said in the above section on presenting talks holds true here. Despite the fact that the "audience" for a poster stands close to it, font size should be large enough to allow the poster to be viewed comfortably from a short distance. Pictures and colors are very effective in this visual set-up.

Students should have ready a brief, under-two-minutes presentation that is a synopsis of their results. Such a speech points out the highlights of the work. At the annual joint winter meetings, the MAA sponsors a student research poster session and cash prizes are awarded for the best posters.

In helping their students prepare a poster, advisers should comment on such things as poor grammar, awkward style or displeasing appearance. Anything you do to help the student learn to communicate mathematics better is important. Emphasize to your students that effectively communicating mathematics is as important as getting good results. Any difficulty in understanding the work should be due to the depth of the mathematics, and not due to the author's exposition.

AFTER THE PRESENTATION

If a student from your own institution gives a talk, presents a poster at a mathematics meeting, or gets a paper published, it is important to publicize this information. Let the campus newspaper know about it. An item in the newspaper will be read by other faculty on campus and will reflect well on your department. Let the Dean, the student's adviser and head of the department know about the presentation or publication. Deans often send congratulatory notes to students, which are appreciated. The student will be pleased with a thank you from you for taking the time and effort to make a professional presentation.

It is our experience that most students enjoy presenting their research at a mathematics meeting and having the opportunity to hear other mathematicians speak. Students are delighted when someone at a meeting asks them about their work. It is not uncommon for a student who has presented at a meeting to present at future meetings. Such students often have an enthusiasm for mathematics meetings that is infectious, and fellow students get caught up in the excitement of undergraduate research and its presentation.

PUBLICATION OUTLETS FOR UNDERGRADUATE RESEARCH

1. Pi Mu Epsilon Journal, http://www.pme-math.org/journal/overview.html
2. The Pentagon, official journal of Kappa Mu Epsilon,
   http://kappamuepsilon.org/pages/pentagon_staff1.html
3. The Missouri Journal of Mathematical Sciences,
   http://www.math-cs.cmsu.edu/~mjms/mjms.html
4. Rose-Hulman Undergraduate Mathematics Journal,
   http://www.rose-hulman.edu/mathjournal/
5. The Journal of Undergraduate Sciences,
   http://www.hcs.harvard.edu/~jus/home.html
6. Journal of Young Investigators, http://www.jyi.org
7. Furman University Electronic Journal of Undergraduate Mathematics,
   http://math.furman.edu/~mwoodard/fuejum/welcome.html

REFERENCES

1. Paul J. Campbell and Kunio Mitsumi, Where else to publish, UMAP Journal 26:2 (Summer 2005), 93-114.
2. Joseph A. Gallian, How to give a good talk, Math Horizons, April 1998 29-30. (Available at http://www.d.umn.edu/~jgallian/advice.pdf)

3. Joseph A. Gallian, Advice on giving a good Powerpoint presentation, Math Horizons, April 2006 25-27.
  (Available at http://www.d.umn.edu/~jgallian/goodPPtalk.pdf)
4. Joseph A. Gallian and Aparna Higgins, Helping students present their research, Proceedings of the Conference on Summer Undergraduate Mathematics Research Programs, Amer. Math. Soc. (2000) 289-295. (Available at http://www.ams.org/employment/REUproceedings.pdf.)
5. David F. Griffiths and Desmond J. Higham, Learning LATEX, SIAM, Philadelphia, 1997.
6. Nicholas J. Higham, Handbook of Writing for the Mathematical Sciences, Second Edition, SIAM, Philadelphia, 1998.
7. Steven G. Krantz, A Primer of Mathematical Writing, AMS, Providence, 1997.

UNIVERSITY OF MINNESOTA DULUTH, DULUTH, MN 55812
E-mail address: jgallian@d.umn.edu

UNIVERSITY OF DAYTON, DAYTON, OHIO, 45469
E-mail address: Aparna.Higgins@notes.udayton.edu

Academic Year Research at Valparaiso University

Richard Gillman and Zsuzsanna Szaniszlo

Introduction

Valparaiso University has a long-standing undergraduate research program in the Department of Mathematics and Computer Science. The program was initiated in approximately its current structure in 1991, and in the past 15 years it has provided successful experiences for both students and faculty.

The VU mathematics faculty have directed undergraduate research for decades. In the early years most of these projects were set up on an ad-hoc basis based on mutual interest between individual faculty members and individual students. The level of organization of this work changed dramatically when James Caristi obtained a NASA Joint Venture grant, part of which involved engaging teams of undergraduates in research projects of interest to NASA. The grant provided scholarships to participating students. After the grant ran out the participating students urged the faculty to continue the research projects even without any real incentives to the students. They were right, and we always find students interested in learning about mathematical research.

Program Goals

It became obvious while the NASA grant was in progress that we needed to identify clear goals for our undergraduate research program in order to communicate its purpose to incoming students and to faculty members. Consensus quickly converged on the following four goals:

1. Model the research experience of mathematicians.
2. Provide a growth experience appropriate for the maturity level of the participating student.
3. Help students build meaningful connections to the faculty and the department.
4. Introduce students early in their studies to the discovery of new mathematics.

Received by the editor December 4, 2006.

These four goals reflect the belief of the department that, while we were providing students with research experiences, the emphasis should be on the learning experience rather than on the production of publishable mathematical results.

Program Structure

The program has two distinctive features that determine the other structural components. First, students commit to working on a project for a full academic year. The participants can receive one credit in each semester for the project, although in our experience, some students do not register even though they participate. Theoretically a student could obtain eight research credits during their studies; however, this almost never happens. Most students do not participate in the program every year, and many who do participate do not register for credit later in their studies. Many of our majors are double majors, spend a semester abroad, or participate in internships. It is quite common that a student participates in research one or two years out of the four years they spend at Valparaiso. Those students who pursue graduate degrees in mathematics or computer science usually participate for at least three years. In the last four years the department directed 13 projects and students registered for 75 credit hours during this period.

The second notable feature is that students work in vertical groups. These groups usually consist of 3-4 students including at least one freshman and one upperclassman. We direct 4-6 groups each year, and approximately 30% of our freshmen majors participate.

Each summer the incoming freshmen mathematics and computer science majors receive a letter of invitation to the program. If interested, the students apply with a personal statement about the nature of their interest and current professional goals. As we are interested in engaging all interested students in research, it is very rare that an application is turned down. When a student is turned down, it is usually due to the student being overcommitted with extracurricular activities (band, theater, sports, etc.) during their first year on campus. We talk to these students and recommend that they reapply after a year if they think they have time for the research project.

Upper classmen are recruited several ways. Some students will have participated in the program earlier and they request to stay in, or to re-enter the program. Individual faculty recruit others based on a class experience with them. And still others approach a faculty member and express interest in participating in a project, and occasionally, an upper classmen has an idea for a project and asks a faculty member to direct the project.

One student, whose path through the program is typical, participated in a project as a freshman, played baseball as a sophomore, participated in another project as a junior and then spent one semester of his senior year abroad and the other one student teaching. One of our recent graduates, who is currently working on her PhD in mathematics, participated in four research groups, and each of these four projects resulted in a refereed publication.

We make an attempt to match the interest of the students to available projects and we try to balance personalities in the groups. We expect the students with

different maturity level to play different roles in the groups. Freshmen and sophomores are responsible for understanding the project, and generating examples and
counterexamples. Sophomores and juniors are responsible for the literature search,
and making conjectures. Juniors and seniors are responsible for generating and
proving conjectures. As our goal is to introduce students to the "trade" of mathematical research, we follow the apprentice model. We explain the students these
different roles and let them know how the expectations are different for the apprentice, journeyman, and tradesman who all work under the direction of the professor,
the master craftsman.

The groups meet weekly for about an hour during which every member presents
their progress for the week. The group evaluates their progress together and sets
new goals for the following week. It is important that every person in the group
has a clearly defined task when (s)he leaves the meeting.

We expect that every project finish with a written report that resembles a
mathematics paper. If the group solves their problem, and/or their progress is worthy of publication the faculty advisor submits the results for publication. However,
producing publications is not an expected outcome of the projects. We encourage
our students to present their results at local, regional and national meetings. At
the minimum, the groups present posters at the Valparaiso Celebration of Undergraduate Research. Our students regularly present at Indiana section meeting of
the MAA, the Butler University Undergraduate Research Conference, and at the
Rose-Hulman Institute of Technology Undergraduate Research Conference. Occasionally, students will present at MathFest, NCUR and NCTM national meetings.

Project Features

Just as with any other undergraduate research project it is important that the
problem can be easily explained to the participants, including the freshmen. Good
projects provide immediate access to examples, either computer generated or pencil
and paper examples that the students can find and play with. It is useful if the
problem has some literature in order for the students to experience literature search,
however, it is also useful if the problem has not been studied extensively. Naturally,
it is unlikely that undergraduates will make progress on the major conjectures of
any field. The solution or study of the research problem should require the study
of new mathematics for every member of the group including the seniors. For a
successful research experience for everyone involved, it is essential that the problem
is interesting for the faculty sponsor of the group.

Many of our projects are resulted from faculty members looking at the world
around us in a mathematical way. Some recent projects included modeling ground
water flow problems, investigating a generalization of the segregation game studied
in social work, studying social mixing, or seating arrangements, creating visual
results for computer information retrieval systems, researching the teaching and
learning of Differential Equations, generating probabilistic word-square puzzles,
exploring colorings,labelings, and pebblings on graphs.

Timeline

As mentioned before, each group works for an academic year. It seems that the projects have very similar life cycles. Usually the first half of the first semester is spent on understanding the problem, conducting literature search and generating basic examples. The second half of the first semester is the time to try out ideas, generate conjectures and hit roadblocks. Christmas break provides a good opportunity to step back from the problem. By the second semester the students are usually eager to get back to working and the first half of the second semester is where breakthroughs happen (if ever). This semester is the most productive in generating new mathematics. The second half of the second semester tends to be very busy for everyone involved, and it is the perfect time to learn about the difficulties involved in writing a mathematical paper. In March and April students present their results at the various conferences.

Benefits

We have no formal instrument for assessing the effectiveness of the program. Based on informal questionnaires and discussions with participants, we believe the program meets its goals as stated earlier in this paper. Students have a different view of the program as the faulty does, and they emphasize different benefits. We often hear that the research project brings the student into the department early on. It provides a connection to a faculty mentor and makes the student feel more at home in the department. Students often state how the view of mathematics they see while working on research projects is different from traditional classroom material.

Students usually do not claim that the research experience radically change their post graduation plans, however, they do say that it has a significant impact on their course of study at the university. Of course it is very difficult to measure the impact of the program on postgraduate plans, as freshmen very rarely know what they plan to do after graduation.

It is interesting to note that students generally do not consider the fact that they get academic credit for the experience important. Many of the students ask for recommendation letters from their former project directors, and the personal connections developed over the year enables faculty to write detailed and meaningful letters of recommendation.

Almost everything mentioned as a benefit for the students can also be considered as a benefit for the participating faculty members. Mentoring students in our profession, building personal connections, conveying the true nature of mathematics are all goals we aim for as teacher-scholars. In the past three years a participating faculty member has received one-hour teaching credit for each year-long project directed. This means that after directing 3 year-long projects the faculty member can receive a course release. This benefit has been available for the last 3 years of the 15 years of the program. Although it is very much appreciated by the participating faculty, we find that it has not been a motivational force for anyone. Faculty members usually volunteer to direct a group when they have a project in mind regardless of the existence of the teaching credit.

Conclusion

The approach that we've taken here at Valparaiso University is somewhat unorthodox, since it is rare that freshmen college students have sufficient mathematical tools to solve research problems. However, as a department that values undergraduate teaching and learning, we realize that our principal goal is not to produce publishable mathematics (a great result when it happens), but rather to produce graduates who are capable and ready to excel both in graduate school and in the workplace.

DEPARTMENT OF MATHEMATICS AND COMPUTER SCIENCE, VALPARAISO UNIVERSITY
E-mail address: Rick.Gillman@valpo.edu, Zsuzsanna.Szaniszlo@valpo.edu

Graduate Students as Mentors in Mathematics REUs

Stephen G. Hartke, Daniel C. Isaksen, Philip Matchett Wood

The purpose of this article is to address the question: How can graduate student mentors improve Research Experiences for Undergraduates (REU) programs? A typical graduate student mentor is in the early stages of graduate study and is preferably an alumnus of the REU. Graduate student mentors often live with the REU students. Their jobs can range from giving mathematical suggestions to the students, to helping people find the grocery store, to generally assisting the REU director. A graduate student mentor requires a relatively modest amount of money (on the order of a few thousand dollars for a summer), so it is highly economical to employ such students. Funding for graduate students can be budgeted as participant support costs, not as senior personnel costs, to avoid overhead and other budgetary restrictions.

Below we will describe how graduate student mentors can benefit the REU students and directors, how being a graduate student mentor benefits the mentors themselves, and how graduate student mentors strengthen REU programs. We also believe that having graduate student mentors helps develop the next generation of REU directors.

The ideas in this article are based on a group discussion about graduate student mentors at the Conference on Promoting Undergraduate Research in Mathematics (PURM) in 2006. We have also relied heavily on our cumulative personal experiences. We have all participated in REUs as both students and graduate mentors, and one of us has also served as an REU director. Our experiences occurred at the REUs at Lafayette College, DIMACS at Rutgers University, the University of Dayton, the University of Minnesota–Duluth, and the University of Notre Dame. All of these programs except Lafayette have utilized graduate student mentors.

The article [1] contains more information on the same topic.

1. Direct benefits to the undergraduate researchers

Having a graduate student mentor gives the REU students another interested person with whom to discuss their math research. Another advisor for the REU students can provide a fresh mathematical viewpoint and thus effectively reduce the student/faculty ratio. Importantly, a graduate student mentor can sometimes be more useful to the REU students than additional faculty. Since the graduate student mentor is only a few years out of college, undergraduate students typically feel more

Received by the editor November 30, 2006.

comfortable asking certain kinds of questions, such as "What is a graph?" or "How do I microwave a potato?" Graduate student mentors are likely to understand which parts of mathematics will be confusing to beginners, and they can also give up-to-date career advice on issues such as taking the Graduate Record Examinations (GRE) and applying to graduate schools. Graduate student mentors provide unique and useful resources to the undergraduate students.

2. Direct benefits to the REU director

Graduate student mentors play a critical role in the scientific program of an REU. They can take responsibility for some of the more mechanical aspects of supervision, thereby freeing the REU director to focus on more substantive issues.

For example, they can help REU students with LaTeX, MATLAB, and other computer systems (depending on the graduate student mentors' specific skills). Even more importantly, graduate student mentors can assist students in the preparation of reports or papers. Graduate student mentors can also help students with their presentation skills by organizing practice sessions before formal presentations. Graduate student mentors can relieve administrative burdens on the director by handling easily delegatable jobs, such as organizing field trips or other social activities, producing a program T-shirt, and maintaining the program website.

3. Direct benefits to the graduate student mentors

Advising undergraduate students is a chance to learn how to mentor, which is an important aspect of being an academic mathematician. Graduate students rarely have other opportunities to develop this skill. Graduate student mentors also learn more about the research process by helping others discover new mathematics. In particular, mentors develop the ability to ask interesting and insightful questions—a good skill for their own future research.

In the summer between undergraduate and graduate school, students are eligible for relatively few mathematically-related work opportunities. Serving as a graduate student mentor in an REU is one notable exception.

4. How graduate student mentors strengthen an REU program

Alumni serving as mentors help preserve institutional memory of successful features in the REU program from previous summers. Program alumni also enhance networking opportunities for REU students.

Graduate student mentors can help undergraduate students adjust to a new environment for the summer. Besides practical matters ("Where is the nearest grocery store?"), the mentors can also facilitate the social life of the program. By living with REU students, the graduate student mentors feel the social pulse of the program in a way that the director cannot. Interpersonal conflicts are more easily resolved when detected early. Smooth social interactions allow everyone to increase their mathematical productivity.

References

[1] D. K. Biss and D. C. Isaksen, *Student mentors in the Duluth mathematics REU*, Council on Undergraduate Research Quarterly **19**, no. 4, 163–167, June 2000.

DEPARTMENT OF MATHEMATICS, UNIVERSITY OF ILLINOIS, URBANA, IL 61801
E-mail address: hartke@math.uiuc.edu

DEPARTMENT OF MATHEMATICS, WAYNE STATE UNIVERSITY, DETROIT, MI 48202
E-mail address: isaksen@math.wayne.edu

DEPARTMENT OF MATHEMATICS, RUTGERS UNIVERSITY, PISCATAWAY, NJ 08854
E-mail address: matchett@math.rutgers.edu

National Research Experiences for Undergraduates Program (NREUP)

William Hawkins, Robert Megginson, and Michael Pearson

1. Introduction

The MAA began its Strengthening Underrepresented Minority Mathematics Achievement (SUMMA) Program in 1990. The goals of SUMMA are to increase the representation of minorities in the fields of mathematics, science, and engineering, and to improve the mathematics education of minorities. A number of factors led SUMMA to concentrate on mathematics-based, pre-college intervention (enrichment) programs in the 1990s. But in Summer 2003, SUMMA inaugurated the National Research Experiences for Undergraduates Program (NREUP) with funds from the National Security Agency (NSA). The goal of NREUP is to provide undergraduate students from underrepresented groups majoring in mathematics or a closely related field with challenging research experiences to increase their interest in obtaining advanced degrees in the mathematical sciences.

SUMMA had applied to the National Science Foundation (NSF) in Fall 2002 for funds to conduct a traditional REU at a university campus for 24 underrepresented minority students. That program was modeled on a successful partnership of the Society for the Advancement of Native Americans and Chicanos in Science (SAC-NAS) and the University of Puerto Rico at Humacao. Their lead researchers were Herbert Medina of Loyola Marymount University and Ivelissa Rubio of University of Puerto Rico. The proposal was not funded by NSF.

2. NREUP

Starting only with funds from NSA in 2003, NREUP was radically restructured to support small research groups of one faculty member and four underrepresented students at multiple sites. The Summer 2003 pilot consisted of 3 students at California State University at Chico, 1 student at Goshen College (IN), and 4 students at Texas Southern University. A new proposal to NSF based on this model of supporting several small sites was submitted in Fall 2003 and funded. The combination of NSF and NSA funds made it possible for NREUP to support six sites

Received by the editor October 5, 2006

I would like to thank Dr. Sangeeta Gad of the University of Houston-Downtown for allowing me to present this paper in her session on ATTRACTING AND RETAINING STUDENTS TO MATHEMATICS PROGRAMS VIA OUTREACH. .

and 25 students in Summer 2004. The Moody's Foundation also provided support so that NREUP was able to support twelve sites and 51 students in Summer 2005. NREUP submitted a new proposal to NSF in Fall 2005 which was funded. Those funds together with funds from NSA and the Moody's Foundation allowed NREUP to support twelve sites and 52 students in Summer 2006.

The following tables show the sites, site directors, research topics, and numbers of students, their ethnicity and gender by year.

	Site	Director	Topic	Student No., Ethnicity Gender
Summer 2003	CA State Univ. Chico	Thomas Mattman	Knot Theory	1 Af. American (f) 1 Latina 1 As. American (m)
	Goshen College	David Housman	Fair Division	1 Af. American (m)
	TX Southern Univ.	Nathaniel Dean	Geom. Graph Theory	4 African Americans (2 m, 2 f)

	Site	Director	Topic	Student No., Ethnicity Gender
Summer 2004	CA Lutheran Univ.	Cynthia Wyels	Graph Pebbling	1 Af. American (f) 1 Latina, 2 Latinos
	CA State Univ. Chico	Thomas Mattman	Knot Theory	2 Latinos 1 As. American (m) 1 Nat Pac Islander(m)

	Site	Director	Topic	Student No., Ethnicity Gender
	East TN State Univ.	Debra Knisley	Comp. Biology	4 African Americans (3 f, 1 m)
	Goshen College	David Housman	Fair Division, Coop. Games	1 As. American (f) 2 African Americans (1 m, 1 f) 1 Latino
	TX Southern Univ.	Nathaniel Dean	Graph, Game Theory	4 African Americans (2 m, 2 f)
	VA State Univ.	Dawit Haile	Graph Theory	4 African Americans (3 m, 1 f) 1 Caucasian (m)

	Site	Director	Topic	Student No., Ethnicity Gender
Summer 2005	Atlanta Metro College	Jack Morrell	Discrete Math.	1 As. American (m) 3 Af. Americans (m) 1 Latino 1 Nat Pac Islander(m)
	CA Lutheran Univ.	Cynthia Wyels	Graph Pebbling	1 Af. American (m) 3 Latinos
	CA State Univ. Chico	Thomas Mattman	Knot Theory	2 African Americans (1 m, 1 f) 2 Latinos
	East TN State Univ.	Debra Knisley	Comp. Biology	4 Af. Americans (f)
	Goshen College	David Housman	Fair Division, Coop. Games	2 Af. Americans (f) 2 Latinos
	Howard Univ.	Louise Raphael Daniel Williams	Bioinformatics	4 Af. Americans (m)

	IN Univ. -Purdue Univ., Indianapolis & St. Mary's College of MD	Benzion Boukai et. al. & Katherine Socha	Theory, Apps. of Math.	4 African Americans (2 m, 2 f)
	Morehouse College	Brett Sims	Modeling Biomagnetism	4 Af. Americans (m)
	Rochester Inst. of Technology	Darren Narayan	Combinatorics	2 African Americans (1 m, 1 f) 2 Latinas
	Spelman College	Jeffrey Ehme	Cryptographic Math.	4 Af. Americans (f)
	TX Southern Univ.	Nathaniel Dean	Game, Graph Theory	2 Af. Americans (m) 2 Latinas
	VA State Univ.	Dawit Haile et. al.	Cevian Algebras	4 African Americans (3 m, 1 f) 1 Caucasian (m)

	Site	Director	Topic	Student No., Ethnicity Gender
Summer 2006	Atlanta Metro. College	Jack Morrell	Discrete Math.	4 Af. Americans (m) 1 As. American (m) 1 Latino
	CA State Univ. Channel Islands	Cynthia Wyels	Graph Labeling	2 Latinas 3 Latinos
	CA State Univ. Chico	Thomas Mattman, Sergei Fomin	Knot Theory, Math. Modeling	3 African Americans (2 m, 1 f) 1 Latino
	Clayton State Univ.	Aprillya Lanz	Diff. Eqs.	2 Af. Americans (f) 1 As. American (f)

	DE State Univ.	Mazen Shahin, Elena Surovyatkina	Nonlinear Dynamics	3 African Americans (2 m, 1 f) 1 Latino
	DeVry Univ. of NJ	Dov Chelst	Network Analysis	4 African Americans (3 m, 1 f) 1 Latino
	Grambling State Univ.	Brett Sims	Modeling Biomagnetism	4 African Americans (3 m, 1 f)
	Spelman College	Monica Stephens	Climate Modeling	4 Af. Americans (f)
	St. Mary's College of MD	Katharine Socha Et. al.	Theory, Apps. of Math.	4 African Americans (2 m, 2 f)
	St. Peter's College	Brian Hopkins	Ramsey Theory on Finite Groups	4 African Americans (2 m, 2 f) 1 Latina
	Univ. of TX Arlington	Tuncay Aktosun Minerva Cordero	Modeling the Vocal Tract	2 Latinas 2 Latinos
	VA State Univ.	Dawit Haile	Cevian Algebras	4 Af. Americans (m)

NREUP has served 136 students (128 from underrepresented groups) since Summer 2003. All of the students have been U.S. citizens or permanent residents. The following table gives the breakdown by ethnicity and gender.

	African Americans	Latinas/ Latinos	Native Pacific Islanders	Asian Americans	Caucasians
Male	53	22	2	4	2
Female	40	11	0	2	0
Total	93	33	2	6	2

3. Outcomes

The success of NREUP must be judged on whether it increases the interest of the participants to pursue graduate studies in the mathematical sciences. If a student is a rising senior when a NREUP participant, then it is easy to track his/her matriculation into graduate school. With younger students, tracking is more difficult for two or more years into the future although planned enrollment can be noted. The site directors can provide contact information but students are less responsive once they have left the program. It might be beneficial for NREUP to focus on rising seniors in its current incarnation with rising juniors and sophomores encouraged to participate in later years. Pre- and post-surveys and faculty/student comments verify the effectiveness of NREUP in increasing student interest in graduate study. Moreover, NREUP improves students' confidence in their ability to do independent research.

The following results have occurred.

- Summer 2003: All seniors (3 students) attended or planned to attend graduate school.
- Summer 2004: 9 out of 25 students attended or planned to attend graduate school. Some seniors entered the workforce or did not respond to the evaluator.
- Summer 2005: Data collection continues in conjunction with Summer 2006 results.

4. Conclusion

The MAA's focus on undergraduate mathematics includes great interest in the Association and among MAA members on undergraduate research. MAA has also focused on encouraging underrepresented minority students to study mathematics and consider careers in mathematics and mathematically intensive fields through a variety of programs under the SUMMA banner. Thus, NREUP is a perfect match of interests with the MAA. Believing that researchers are made not born, SUMMA has established NREUP as a means of addressing the dearth of minorities pursuing graduate studies in the mathematical sciences. Significant numbers of students are benefiting from NREUP and should be attending graduate school in the near term. Other efforts are underway to address these concerns but NREUP can make an enormous impact in the next few years.

MATHEMATICAL ASSOCIATION OF AMERICA, 1529 18TH ST. NW, WASHINGTON, DC 20036
E-mail address: bhawkins@maa.org

UNIVERSITY OF MICHIGAN, DEPARTMENT OF MATHEMATICS, 2074 EAST HALL, 530 CHURCH ST., ANN ARBOR, MICHIGAN 48109
E-mail address: meggin@umich.edu

MATHEMATICAL ASSOCIATION OF AMERICA, 1529 18TH ST. NW, WASHINGTON, DC 20036
E-mail address: pearson@maa.org

Undergraduate Research during the Academic Year

Aparna W. Higgins

I am an unabashed proponent of undergraduate research in mathematics. I want to share with my undergraduate students what I find so joyful about mathematics. The NSF-sponsored Research Experiences for Undergraduates are a wonderful way of providing a focused immersion experience in undergraduate research. The summer REUs, however, serve a very small number of those mathematics majors who wish to explore mathematical questions and produce original results in mathematics. In addition to the national REUs, I advocate undergraduate research done at our own institutions with our own students during the academic year.

1. Goals of undergraduate research

1.1. Graduate studies. There is substantial agreement that one of the goals of undergraduate research in mathematics is to encourage students to pursue graduate studies. The University of Dayton has long been one of the higher-ranked schools for mathematics in its category in the Franklin & Marshall Baccalaureate Origins of Doctoral Recipients. Yet, the University of Dayton was an REU Site for mathematics for only a handful of years in the early 1990s. In fact, very few of the top-ranking 4-year colleges and Master's-granting institutions mentioned in this report have been NSF-REU Sites. Although it is possible that the students from these schools who go on to get Ph. D. degrees in mathematics all went to summer REUs, I believe that the more likely explanation is that most of these schools have means in place to support students who wish to undertake serious investigations in mathematics beyond the setting of a course.

There are many formats for undergraduate research established at various schools. Some are formal, such as University-wide Honors or Scholars programs, or a senior thesis or project requirement. Then there is the far less formal one-on-one interaction between a student and a faculty member that leads to exploring some mathematical question. Since the summer research programs available to mathematics undergraduates are few, they are highly competitive, and the students who come to them have likely already considered advanced study in mathematics. In contrast, one of the advantages of undergraduate research experiences in one's own institution is that it can reach students who would not have considered their own

Received by the editor December 27, 2006

This article is based on the opening address I gave at this conference. I had the honor of sharing the opening address with Joe Gallian.

talent and ability sufficient for applying to mathematics graduate school. Such students build up confidence in their mathematics abilities through their academic-year research experiences and often end up applying to mathematics graduate schools.

1.2. Introduction to our profession. It is generally accepted that goals for doing research with undergraduates include introducing undergraduates to our profession, and advancing knowledge in a particular field of mathematics. These goals are often met by undergraduate research efforts in our own institutions. Such research definitely introduces students to our profession, but it may or may not lead to new and publishable results. Whether or not a publication results from the experience, I believe strongly that the experience is valuable for the student and for the mathematics community.

The value to the student's mathematical and intellectual growth is undisputed. The value of undergraduate research to the mathematics community lies in a new member of our community seeking advanced preparation in mathematics, or in gaining an addition to the workforce who is knowledgeable about the workings of mathematicians and about the potential of mathematics.

2. The Time Factor

At the University of Dayton, I have had the experience of conducting year-long academic research in addition to having co-directed an NSF-sponsored REU. I think that the luxury of time is a major benefit of working with students on research during the academic year. The pace is far more leisurely than in an intensive summer program, and hence the student and I can spend some more time on various important aspects of undergraduate research. Typically, I advise students after they have heard from me about specific open questions, and after they have expressed interest in at least one of those. We spend some time getting to know the background of the problem, and this allows us to adjust the question so that this particular student can make some progress on it. I suggest readings to the student that provide more background in the subject as a whole (not just on the particular problem). In this manner, the student is always learning, even during the times when the problem doesn't seem to be getting solved. Converging to a problem that is interesting to the student, and that provides the student with a challenge while still not immobilizing her/his creativity, is facilitated by the year-long setting. All the background learning and the fine-tuning of the problem in question usually results in us being able to carve out a niche in mathematics in which the student is an expert.

Academic year research has a feature that is both an advantage and a challenge for the student. Students have to learn to manage their time so that they make sustained progress on the research question while taking a full load of courses. This is a more natural state of affairs than that of devoting an entire summer exclusively to mathematics research.

3. Formats of academic year undergraduate research

From an institutional point of view, a selling point of year-long academic research is that no (or minimal) funding is required, and hence no (or minimal) funders' expectations need to be met. Let me hasten to add that I am in favor of finding funding so that the student can engage in research to earn money to defray expenses. Funding for faculty release time, or as added compensation to attend a meeting, perhaps, is also desirable and welcome. My point here is to emphasize that funding is not a necessary condition for academic-year undergraduate research.

Academic year undergraduate research in mathematics takes place in many institutions. The Fall 2000 CBMS Survey (Table Ar. 12, Statistical Abstract of Undergraduate Programs in the Mathematical Sciences in the United States, by David J. Lutzer, James W. Maxwell, Stephen B. Rodi, AMS website) provides the percentage of mathematics programs that report having undergraduate research opportunities: 84% of Ph. D. granting institutions, 67% of Master's degree granting institutions and 52% of B.A. granting institutions.

At this conference, many formats and rationales of in-house undergraduate research experiences are represented. I provide examples of some of those whose work I know about, and urge you to contact these faculty if you are interested in knowing more about their versions of undergraduate research.

3.1. Communicating mathematics in paper and poster sessions. Some
faculty help students with their research and encourage them to present their work at regional or national conferences. These faculty are involved in providing much-needed guidance to inexperienced students in communicating mathematics effectively. Most students benefit from such attention and their presentations are often noticeably more interesting than those who did not receive faculty help. Some of the faculty who are known to work with students on presentations and who are in attendance at this conference are Ermelinda Delavina of University of Houston-Downtown, Doug Faires of Youngstown State University, Lew Ludwig of Denison University, James Sellers of Penn State University, Dan Schaal of South Dakota State University and me.

3.2. Directing undergraduate research after participating in it. Some
of the more recent Ph. D.s in our profession are now directing undergraduate research because they felt it was an interesting and valuable part of their own undergraduate experience. At this conference, some such faculty members are Joel Foisy of State University of New York Potsdam, Stephen Hartke of the University of Illinois Urbana-Champaign, Dan Isaksen of Wayne State University and Darren Narayan of Rochester Institute of Technology.

3.3. Starting locally and building to a funded program. Some faculty or
institutions built up a record of working with undergraduates and have now received grant monies to continue on a larger scale, either nationally or with their own students. At this conference, some of those institutions and their representatives are Brigham Young University (Michael Dorff), California State University Channel Islands (Cindy Wyels), Rochester Institute of Technology (Darren Narayan) and Valparaiso University (Zsuzsanna Szaniszlo).

4. Undergraduate research is "hot"

There is a trend in undergraduate mathematics departments to do, or consider doing, undergraduate research. You can find a historical perspective in "Research by Undergraduates is Hot!" by Gallian and Higgins, MAA FOCUS 2002, March Vol 22, #3, 16-17. Evidence of this trend abounds.

- For the past decade or so, Joe Gallian and I have presented an MAA minicourse at the January Joint Mathematics Meetings on getting your students involved in undergraduate research. Each year, we have about forty participants.

- For the last twelve summers, I have presented a course at the Project NExT workshop. (Project NExT - New Experiences in Teaching - is a professional development program of the MAA for new and recent Ph. D.s in the mathematical sciences who are in the first two years of a full-time, post-Ph. D. college or university position.) The four-hour course is entitled "Undergraduate Research: How to make it work." It is capped at twenty-five participants in order to allow for interaction, but that is inadequate to allow everyone who wants it to take it. The seventy or so incoming Project NExT Fellows are asked to rank order six Project NExT courses by interest. About 35% designate the undergraduate research course their first choice, and another 35% request it as their second choice.

- Both Joe Gallian and I are frequently invited by colleges or universities and MAA Sections to provide a workshop on getting started in undergraduate research.

5. Undergraduate research as a pedagogical enterprise

Why is undergraduate research so popular these days? In my opinion, a large part of the phenomenon is due to the broadening of the understanding of the term "undergraduate research." The term no longer refers exclusively to work done by undergraduates that increases the current state of mathematical knowledge in a way that the research community might consider significant. In fact, undergraduate research is now valued as a pedagogical enterprise, with all the ensuing benefits and challenges.

5.1. Institutional point of view. From an institutional point of view, undergraduate research often serves as a recruiting and retaining tool. Several schools have institutional initiatives where first-year students join faculty in research. In mathematics, I am aware of such initiatives at The University of Louisiana at Monroe and at Valparaiso University. Pairing up students with individual faculty mentors in research is often used in conjunction with other retention efforts for minority students.

5.2. Departmental point of view. From a departmental point of view, undergraduate research allows faculty members to stay professionally active at schools with large teaching loads, helping to accomplish the increasing demand of "scholarly activity" at these schools. Curricularly, some interesting capstone experiences can be created for students based on undergraduate research. The department faculty may also be able to bring in small grants for their work with students.

5.3. Showcasing results. In addition to traditional mainstream research outlets, there are many more venues for undergraduates to present the results of their research. Some of these are listed below.

- **Mathfest** (the summer meeting of the MAA) At the Knoxville Mathfest 2006, there were forty-one papers presented in the Pi Mu Epsilon student paper sessions and fifty-one papers in the MAA Student Paper Sessions.

For many years, student paper presenters have accounted for about ten percent of total number of attendees at Mathfest.

- **Joint Mathematics Meetings** (in the winter) There is a poster session on research by undergraduates at the winter meetings. There were over one hundred fifty posters last year, and the number of posters is limited only by the size of the room. More information on the poster session can be found in articles by Mario Martelli on the MAA website. Also, there are paper sessions dedicated to research by undergraduates.
- **MAA Regional Undergraduate Mathematics Conferences** The MAA received a grant from the NSF to partially fund regional undergraduate mathematics conferences. The MAA website has more information on these, including links to the funded conferences.
- **Other Undergraduate Conferences** Many colleges and universities host annual undergraduate conferences in mathematics. Several universities with NSF-VIGRE grants also host such conferences.
- **Conferences for Women** Conferences that are targeted to supporting women in undergraduate mathematics are funded by the MAA Tensor Grants.

5.4. Discovery Learning. A case can be made that some current trends in undergraduate mathematics education favor the implementation of undergraduate research. One such trend is that of discovery learning. This form of learning is championed in K-16 mathematics education. Many undergraduate students are being taught in ways different from lectures. They are engaging in active learning, and some are being taught using modified Moore methods. This could mean that our current students are more attuned to independent learning and discovery than those of past years.

Many departments have instituted an undergraduate research or senior project requirement. When undergraduate research is part of the required curriculum, the faculty at these schools are advocating that every mathematics major is capable of undertaking some measure of research. Consider that most undergraduate mathematics have one of three career plans: being a K-12 teacher, entering the workforce, or going to graduate school in a mathematical science. For pre-service teachers, going through guided discovery is very useful so that they can use that experience to guide their own students through discovery. Learning what mathematics research is also helps them answer questions regarding what mathematicians do. For those majors joining the workforce after graduation, engaging in mathematics research provides them with a heightened ability as problem-solvers. The benefits of doing undergraduate research are very obvious for those majors who intend to go to graduate school.

6. Promoting Undergraduate Research

As we open this conference, we try to state its purpose. Although it is clear that all attendees of this conference are engaging students in undergraduate research, the organizing committee wanted to bring all of us together so we could become aware of the diversity of ways and formats in which undergraduate research in mathematics is conducted. So, the hope of the organizing committee, the hosts and the funders is that we learn from each other about different ways in which we can engage our students in undergraduate research.

7. Thanks

Thanks go to the National Security Agency, and to Barbara Deuink in particular, for funding PURM in both 1999 and 2006, to the AMS (John Ewing, Ellen Maycock and all the staff) for making all the abstract arrangements concrete, and to Joe Gallian for agreeing to edit the Proceedings. I feel immense gratitude to each of you for being willing to share your successes (and not-quite-successes), and for coming to this conference to increase your awareness of the diversity of mathematics undergraduate research.

University of Dayton, Department of Mathematics, Dayton, Ohio 45469-2316
E-mail address: Aparna.Higgins@notes.udayton.edu

IDeA Labs: BYU's NSF CSUMS Program

Jeffrey Humpherys and Sean Warnick

1. Program Overview

The CSUMS program at BYU furthers the development of a network of interdisciplinary student-focused research laboratories that explore algorithmic decision processes from a common mathematical framework of information, dynamics, and control. These "Information and Decision Algorithm Laboratories," or "IDeA Labs," form a research model designed to demonstrate the power of mathematics to students by allowing them to abstract fundamental concepts such as approximation, optimization, and control from a diverse set of application domains.

IDeA Labs consists of four distinct laboratories, which are thematic centers for the study of algorithmic decision processes arising in various disciplines. They are:

- The Computational Biology and Environmental Systems Lab (CBES),
- The Computational Economics and Financial Systems Lab (CEFS),
- The Operations Research and Engineered Systems Lab (ORES),
- The Policy Sciences and Human Systems Lab (PSHS).

Each of these laboratories house research projects, invite collaborations, facilitate interaction with other research groups, and encourage partnerships with industry. Nevertheless, all the projects are focused abstractly on algorithmic decision processes and introduce students to the common mathematical themes underlying these problem areas.

For many undergraduates, the research focus on specific applications is important. Not only does it give them concrete examples that align with their individual interests and intuition, but it also motivates the need for deeper and more abstract mathematics. Moreover, through the use of numerical computation, we have found that students can be quickly trained to begin research projects and as they develop computational expertise, they find themselves well-equipped to meaningfully contribute to the overall research program.

The aim of IDeA Labs, however, is more than simply using applications as specific examples for the mathematics of information, dynamics, and control. The real impact emerges when several application areas are engaged simultaneously. Projects from IDeA Labs feed our undergraduate and graduate mathematical sciences curriculum with examples that cut through application-specific jargon and

Received by the editor December 1, 2006.

focus attention on the common mathematical themes underlying seemingly disparate problems. Mathematical science students with interests in computational biology, ecological dynamics, finance, economics, operations research, engineering, and the policy sciences all take a common core of courses that focus on the mathematical issues surrounding the formulation and computation of decision processes; examples include courses in dynamical systems, modeling & simulation, numerical analysis, and probability & statistics. Introducing application-specific laboratories for student researchers allows us to keep the curriculum focused on the central theory at the intersection of all of these applications, and does so without sacrificing the pedagogical advantage of concrete examples used to feed a wide variety of student interests. Moreover, engaging all of these application areas simultaneously keeps the curriculum focused on the underlying theoretical principles rather than allowing it to become dominated by any one particular application.

2. Central Theme: Algorithmic Decision Processes

Research in algorithmic decision processes is the study of the dynamics of systems that process information to make decisions. Decision systems are ubiquitous, arising everywhere from the strategies deployed by sports teams to investment decisions made by fund managers, and from regulatory signals deployed in metabolic networks to policy decisions made by governments. These processes become algorithmic when an effort is made to make the decisions scientifically, that is, based on data.

Such processes historically have been decomposed into a sequence of stages: modeling, parameter estimation, control, and verification. Although various disciplines have different names for these stages, virtually all scientific decision processes generate a model, parametrize this model using part of the available observation data, use this model to select a decision that is acceptable or optimal according to some objective or measure, and then use the remaining data to verify that the model and/or decision is adequate. As students strip away the jargon of various fields and focus on these scientific decision process, they will discover the universality and importance of abstraction in the mathematical sciences. Through the algorithmic development of decision processes, students will discover the fundamental relationships between information, uncertainty, and complexity that govern the transformation of data into useful decisions.

3. Schedule, Goals, and Activities

Our CSUMS program commences on the first day of the summer term and concludes the following year at the end of the spring term. This allows for intensive 8-week blocks of time at both the beginning and the end of the program so that the students will have time to focus both on (i) learning how to do research at the beginning of the program and (ii) finalizing their projects at the end. It also allows students to take courses in the spring term, before the program begins, if desired or necessary.

The program schedule is broken down into the following periods:

- Summer Workshop (4 weeks)
- Summer Research Immersion (4 weeks)
- Fall Semester (15 weeks)
- Winter Semester (15 weeks)

- Spring Wrap-up (8 weeks)

During each period there are goals and activities that are in place to help insure that students are learning and progressing at each stage in their research, as well as developing good working relationships with those in the group. At the end of each period, we review each student's progress, and meet with them to discuss the review and have an open discussion with the student about their experience in the program.

In order to facilitate positive and open interaction we also have social activities, approximately one activity per period. For example, we kick-off the program with an opening social so that students can get to know us and each other. Then throughout the program we have other activities to further strengthen the overall group dynamic.

Below we outline the goals and activities of each period in the program.

3.1. Summer Workshop (4 weeks). The purpose of the Summer Workshop is to get the students ramped up to do research. Activities are broken up into morning classroom instruction and afternoon computing assignments. Class instruction is given by PI's, senior personnel, graduate students, and guest lecturers. Afternoon computer assignments then provide time for the students to develop computational skill and to also work on their own and in groups.

Classroom instruction starts with more advanced topics from linear algebra, probability and statistics. Care is taken to point out how interconnected these three subjects are. For example, regression is reviewed as an application of least squares, principal component analysis is presented as an application of the singular-value decomposition. Next, students will explore dynamical systems and control, but from a perspective that combines together discrete and continuous dynamics, as well as both deterministic and stochastic modeling. The summer workshop then concludes with an exploration into estimating model parameters from data through state estimation (e.g., Kalman filters), Bayesian methods, time series, and system identification.

In conjunction with BYU's REU program, the summer workshop also has a joint career development seminar, which includes panel discussions and talks from guest speakers on "Misconceptions about graduate school," "How to write a paper," "How to use LaTeX," "How to read a research paper," and "How to use the library and Internet to do research." We have found such career developing experiences to be helpful to students starting research program and it better prepares them for graduate school.

3.2. Summer Research Immersion (4 weeks). The purpose of the Summer Research Immersion is to get students to the point where they are making progress on their research before the academic year begins. During this period, we formally lay the groundwork for student projects. Each student is assigned two projects, according to their interest and input, one where they are the lead investigator and another where they are supporting another student's project. The goal is for each student to be working on projects in both emphasis areas, and to also build a better group environment.

Once projects are selected, we have regular meetings at the project level and at the cohort level. At the start, meetings are held on a daily basis, and as students progress through this period, meetings drop down to a weekly basis. It is hard to have more than one cohort meeting per week once classes begin in the fall, and so it is important for students to develop some independence during the Immersion period so that they are able to make good use of their time during the school year.

It is our philosophy that every meeting should close with a discussion of the goals and deliverables for the next meeting so that students have something on which to focus. For example, at every cohort meeting we will expect one of the students to present something to the entire group.

During the Immersion, students begin outlining their projects doing literature searches. They start modeling and simulating simplified examples and toy models related to their research project so that they build intuition. We have daily project and cohort meetings at this early stage.

As the Immersion period evolves, meeting frequency gradually shifts from daily to a couple meetings per week. The final meeting of the period will conclude with a discussion of project and cohort goals for the rest of the program.

3.3. Fall and Winter Periods (15 weeks each). The goals for the Fall period are to develop the core research methodologies, create an algorithmic decision process, and apply it to a research problem. At the end of the Fall period, each student presents a 15-minute talk to the cohort on their main project.

The goals of the Winter period are for the students to generalize their work as far as possible and to give a talk on their work at our college's Spring Research Conference in mid-March. Students then investigate the implications of weakening constraints and/or hypotheses, and they attempt to connect their work to the broader literature.

Throughout the Fall and Winter periods, we continue have weekly project and cohort meetings so that students can get the feedback that they need, as well as continue to develop the broader familiarity with algorithmic decision processes by exposing them to the other projects in the cohort.

The Spring Research Conference is a great opportunity for students to give a talk and get good feedback from faculty judges and organizers. For many students in the cohort, it is their first talk. One of the purposes of having students present to the entire cohort at our weekly meetings is to give them some experience speaking about their research. The Spring Research Conference extends that to a more general audience. Students are then be able to make improvements to their talks and give their "final" program talk at a professional meeting, e.g, the American Controls Conference, in the summer.

3.4. Spring Immersion and Wrap-up (8 weeks). The purpose of the Spring Immersion and Wrap-up period is to give students time to focus intensely on their projects at the end of the school year, and to wrap up their work into a final report or journal paper. This is a hard process for many students and so much guidance will be needed. After extensive editing and input from faculty, students are expected to submit their work in an appropriate refereed journal or conference venue. An additional benefit of having students work on two projects at once is that some research directions can hit a dead end, or the student might need more time to strengthen their results. This way, they are still able to participate in a conference and/or have a publication in progress at the end of the program year.

4. Student Recruitment and Selection

We are primarily interested in recruiting students who are nearing the end of their sophomore year (or those who have two more years left in their undergraduate education). Our reasoning is that we want students to complete their experience in the cohort before solidifying graduate school and/or career plans. This allows for the full advantage of the experience to have taken effect.

Students apply for the program at the beginning of each calendar year. Application information and a description of the program can be found on the IDeA Labs website (see `idealabs.byu.edu`). To keep the applicant pool high, we also advertise aggressively. In conjunction with other efforts in the math department to bolster our recruitment of students, including women and minorities, we also email all incoming freshman who score highly on the SAT and ACT exams describing the program and outlining the courses that should be taken by the end of their second year. Along these lines, we visit the "Seminar for Women in Math, Science and Engineering" and talk about IDeA Labs in an effort to recruit women into the math and statistics departments, and then into IDeA Labs. Finally, mathematical sciences students from underrepresented groups, including women, will be personally contacted through email and invited to apply.

Students selection is based on (i) course-work preparation, (ii) potential for success, (iii) evidence of work ethic, (iv) curiosity, (v) attitude, (vi) citizenship, and (vii) diversity.

5. Project Evaluation and Reporting

5.1. Quality Control. We have organized a network of 18 advisers and reviewers who will help to insure that our program running well. With each person, we have discussed the scope and vision of IDeA Labs and have received warm responses, good feedback, and an eager willingness to assist.

We have recruited two BYU faculty members to be overall program reviewers to help us with the recruitment, selection, and mentoring of women. Both have agreed to annually review our program provide us with helpful feedback.

In an effort to insure quality of our work at IDeA Labs, we have developed a team of research area advisers and reviewers. Each lab will have (i) an internal adviser, (ii) an internal reviewer, (iii) an external reviewer, and (iv) an industrial reviewer.

The purpose of the internal adviser is to provide the PIs with subject-matter expertise in each particular field. Our interactions with them help us to better guide students into projects that are current with the mainstream thought and the cutting-edge ideas. In fact all four advisers have expressed an interest in working with us on projects. In fact, one remarked to us after reviewing our students' work in the area that our undergraduates were more capable of doing research in his field than his own masters students because they lacked the mathematical and computational skill necessary to help carry out the work.

The purpose of the three reviewers is to biennially visit the labs and to write-up a formal review with recommendations. Reviewers will also be given a questionnaire to help solicit specific feedback about our program. By design, each reviewer comes from a different vantage point. Internal reviewers are able to uniquely offer advice that's coupled with institutional knowledge of BYU, external reviewers add intellectual diversity into their feedback, and industrial reviewers are able to provide feedback that will help us to strengthen our industrial partners program. In addition to good evaluation and reporting, an additional benefit of having a team of advisers and reviewers is that they are all be well-positioned to write letters of recommendation on behalf of our students when they apply to graduate schools.

5.2. Key Performance Indicators. Key performance indicators are collected and reported in an effort to evaluate the output of this program and its overall benefit. For example demographic data is collected and broken down, research activities summarized, and class enrollments tabulated. Quantitative measures of undergraduate research is also collected, including numbers of participants, academic presentations, publications in refereed venues, and additional support through grants and gifts are noted.

Following the lead of [1], surveys are administered at the beginning of the program, at the end of the program, and the following two years after the project concludes. Questions gauge courses taken, future courses planned, and hours per week spend doing mathematics outside of coursework. Questions regarding graduate school and career interests are included.

As described above, students are reviewed 5 times during the cohort year, and the review is shared with them so that they can improve. Through that process, we also get qualitative feedback from students so that we can continue to improve the overall CSUMS mentoring experience.

References

[1] A. Adhikari and D. Nolan, "But what if good came at last" How to asses the value of undergraduate research. Notices of the AMS 49:1252-1257, 2002.

DEPARTMENT OF MATHEMATICS, BRIGHAM YOUNG UNIVERSITY, PROVO, UT 84602
E-mail address: jeffh@math.byu.edu, sean@cs.byu.edu

Assessment Methods for Undergraduate Research Programs

Daniel C. Isaksen

Undergraduate research programs in mathematics are more and more widely acknowledged as a valuable part of the total mathematics education and research community. In order to secure long-term funding for such projects and to increase even further the support for such programs in the mathematics community, directors of undergraduate research programs bear a responsibility to provide concrete evidence of the effectiveness of their programs.

This article summarizes some of the methods used by various programs to assess their own effectiveness. We have collected information not just on undergraduate research programs but also on related programs for undergraduates.

Participant databases. The most common assessment method is to maintain an on-going database of former participants. Typically, the database tracks undergraduate and graduate degrees as well as employment history. It can also include more specific information such as scholarships and fellowships received. Over time, these databases provide a good summary of what happens to students after they participate in a research program. One example of such a student database is [1].

Maintaining contacts with many former participants can be very difficult and time-consuming. Some program directors choose to stop tracking former participants when they leave academics for another type of career (but to record what career they adopted). For those former participants who end up in academic careers, it probably makes sense to never stop tracking.

Encouraging networking amongst former participants is valuable in many ways, including for assessment. Maintaining contacts with former participants becomes much easier. Direct conversations also provide additional anecdotal evidence of the effectiveness of undergraduate research programs. As one example, some programs organize reunions annually at the Joint Mathematics Meetings.

Evaluation forms and surveys. Some programs ask participants to complete evaluation forms immediately at the end of their active participation in the program. A few programs use a coordinated set of surveys, one of which occurs at the beginning of the program or even before it starts. This approach shows how students' perspectives change over the course of a research experience.

Received by the editor October 24, 2006.

Program directors can create evaluation forms on their own. However, it is also possible to pay for formal evaluations that are designed and administered by outside professionals. Depending on the anticipated value of the data and also depending on budgets, the extra cost may be well worth it. Outside evaluators are experts at phrasing questions that encourage useful responses.

Evaluating the mathematics. The methods described above are mostly intended to assess the social impact of undergraduate research programs. However, program directors should also pay attention to assessing more directly the mathematical output of their programs.

One way to measure this is to consider student activities that are directly linked to a recent research experience. These include writing or publishing an article, giving a talk at a conference, giving a colloquium lecture at the student's home institution, or presenting a poster in a poster session. Another example is applying for or receiving funding for research activities; many institutions have funds to support their own local students. Some programs require that students perform some or all of these activities. In addition to providing a chance to hone communication skills, such activities also provide some validation of the quality of the mathematical research.

It is difficult for traditional summer program directors to enforce requirements of this nature. One possible solution is to solicit the cooperation of a faculty member from a student's home institution. For local students working on year-round projects, these requirements are easily enforced.

Summer programs compared to year-round programs. Although it is impossible to fit all types of undergraduate research activities into clean categories, most such activities either are part of a summer program involving multiple students (e.g., NSF-funded REUs) or are year-round activities involving at most a few local students and a local faculty member. These two types of research activities require different approaches to assessment.

The document [2] provides reporting guidelines for NSF-funded REUs; this serves as a guide for data collection for all summer programs. The basic idea is to provide evidence of progress towards the goal of a "diverse, internationally-competitive, and globally-engaged workforce".

The NSF requests information on each student participant, including year of schooling completed, home institution, gender, ethnicity, race, disability status, and citizenship. Also requested is information on research, such as project titles, specific examples of research results, publications, presentations, and participation in conferences. NSF is also interested in other kinds of information, such as the impact and nature of mentoring; the impact of the REU experience on career choices; findings from formal evaluations; and details of the recruitment and selection process, especially with respect to diversity.

Assessment for year-round research is much more flexible. Since the number of students is so low, formal evaluations are probably not relevant. Still, research mentors can meaningfully measure whether or not a project was mathematically successful; they can also maintain a database of former students as described above.

Sample evaluation questions. To give program directors an idea of suitable evaluation questions, we provide the following samples. They are suggested by questions from actual surveys of the Summer Institute in Mathematics for Undergraduates (SIMU) at the University of Puerto Rico – Humacao.

The first set of questions is designed for students who participated in a research program in the previous year.

1. Before our program, had you seriously thought about attending graduate school?

2. Did your experience in our program increase or decrease your desire to attend graduate school?

3. Before our program, had you worked on an undergraduate research project in mathematics? After our program, have you worked on an undergraduate research project in mathematics? Will you participate in a research program this year? If so, where?

4. Since the end of the program, with which other students and program staff have you communicated?

5. After our program, have your grades in mathematics courses changed significantly? How?

6. List any conferences at which you have presented a poster based on your work in our program. Also, list all talks which you given based on your work in our program. Have you won any awards for these presentations?

7. List the titles of any papers that you have written based on your work in our program. Have you submitted any for publication? If so, to which journals?

8. Have you already applied to graduate schools? If so, list the universities and departments to which you applied. Do you feel that your participation in our program helped you in the application process?

The following questions are intended for former students who participated in a research program several years ago.

9. Are you currently enrolled in a graduate program? If so, at which university and in which field? When did you start this program? What graduate degree are you currently pursuing? Do you hold a national fellowship? Do you feel that your participation in our program helped you to be accepted?

10. Have you completed a graduate degree? If so, at which university and in which field? What degree did you receive, and when did you receive it? Do you feel that your participation in our program helped you to succeed in graduate school? If you are employed, describe your current position.

11. Did you start a graduate program but subsequently leave without obtaining a degree? What university did you attend? Why did you leave? If you are employed, describe your current position.

12. Have you never attended graduate school? Why not? Did you apply to graduate school, or do you plan to apply in the future?

Acknowledgement. We acknowledge the assistance of Deanna Haunsperger, Aparna Higgins, Ive Rubio, Zsuzsanna Szaniszlo, and Judy Walker.

References

[1] http://www.d.umn.edu/~jgallian/partnew.html
[2] http://www.nsf.gov/pubs/2001/nsf01124/nsf01124.html

DEPARTMENT OF MATHEMATICS, WAYNE STATE UNIVERSITY, DETROIT, MI 48202
E-mail address: isaksen@math.wayne.edu

My Experiences Researching With Undergraduate Mathematicians: The Collaboration Model

Joshua D. Laison

During the past six years as a Visiting Assistant Professor at Kenyon College, Colorado College, and St. Olaf College, I collaborated with 15 undergraduate students on 11 research projects, both singly and in groups of two or three. I say "collaborated," rather than "supervised," since my approach to undergraduate research is to model my work with students on my collaborations with other Ph.D. mathematicians. In my experiences speaking with participants at the Conference on Promoting Undergraduate Research in Mathematics, it seems that this approach is unusual enough that others might benefit from a description of my experiences, ideas, and results. This article summarizes what I believe to be the most significantly different aspects of my approach, and what I feel were the most important ingredients of my successful research experiences.

1. Finding a topic

There are two criteria that I look for when I select a research problem for collaboration with students.

The problem is new to me and to the student. In order to be able to collaborate on a problem as equal contributors, it's necessary that my students and I start at roughly the same point in the research, in knowledge of the relevant background information, and experience with techniques in the area. As a rule, I try not to work on problems with students that I have previously studied and published results on. An exception to this rule is a problem that I have worked on with other students previously and left unfinished, since usually the first step a new collaborator takes on a problem is to change the direction of inquiry.

This is frequently a difficult requirement to fulfill, since I have more knowledge of my research areas than my students, and I guide the choice of area we'll be working in. Certainly the field of graph theory helps in this regard. A distinctive feature of the field is that there are hundreds of distinct subfields with different kinds of research problems that require different techniques of solution. It is relatively easy to find a new corner of a particular small subfield which is unexplored by anyone, and in particular by me. Graph theory is not unique in this respect, but perhaps in the minority among fields of mathematics. It also helps greatly to work

---

Received by the editor December 1, 2006.

in an area in which the problems are easily accessible and don't take a lot of startup time.

The student plays a central role in the selection of the problem. I never have a specific problem picked out before our research begins. Before the collaboration, I select one or several possible narrow fields of study. Ideally the field will have at least three papers published in the area, but usually less than twenty. With a few papers to read, we can get a good idea of what's been done in the field, and have a starting point to begin investigating problems in the area. On the other hand, with more than 20 papers already published, the problems start to become more difficult to find, and perhaps more importantly, the time it takes to read and absorb the required background material is prohibitively large (often for both me and my students).

I obtain these papers and skim through them, to determine what type of techniques are used in the proofs, and the range of possible unanswered or unasked questions. The project works much more smoothly when the techniques used in past research are accessible to the student, given his or her mathematical experience thus far. Often the topic of the project changes significantly enough in the first few weeks of work that the resources I have found become irrelevant, and we must return to perform a literature search again. This time the student takes an active role in the search, and gets to learn about mathscinet and requesting articles through interlibrary loan.

I believe that an undergraduate student's biggest strength as a researcher is that she or he is able to ask creative questions in a way that no other researcher in the field has done. Since they are new to the particular research area, undergraduate students are not familiar with the standard form of research problems in a given field, and often their questions are ones I would never have thought to ask myself. Indeed, since all of the problems I've worked on with undergraduate students have been new problems, no other mathematician has ever thought of them either. I'll give a few examples of this process of finding a problem from my past work with students.

Critical Pebbling Numbers of Graphs. Given a set of pebbles distributed on the vertices of a graph, a *pebbling step* takes two pebbles from one vertex and replaces one at an adjacent vertex. A distribution D of pebbles is *solvable* if, starting from D, a pebble can be moved to any specified vertex by a sequence of pebbling steps. The *pebbling number* $p(G)$ of a connected graph is the smallest number of pebbles such that every distribution with $p(G)$ pebbles is solvable. The *(global) critical pebbling number* $c(G)$ is the largest size of a minimally solvable distribution [1].

This graph parameter was defined by Courtney Gibbons, Erick Paul and myself in a research project in the summer of 2004. I started by suggesting a few possible research areas, including graph pebbling. I had seen a few talks on graph pebbling, found them interesting, and given a talk about pebbling in my department colloquium series. I used a survey paper on graph pebbling to write my talk, and I gave this paper to Courtney and Erick to study [2]. After reading through the paper, they decided they were interested in looking more closely at greedy pebbling and greedy graphs. We worked through some examples, and discovered that some of our initial assumptions about greedy pebbling were false, corresponding to a modified version of the definition of pebbling. We reworked this new definition for a few weeks, until we obtained the definitions of critical pebbling numbers in our paper.

Obstacle Numbers of Graphs. Given a set of vertices in the plane, and a set of (possibly non-convex) disjoint polygons in the plane called *obstacles*, we put a straight-line edge between two vertices if it does not intersect any obstacles. Given a graph G, the smallest number of obstacles required for such a representation of G is called the *obstacle number* of G.

This graph parameter was defined by Arden Brookstein, Richard Marcus, Andrew Ravits and myself during a mathematics research course I taught in the spring of 2006. I had eight students in this course, and at the beginning of the course I selected seven possible topics for research. I pitched each of the topics to the class on the first day, and they formed groups around the topics they were interested in. This particular research group selected the field of visibility graphs (somewhat broad considering my constraints), and started by looking at five papers on bar visibility graphs and rectangle visibility graphs. During a discussion about one of these papers, Richard remarked that lines of sight in a visibility representation reminded him of lines of sight in a video game, in which obstacles block visibility between opposing players. This led the team to the definition of obstacle numbers of graphs. The problem of finding a graph with large obstacle number carried the group through the course, and I have started investigating related problems with research colleagues since the course ended.

There are a couple big advantages to these requirements for the selection of a problem. First, as I discuss below, having been placed on roughly equal footing, I am free to involve myself as a full member of the research team. I don't feel the need to withhold information from my students to insure that they get the complete research experience. This means that we can accomplish more than we would otherwise. Second, with a problem which has never been studied, any progress is new, and it is much easier to find new results if the questions have not already been asked by other researchers.

2. The Research

Goals for a Successful Project. There are many measurements of the success of a student research project. I believe that if the student learns some mathematics, changes her or his attitude towards doing research in mathematics, or just enjoys working on a mathematics problem, then the project is successful. By this measure, all of my projects have been successful. Unfortunately, oftentimes a more objective measure of success is needed. Many attendees at the Conference on Promoting Undergraduate Research in Mathematics spoke of an increase in percentages of students enrolling in graduate school in mathematics as being a useful measurement of the success of a Research Experience for Undergraduates program. My ideal goal for a research project with an undergraduate student is that my student and I obtain original results and write an article for publication in a journal. This is a somewhat selfish goal on my part, since by this measure, a successful project means a publication for me, but it is one that every student I have talked to so far has been happy to embrace as his or her own as well. Indeed, many students are much more motivated to work hard with this goal in mind. Using this measure, about half of my student research projects have been successful so far.

Commitment to time. One clear pattern has emerged in my research with students. No other factor (level of student experience, difficulty of the research problem, area of research) matters more in predicting the success of the project than the amount of time the student commits to the project. Due to unpredictable variations in the difficulty of the problem, my projects have taken widely varying amounts of time, but a crude estimate is that all my student projects which produce a submitted article have resulted from at least 100 hours of work on the part of both my student and myself, and the other projects have generally taken less time. A student I have written two articles with estimates that we worked on each of our projects 20 hours a week for most of the summer, followed by additional time during the school year. Again, while this may seem like a big time commitment, it is time that I see myself investing not only in the quality of my students' education, but also in my own research program.

Since I have determined this rough guideline, I tell all my research students about it at the beginning of our project. Each student I talk to makes a decision about their commitment to time at the beginning of our collaboration, and most of those who decide they are willing to commit the time to achieve original results and a paper accomplish that goal.

Working as a team. My goal in research with students is that we work together as a research team, with each team member having equal control over the direction of the team. Most or all of the work is done together, when all team members are thinking about the problem in the same room. While I often assign tasks for students to complete between meetings, sometimes quite ambitious tasks, most of the time students use these tasks as a way to clarify and articulate their ideas and frustrations to be considered at the next meeting. The time I spend outside meetings thinking about the problem varies depending on how much progress we're making.

In all of my collaborations, both with students and colleagues, it's difficult or impossible to say how much each member of the collaboration contributed to the results. In a good collaboration, the end product would have been impossible without the work of any one of the team members. This has been true of most of my student research collaborations. While different members of the team had different strengths, the results would not have been possible without the participation of all collaborators.

Follow through. Any type of research with undergraduate students has the additional difficulty that the student will not be a student for a very long time. For example, if the research takes place over the summer before a student's senior year, and is submitted for publication in September of their senior year, it is very likely that the paper will not make it through the refereeing process until after the student graduates and probably moves out of town. After the student moves, regular meetings are impossible, and the remaining communication must be done through e-mail. Just as with any undergraduate research, if the team wants to submit an article, the professor must take on more responsibilities, including more of the revising of the article based on a referee's report. When I start a project, I commit to that extra work possibly some years down the road, but with the extra commitment comes the extra reward of an additional publication.

3. Results

I am very happy with the outcomes of my research collaborations with students. It is true that research with students takes a large amount of my time, and sometimes it doesn't work out. A student might feel pressures from other obligations, and run out of time to get anywhere on a research project. On the other hand, the model that I have been following for the past six years for research collaborations with students has many advantages. First, about half of my research collaborations with students have produced results significant enough to submit for publication, and these submitted papers constitute about half of my submitted research articles so far. Second, while some students and faculty might feel that undergraduate research is only an approximation of real mathematical research, by its nature there is essentially no difference between the research students do under my model and the research done by professional mathematicians. Lastly, and perhaps most importantly, the research I have done with students has been a lot of fun! It is difficult to find a collaborator who is more excited just to be involved in a research experience than the typical junior mathematics major thinking about a career in mathematics, and that excitement is infectious. I believe that I've found a model that works for me, and I plan to continue doing it for as long as I get the chance.

References

[1] Courtney R. Gibbons, Joshua D. Laison, and Erick J. Paul. Critical pebbling numbers of graphs. Submitted for publication.

[2] Glenn H. Hurlbert. A survey of graph pebbling. *Congr. Numer.*, 139:41–64, 1999.

MATHEMATICS, STATISTICS, AND COMPUTER SCIENCE DEPARTMENT, ST. OLAF COLLEGE, 1520 ST. OLAF AVENUE, NORTHFIELD, MN 55057

E-mail address: laison@stolaf.edu

Pi Mu Epsilon Student Presentations and Mathfest

Terry Jo Leiterman and Rick Poss

1. History

Pi Mu Epsilon (PME) is the original national honorary mathematics society for college and university students. Although PME was established in 1914, it didn't sponsor student paper sessions at the summer national meetings until the joint MAA/AMS meeting at Michigan State University in 1952. J. Sutherland Frame was just beginning his tenure as an councilor of PME and he was also on the faculty at Michigan State. In that first year, there were six PME student presentations. There were no PME summer meetings in 1953 or 1954, but there were eight student presentations when the national summer meetings were held at the University of Michigan in 1955. Again, in 1956, there was no national PME summer meeting but in 1957, when the meeting was held at Pennsylvania State University, there were five student presentations.

The next student sessions were at Michigan State University in 1960 (eight presentations) and Oklahoma State University in 1961 (4 presentations). In 1962 the MAA/AMS summer meeting was in Canada, so there was no PME national meeting. Starting with the Boulder meeting in 1963, (9 student presentations), there has been a PME student session at every year's summer meeting, except for 1974. For a more detailed account of PME's first seventy-five years and a listing of all PME student presentations up to 1988, please see the article by J. Sutherland Frame in the spring 1989 issue of the Pi Mu Epsilon Journal.

The PME summer meetings were held in conjunction with the summer joint meetings of the AMS and MAA through 1991. In 1992, because of the joint meetings being held with ICME in Quebec, Canada, PME held a joint meeting with the MAA Student Sections at Miami University, in Ohio. Starting again in 1993, PME held its summer meetings with the AMS and the MAA. Since 1997, the AMS has not held summer national meetings and the PME summer meeting has been meeting with the MAA as part of Mathfest.

2. Growth

The number of student presentations at summer meetings has shown a pattern of growth. Although the numbers might go up or down in any particular year - primarily, perhaps, due to the location of the summer meetings - the trend has been

Received by the editor December 28, 2006.

upward. The average number of student presentations from 1952 through 1969 was 7.9 per year. From 1970 through 1979 the average was 15.3. During the 1980's, the average was 25.1; during the 1990's the average was 32.8. Since 2000, the average has been 36.4.

The reasons for the growth are many. The financial support from the NSA and the prize money given by the AMS have certainly been factors. In fact, the largest number of talks in a single year was 46 in 1989, the year that the AMS and the NSA began their support. 1989, however, was also the diamond jubilee year for PME, and the large number of papers might have been a result of the increased publicity.

The watershed year appears to have been 1985. The number of speakers in 1984 was 13; the number jumped to 34 in 1985. The largest number before 1985 was 22; the smallest number after 1985 was 22. No one event in 1985 would explain the large increase in student presentations. At about that time, however, more chapters started taking advantage of the fact that PME would support (at some level) more than one speaker from the same chapter. The number of chapters sending multiple speakers has been increasing over time.

The presence of the NSF REUs has certainly been a factor in maintaining and increasing the number of student presentations. Quantifying this effect, however, is difficult. Although students who present the results of their REUs give proper credit to the NSF and their REU advisors, there are no records kept about which talks are REU-based and which are not. It is reasonable to assume that the increase of a bit more than ten presentations per year that has occurred since the inception of the REU program in 1987 can be attributed to this program.

The MAA Student Chapter program was initially seen as a problem for PME. There was a feeling that there would be competition between the two groups for the same student speakers. That has not turned out to be true. The Student Chapter program has provided opportunities for students from schools without PME chapters. Additionally, some schools are able to send speakers to both the PME and the MAA student session, providing funding for more PME students. (Note that no student can receive travel support from both PME and the MAA for the same meeting.) The presence of the MAA student speakers has also increased the total student visibility at the summer meetings, making the meetings more comfortable for all students in attendance. The MAA Student Paper sessions have also been especially effective in providing a venue for presentations based on REUs.

3. Prizes and Judging

In 1989 the AMS gave Pi Mu Epsilon a grant of $1000 to use for prizes for outstanding student presentations at Mathfest. Each year, about seven prizes of $150 each have been awarded to student speakers. Student presentations are judged on various criteria, including: whether the level of topic is appropriate for the audience; quality of research (for research presentations) or interest of subject (in the case of expository talks); appropriate use of media; and ability to answer questions. One overriding criterion is the delivery of the material. Student presentations are judged on the effectiveness in communicating the mathematics to the audience.

A panel of three judges each student presentation. The judges are current or former Pi Mu Epsilon councilors or other mathematicians with ties to the organization. Since there are parallel sessions and no one judge can attend all of them, each judge assigns a score to the presentations that he or she attends. At the conclusion of all of the student presentation sessions, the judges' scores are tallied and award decisions made.

Since 2003 CUR (the Council on Undergraduate Research) has been awarding one prize each year for an outstanding research talk during the PME sessions. The decision on the recipient of this prize is made at the same time that the decisions on PME awards are made.

In some years, there are special prizes awarded by SIAM and by the SIGMAA on Environmental Mathematics for presentations that are applied in nature or else deal with environmental topics. The respective groups make decisions about these prizes, with the concurrence of the PME panel of judges. Each of the prizes is for $150.

4. Funding

Ever since the first PME sessions at the summer national meetings, the organization has provided travel support for student speakers. PME currently reimburses student speakers for their transportation expenses up to a maximum of $600 for one student speaker per chapter. There is significant additional support when there are multiple speakers from a single chapter. In 2006, over $11,000 was given by PME to student speakers to support their transportation to Mathfest. Support for student speakers at Mathfest continues to be the single largest expenditure in Pi Mu Epsilon's annual budget.

The National Security Agency (NSA) has also supported the PME student speakers. Each year since 1989, the NSA has awarded PME a grant of $5000 to be divided among student speakers to be used as a subsistence allowance. In recent years, PME has supplemented this grant so that each student speaker receives about $150 to defray the costs of registration, meals, and housing at Mathfest. To help cover the costs of additional speakers and also deal with inflation, the NSA has recently agreed to increase its grant to $7000 per year.

5. Conclusion

The stated purpose of Pi Mu Epsilon is the promotion and recognition of scholarly activity in the mathematical sciences among its student members. One way in which it has achieved this purpose is by organizing and funding its sessions for student presentations at each summer's Mathfest. The financial support of the AMS and the NSA, the research opportunities funded by REUs, and the welcoming atmosphere provided by the MAA have all contributed to the increasing success of these summer student presentation sessions.

DEPARTMENT OF MATHEMATICS, ST. NORBERT COLLEGE, DEPERE, WI 54115
E-mail address: terryjo.leiterman@snc.edu, rick.poss@snc.edu

On the path to scholarship

Lewis D. Ludwig

1. Introduction

When trying to promote undergraduate research, we should be mindful of the many steps needed to achieve a successful research experience for our students. It often seems that posing that perfect open problem to the student is the start of the undergraduate research experience. However, we feel that a good deal of preparation is needed to lead students to this point. In the following, we discuss the evolution of a course in our mathematics and computer science curriculum at Denison University that sets the students on the pathway to undergraduate research.

2. The evolution of a course

We are fortunate at Denison University to have an active campus-wide undergraduate research program. Every year, approximately 100 of Denison's 2000 students receive financial support to spend their summer working with Denison faculty members on a variety of research projects [1]. Although our joint department of mathematics and computer science had a number of summer scholars over the years, a surprising few presented their original research at regional or national meetings. In addition, virtually none of our students participated in research activities outside the department such as REUs. Nonetheless, these are two major steps leading to a successful undergraduate research experience.

In an attempt to address the first concern, in the spring of 2003, I convinced a Denison student to present his previous summer's work at the Ohio Sectional Meeting of the MAA. The student did a fine job in the end, but the process leading to this was trying and challenging for both of us. Upon reflection, I realized that our current mathematics and computer science curriculum was not preparing our students to present their technical work to a general audience. After discussing this with colleagues, it was clear we needed to make changes in our curriculum to address this issue. Initially, we decided to augment our sophomore level Introduction to

Received by the editor November 28, 2006.

Partial support for this work was provided by the National Science Foundation's Course, Curriculum, and Laboratory Improvement (CCLI) program under Award No. 0632804. Any opinions, findings, and conclusions or recommendations expressed in this material are those of the author(s) and do not necessarily reflect the views of the National Science Foundation.

Week	Topic(s)
1	Introduction, talk logistics, and LaTeX
2	Critiquing talks and more LaTeX
3–5	Students give a round of 5 minute talks
6	Review first round of talks; prepare for second round; Maple
7–9	Students give a round of 7 minute talks
10	Review second round of talks; prepare for the last round
11–13	Students give final round of 10 minute talks
14	Wrap-up, awards for best talks

FIGURE 1. Timeline for the oral communication lab.

Proofs course. This seemed the ideal time to instruct students on effective oral presentations skills. They were well into the major, but had adequate time left at Denison to use these skills.

After three evolutionary years, our Proofs course now includes a "lab component" which meets jointly with the computer science counterpart, Data Structures and Algorithm Analysis. The traditional courses meet three days a week with an additional two hour lab on Tuesdays. In the lab, students are instructed on orally communicating mathematical and computer science concepts at varying levels, how to use Waterloo Maple, and how to use LaTeX. Due to a lack of pedagogical materials, the author developed a series of pilot video vignettes for the course that compare and contrast various presentation techniques. A working version of these videos will be available for general use by the fall of 2007 [3].

Over the 14 week semester, the students obtain a thorough introduction to technical speaking without adding an additional course to the already full curriculum. This was one of the first courses at Denison, outside of the Communication Department, to fulfill the oral communication requirement for the general education package. Figure 1 provides an overview of how the 14 weeks of the semester are allocated. A separate article detailing the logistics of the course is available from the author upon request [2] .

The lab consists of five main components.

(1) Each student presents three talks on various topics in mathematics or computer science.
(2) Each student talk is peer reviewed by four members of the class as well as by the two instructors.
(3) Each student writes a review of each of his or her own talk after viewing a recording of the talk and reading the peer evaluations.
(4) Each student critiques five outside talks, such as department or campus-wide lectures, over the course of the semester.
(5) Each student is introduced to typesetting technical documents with LaTeX and performing basic tasks in Waterloo Maple.

3. Emerging scholars

Since the inception of this course in 2003, we have witnessed two trends in our undergraduates. First, the number of Denison students presenting their work at regional and national meetings has greatly increased. Second, there has been a marked increase in the number of Denison students attending REUs or other national programs.

As noted, prior to this course, it was rare to have a student present his or her work to an external audience. Now we average seven student presentations each spring at the Ohio Sectional Meeting of the MAA. In the past three years, nine talks by Denison students were delivered at national meetings, two winning awards based on quality of research content and delivery. In addition to external presentations, students share their work with the department and peers at our biweekly "FASt Talks" (Faculty and Student Talks). A survey of the students who completed this course in the past two years showed that of the 34 respondents, 20 were more willing to present a mathematical or computer science talk outside the department after taking this course. Upon completing his presentation at a sectional meeting, one student commented "...after Proofs and the FAsT Talk, the section talk was a breeze." Clearly, preparing the students in the art of oral presentation has given them the confidence to successfully present their work to external audiences.

In the last two years, we have had three students participate in REUs and three conduct research internships with NASA. While these numbers are not staggering, they are meaningful when we had only one student participate in an REU in the three years prior to this course. This trend seems to be a corollary of the first. As students presented at various conferences, they were exposed to other undergraduates presenting their REU findings. This encouraged our students and gave them the confidence that they could do the same. In the survey noted above, 14 students reported that they are now more likely to pursue a research project after taking this course. After presenting at a national meeting, one of our students noted "I never knew there was so much math out there! I saw really cool stuff that wasn't in any of my classes. It was brand new mathematics discovered by students like me!" She ultimately attended a strong REU, presented her work at MathFest, and is now in graduate school.

4. Concluding remarks

We hope our experiment in setting students on the pathway to undergraduate research inspires others to do the same. Not all schools have REUs or research programs like Denison. Nonetheless, we need to think of creative ways to expose students to interesting mathematics and computer science topics that will excite them and encourage them down this path.

The author would like to thank Jessen Havill, Matt Kretchmar, Matt Neal and Todd Feil for their help in developing and teaching this course.

References

[1] Boone, C. Keith, *Entering the Community of Scholars: Summer Research at Denison*, Council on Undergraduate Research Quarterly, March 2003, 113-115.
[2] Havill, Jessen T., Ludwig, Lewis D., *Technically Speaking: Fostering the Communication Skills of Computer Science and Mathematics Students*, to appear in the Proceedings of the SIGSCE Technical Symposium on Computer Science Education, 2007.

[3] Ludwig, Lewis D., *Technically Speaking* ..., http://www.denison.edu/mathcs/techspeaking, supported by NSF award DUE-0632804.

DENISON UNIVERSITY, GRANVILLE, OH 43023
E-mail address: LUDWIGL@DENISON.EDU

Mathematics Research by Undergraduates: Costs and Benefits to Faculty and the Institution

MAA CUPM Subcommittee on Research by Undergraduates

Opportunities for undergraduates to do research in mathematics have increased dramatically over the past twenty years. The National Science Foundation sponsors programs such as Research Experiences for Undergraduates (REUs) to encourage involvement, and individual schools have supported local efforts. Administrators have seized on the idea of undergraduate research as a way to engage undergraduates in meaningful intellectual experiences, and departments are striving to increase the involvement of faculty and students in research activities. It is important to note that there are significant differences between undergraduate research in mathematics and undergraduate research in other disciplines.

This document seeks to address the following questions: What is mathematical research by undergraduates? What are the costs and what are the benefits? How is mathematical research by undergraduates different from research by undergraduates in other fields? Our goal is to provide a framework for students, faculty, and administrators to follow as they seek to participate in undergraduate research activities in mathematics.

What is mathematical research by undergraduates?

For the purposes of this document, mathematical research by undergraduates is characterized by the following:

- The student is engaged in original work in pure or applied mathematics.
- The student understands and works on a problem of current research interest.
- The activity simulates publishable mathematical work even if the outcome is not publishable.
- The topic addressed is significantly beyond the standard undergraduate curriculum.

Undergraduate research in mathematics usually does not lead to publishable work in standard mathematical research journals. Publications are not a universal expectation for undergraduate research projects.

Received by the editor December 1, 2006.

Undergraduate research is an opportunity for students to explore open questions and to gain experience in making and testing conjectures or in building or testing models. It goes beyond learning what others have done.

Undergraduate research can be performed in pure mathematics, where the goal is to develop new mathematics and prove new theorems. It can also be essentially applied, where mathematics is used to solve problems of practical interest.

Excellent undergraduate research in mathematics draws the student close to the frontiers of mathematics, enabling the student to realize that mathematical research is a vibrant pursuit. This often includes reading recent journal or survey articles, doing computer searches for small examples, doing extensive background reading on the mathematics required to understand the problem, and writing the results in proper mathematical style. Undergraduates can and should present their work at regional or national mathematical meetings either in poster form or by giving a talk. There is a large Undergraduate Poster Session at the Joint Mathematics Meetings each year with many posters.

In summary, the activities by the undergraduates should simulate the activities of a mathematician, going through all the stages of a research project, with the expectations of the outcome generally more modest than for that of a professional mathematician. The stages include formulating and solving a problem, writing the paper, communicating results in a talk or poster, and possibly publishing a paper.

How is mathematical research by undergraduates different from research by undergraduates in other fields?

Mathematical research by undergraduates should be distinguished from research by undergraduates in fields such as the natural sciences and the social sciences. Research in the natural and social sciences require many people to be involved, and undergraduates in those fields can participate in valuable ways toward the success of a project. Their names will rightly be included on publications based on their contribution.

Most areas of mathematics do not naturally fit into either of these models: we are not doing physical experiments, and we are not usually collecting data. Research in mathematics does involve much quiet thought, and communication so that ideas may be tested. Researchers in pure mathematics try to answer open-ended questions; these questions are in many ways similar to puzzles (like the Car Talk puzzler you might hear on the radio). But the "puzzles" mathematicians seek to solve, address deep questions about the structure of mathematical systems. Mathematics research is often misunderstood because of its introspective nature, and because some do not see any immediate application of the results. Mathematicians see their open questions, and their solutions to open questions, as beautiful and interesting in themselves, and occasionally applications do come about from their study.

In some applied areas of mathematics, mathematicians do construct models and run numerical simulations (frequently with real data.) In both pure and applied areas of mathematics, the work of undergraduate researchers may not contribute in significant ways to publishable work. But in some projects, publications do result with significant contributions from the undergraduates involved.

Furthermore, while undergraduate researchers in other disciplines often assist the project mentor in his/her own research projects, undergraduate research projects in mathematics are often created specifically for undergraduates. Thus, time spent on leading an undergraduate research project is often time taken away from the mentor's own research endeavors. While undergraduate research is extremely beneficial to the student and to the department, it does not always advance the faculty member's own research program.

What are the costs and what are the benefits?

The "costs" of undergraduate research in mathematics are primarily measured in time. Mathematics does not require expensive lab equipment, and most universities have access to adequate computing facilities and software. Time is a somewhat less tangible resource, but it cannot be underestimated as an obstacle to research activities. One typical format for undergraduate research is a year-long independent study in the senior year culminating in a thesis. Our estimates indicate that the cost in terms of faculty time averages about 3 hours per week for the year, including actual meeting time, time spent thinking about what direction to take the research, and time spent reading and correcting drafts of the thesis. It takes creative energy to formulate good projects for students. Another format for research is an 8 or 10 week summer program. In this case, the supervisor will spend between 5 and 15 hours per week per project. The number of hours can vary quite a bit depending on the backgrounds of the students and the difficulty of the project. Thus, the cost in terms of time to faculty is quite high.

Students receive tremendous benefit from this activity. Students get to be involved in a significant mathematics project under close supervision by a professor. They gain experience with independent learning, a skill that will prepare them for research in graduate school as well as prepare them to be productive members of a company. They get control over their education in ways that are impossible to duplicate in the classroom environment. Students come out of this experience significantly enriched in their understanding of modern mathematics. Presentation of the results in written and oral formats improves the communication skills of the student.

Professors, on the other hand, have less to point to as a benefit. Certainly there is personal satisfaction in working closely with students. Sometimes, these research experiences do not assist the professor with his/her own research agenda. Faculty who choose to participate may have less time to pursue research leading to publication. This can be overwhelming, particularly for young faculty who are in the early stages of establishing their research credentials. In fact, unless there is an institutional commitment to reward supervision of undergraduate research in mathematics, faculty will understandably be reluctant to agree to spend time working with undergraduates when they can see direct benefit from spending their time on other activities.

The mathematics departments of many colleges and universities provide research experiences for undergraduates. Some programs are during the summer and others occur during the semester. Some programs involve students from other universities and some involve only students at their own institutions. Some programs are supported by the National Science Foundation and the National Security Agency, while others receive support from their own administrations. [1, 2] See the Proceedings of the Conference on Summer Undergraduate Mathematics Research

Programs for a variety of types of programs. [3] See also the website of the Council for Undergraduate Research.[4]

We conclude that institutional support is a critical component for a sustainable undergraduate research program in mathematics.

Websites with further information.
1. `www.nsf.gov/crssprgm/reu/list_result.cfm?unitid=5044`
2. `www.nsa.gov`
3. `www.ams.org/employment/REUproceedings.html`
4. `www.cur.org/`

This document was written by the MAA CUPM (Committee on the Undergraduate Program in Mathematics) Subcommittee on Research by Undergraduates and was approved by CUPM. It is posted at `www.maa.org/cupm/CUPM-UG-research.pdf`.

SUBMITTED TO THIS PROCEEDINGS BY SUZANNE LENHART
E-mail address: `lenhart@math.utk.edu`

Reflections on Undergraduate Research

Mario Martelli

1. How it started

In 1987 I was hired as professor of Mathematics at California State University Fullerton, a predominantly undergraduate institution. The Administrators, from the President of the University to the Dean of the School of Art and Science and the Chair of the Mathematics Department, were placing a very strong emphasis in a combination of research and excellence in teaching. Financial support was available in the form of release time and paid participation to national and international meetings. I realized that the following two activities could provide a winning combination to achieve the goals supported by the university.

1. Publish simple and easily readable papers on topics of interests to our students.
2. Do research with our best undergraduates on topics that were accessible to them and of interest to me.

I realized that these two activities were going to absorb my energy and to affect negatively my traditional professional research, but I decided that they were worthy to pursue and I embarked in them fully aware of the potential risks and rewards. By the time I left California State University Fullerton in the year 2000 I had published collaborative work with six students: David Marshall [16], Mai Dang [3, 4], Tania Seph [4], Bethany Johnston [14], Gary Michaelian and Suzanne Sindi [17]. David and Suzanne would go on to earn a Ph.D. in Mathematics. David's thesis advisor was Prof. William McCallum from the University of Arizona in Tucson, and Suzanne's advisor was Prof. Jim Yorke from the University of Maryland, College Park. Gary earned a Ph.D. in Physics from UC Irvine. The others are teaching mathematics in community colleges or high schools in California.

I had also published several papers to meet the first goal of my agenda. My co-authors were Gerald Gannon ([9, 10, 11, 12, 13]), William Gearhart [8] and Harris Schultz [18].

In the fall of 2000 I arrived at Claremont McKenna College (CMC), following a special invitation from the administration. Since CMC is an undergraduate institution, I continued with my research collaboration with students. I collaborated with Carrie Staples [7] and with three undergraduate students: Adam Cox, Christopher

Received by the editor January 3, 2007.

Jones, and Alison Westfahl [1]. All these students are or are going to be graduate students in Mathematics, or Economics, or Law.

Am I done collaborating with undergraduates? Not at all. In my fall 2006 classes in Multivariable Calculus and Differential Equations I mentioned open problems I would be happy to solve in collaboration with some of my students. One of them, from my Multivariable Calculus, is very interested in the problem I presented and want to work with me on its solution. The future is open and promising!

Let me share with you a bit of my life as a mathematician. I received my degree from the University of Firenze in 1966 and my thesis advisor was Prof. Roberto Conti, who passed away about a month ago. I wrote my thesis on an inequality proved by De La Vallée Poussin and the result I obtained was published [15]. I became Assistant Professor at the University of Firenze in 1967 and Full Professor in 1976. I came to the USA for family reasons since my wife is from California. I was first Professor at Bryn Mawr College from 1979 to 1987 and in the fall of that year I arrived at California State University Fullerton (CSUF). To make my transition easier the CSUF Administration granted me tenure upon entrance.

Many colleagues regarded my move from Bryn Mawr College to CSUF as a step down, but I did not perceived it that way. I knew that at Fullerton my influence in the life of the students taking my classes could be far greater than anything I could ever accomplished at Bryn Mawr. In Italy private schools are the natural venue for students with lower than average skills and rich parents. University professors regard with a degree of skepticism every student coming from a private school. Hence, for me, teaching in a public university, was like "going back home." I had always some degree of uneasiness about investing my energies and my knowledge with students coming from financially able families. I felt that Fullerton was the right place for me, and I still remember, with great emotion and sense of accomplishments, what David Marshall's father told me the day his son graduated. "You have turned David's life around. Thank you."

Many readers would be curious to know how I found the time to work with my students, since the teaching load at Fullerton is four courses per semester. First, I am happy to credit Prof. Jim Friel, who at that time was Chair of the Mathematics Department, for repeatedly granting me release time so that I was required to teach four courses only a couple of times and every remaining semester I taught three (or even two) courses. Having said that, I must recognize that I met and studied with my students mostly during the summer. They were not paid for working with me, and I was not paid for working with them, but we studied together, sometimes at the department, other times at my house. Now and then I provided lunch. We all loved to do mathematics and we greatly enjoyed each other company.

This arrangement is impossible at a private institution, since the students go back home during the summer, unless they are provided an income that justifies their presence at the college where they study during the regular academic period. I learned this unpleasant fact the very first summer at Claremont McKenna College. All students went back home and I was left with no one to work with. I had problems to propose, but no one to listen. In Fullerton, it was different. The students lived at home, they usually took a summer job, but they set aside time to do research with me and to talk about mathematics. I really appreciated their willingness to come to my house in Claremont, so that I could avoid the trip of going to Fullerton, 22 miles away from my residence.

2. Topics

How did I select the topics of research I proposed? I did not have only one strategy. First, I did read a lot of mathematical journals, including, but not limited to, the American Mathematical Monthly, the College Mathematics Journal, Mathematics Magazine, Applied Mathematics Letters, etc. I paid particular attention to those papers that had open questions at the end and I asked myself if we could solve some of the questions left unanswered by the writer(s). I asked friends and colleagues to tell me of problems they found and they considered suitable for our students. When I was teaching, I always asked myself if some theorems could be generalized, or proved differently, or established for a different class of functions. I am happy to confess that I was never short of problems. In fact, I had abundance of them, many more than I could assign.

Before presenting a topic of research I did some preliminary investigation to check if the proposed question could be solved. I had the feeling that my students would be frustrated if their work would have come to an impasse. However, I was never sure, when we started, if we were going to be succesful. In some cases, in fact, the problem proved to be more difficult than I had anticipated, and we could not write a paper with our results. I am still puzzled by some of the problems and I do not have an answer, although I believe that the result we wanted to prove is indeed true. Let me give you an example.

In the book by R. Boas [2] I had found the celebrated universal chord theorem.

THEOREM 2.1. *Let $f : [0, 1] \rightarrow [0, 1]$ be continuous and such that $f(0) = f(1)$. Then for every positive integer n there are two points $0 \leq x_1 < x_2 \leq 1$ such that $f(x_1) = f(x_2)$ and $x_2 - x_1 = \frac{1}{n}$.*

The segment joining $(x_1, f(x_1))$ with $(x_2, f(x_2))$ is appropriately called a *horizontal chord* of f. An interesting complement to 2.1 is the following result

THEOREM 2.2. *Let $f : [0, 1] \rightarrow [0, 1]$ be continuous and such that $f(0) = f(1)$. Then for every $a \in (0, 1)$ the function has either one horizontal chord of length a or two different horizontal chords of length $1 - a$.*

Famous mathematicians have worked on these problems and published their discoveries either about proofs different from the existing ones or about different families of functions [5, 6, 19, 20].

I had the strong feeling that Theorem 2.2 could be established in higher dimension if the statement was modified in a suitable manner. In fact, the simple example of the function $f(t) = (\cos 2\pi t, \sin 2\pi t)$, $t \in [0, 1]$ shows that some adjustments are needed. For example, *horizontal* may now be interpreted as *parallel* to the plane $z = 0$. I proposed the problem to Pilar Mata, a very talented student who did not want to pursue her Ph.D. in mathematics despite my encouragement to do so. Pilar and I worked on it for several weeks, and we did establish the validity of a higher dimensional version of Theorem 2.2 in many cases. However, a full proof of the result eluded us, and, in the end, we had to give up since we could not surmount the difficulties presented by some cases.

There was always some preliminary work to do with my students to bring them up to speed. For example, when Suzanne Sindi, Gary Michaelian and I were working on establishing necessary and sufficient conditions for pitchfork, transcritical, fold, and period doubling bifurcation [17], we had to review together the appropriate analytic conditions required by simple and double points of planar curves, including the exceptional case in which the tangent lines are vertical. I had to explain to them that given the set $S = \{(x, y) : f(x, y) = 0\}$ where f is C^n with n sufficiently large, and given a point $P \in S$ we can define the multiplicity of P as a solution of the equation $f(x, y) = 0$ by using the partial derivatives of f. In particular, P will be a double point if

$$\frac{\partial f}{\partial x}(P) = \frac{\partial f}{\partial y}(P) = 0,$$

but not all second partials are 0 at P. We had also to understand why the slopes m of the two tangent lines to S at P are found by solving the quadratic equation

$$am^2 + 2bm + c = 0$$

where

$$a = \frac{\partial^2 f}{\partial x^2}(P), \ b = \frac{\partial^2 f}{\partial x \partial y}(P), \ c = \frac{\partial^2 f}{\partial y^2}(P).$$

Suzanne, Gary and I invested many hours in this problem. I purposely avoided to let the students know what I believed we were going to find. They discovered many, but not all, results on their own.

They also wanted to explore the situation in a higher dimensional setting, but, unfortunately, there was not time for this study. Both students left Fullerton to pursue their graduate studies, Gary at UCI and Suzanne at the University of Maryland, College Park.

Here is perhaps the most appropriate moment to underline that the collaboration with undergraduates is necessarily constrained by their four years training. After they go on to graduate studies, or to other activities, it is extremely difficult to establish a meaningful collaboration with them. The distance, the different interests, the pressure of the graduate program are all conspiring against any plan to continue a research program centered on problems more suitable for an undergraduate than for a graduate student. Therefore, the wisest move, when possible, is to recruit juniors and possibly even sophomores. In this case the instructor can be sure that the collaborative work will have the necessary continuity.

3. Recruiting

Hence, we naturally come to the question of how to find students who are willing and capable of collaborating with you. The "willing and capable" is an important combination. I have selected students from my own classes, I have asked the advice of other instructors, I have looked at the high school record, and at the record of courses taken by the student after their high school graduation. I realized, however, that nothing can replace direct talks with the undergraduates.

At least in one case, I made the choice simply because I had the strong feeling that the individual was very capable and the previous experiences were not representative of the student's real capabilities. Hence, for example, I invited David Marshall to work with me and I found out very soon that I had made the correct decision, even though his previous grades were average at best. I was probably influenced by my experience back in Italy. One of my classmates and later my collaborator was Massimo Furi, who did not shine in high school (Istituto Tecnico Industriale) simply because the mathematics he had to learn was uninspiring and deprived of ideas. At the university Massimo flourished and revealed an amazing potential.

Did I ever selected the "wrong" students? Yes, in two cases. One based on my own experience, and the other based on the advice of a colleague. It can happen. The undergraduates did not measured-up to my expectations and I worked for about a year with no tangible accomplishments.

In two cases I selected problems that were too difficult. We were not able to solve them and the results we obtained were too partial. It was a great disappointment for me, and an even greater disappointment for the students. Unfortunately, there is no way to find out, at the outset, if the result you want to establish is true or not. Moreover, even if we assume that the result is true, it could be very difficult to prove. I found out that missteps can be minimized but not completely avoided.

My name has always appeared among the list of authors of the paper written in collaboration with my undergraduates. Several reasons have dictated this strategy. I will mention only two. First, I felt that the inclusion of my name was more representative of the common efforts that went into the collaborative work. Second, I felt that adding my name would give more visibility to the papers and consequently, would be more beneficial to the students.

I encouraged the student's participation in regional and national meeting. I obtained from the administrations at Fullerton and at CMC the necessary funds to pay for transportation, registration, and lodging. I started an Undergraduate Student Poster Session at the spring meeting of the Southern California-Nevada Section of the MAA. This is now a well established tradition and the activity usually attracts thirty teams of undergraduates. The section pays for their registration and lunch.

For several years I organized a national Undergraduate Student Poster Session in conjunction with the annual joint meetings of the AMS and MAA in January. From humble origins the session had recently become one of the most visible and well attended activities of the joint meetings. I recently relinquished to Dr. Diana Thomas, of Montclair State University, the task of organizing the poster session and I understand that the Executive Committee of the MAA would like to open the participation to 250 teams of undergraduates. The amount of work required to reach this goal with a flawless organization is definitely challenging, but not impossible.

Let me close these remarks by repeating that my collaboration with undergraduates has been very rewarding and I would do it again. I feel that it should not be the primary research activity of a young faculty, since it requires a degree of experience and an investment of time that may prove to be prohibitive for a person at the beginning of the academic career. However, working with these young kids is fun and interesting. The influence a teacher can have in their future will probably last a lifetime.

References

[1] Aarao J.,Cox A., Jones C., Martelli M., and Westfahl A., *A non-smooth band around a non-convex region*, College Mathematics Journal, **34**, (2006) 269–278.

[2] Boas P.R., *A Primer of Real Functions*, The Carus Mathematical Monographs, Number 13, 1996.

[3] Dang M., and Martelli M., *On the derivative of a non-constant function*, Applied Mathematics Letters, **7**, (1994) 81–84.

[4] Dang M., Martelli M., and Seph T., *Defining chaos*, Mathematics Magazine, **71**, (1998) 112–122.

[5] Diaz J.B., and Metcalf F.T., *A continuous periodic function has every chord twice*, Am. Math. Monthly, **74**, (1967) 833-835.

[6] Gillespie D.C., *A property of continuity*, Bulletin of the American Mathematical Society, **22**, (1922) 245-250.

[7] Furi M., Martelli M., O'Neill M., and Staples C., *Chaotic orbits of a pendulum with variable length*, EJDE, **2004**, (2004) 1–14.

[8] Gearhart W., and Martelli M., *A blood cell population model, dynamical diseases, and chaos*, UMAP, **11**, (1990) 309–339.

[9] Gannon G., and Martelli M., *The farmer and the goose: a generalization*, Math. Teacher, **86**, (1993) 202–204.

[10] Gannon G., and Martelli M., *Cutting a chain: a problem made to be generalized*, Math. Teacher, **89**, (1996) 292–293.

[11] Gannon G., and Martelli M., *Weighing coins: divide and conquer to detect a counterfeit*, College Mathematics Journal, **28**, (1997) 365–367.

[12] Gannon G., and Martelli M., *The prisoner's problem: a generalization*, Math. Teacher, **93**, (2000) 192–193.

[13] Gannon G., and Martelli M., *Discrete dynamical systems meet the classic monkey-and-the-bananas problem*, Math. Teacher, **94**, (2001) 299–301.

[14] Johnston B., and Martelli M., *Global attractivity and forward neural networks*, Applied Mathematics Letters, **9**, (1996) 77–83.

[15] Martelli M., *Sul criterio di unicità di De La Vallée Poussin*, Atti dell'Accademia Nazional dei Lincei, **45**, (1968) 7–12.

[16] Marshall D., and Martelli M., *Stability and attractivity in discrete dynamical systems*, Mathematical Biosciences, **496**, (1995) 1–9.

[17] Martelli M., Mikaelian G., and Sindi S., *A geometric approach to transversality conditions for bifurcation*, Recent Trends in Nonlinear Analysis, J. Appel Ed., Birkauser, March 2000, 205–215.

[18] Martelli M., and Schultz H., *Geometry and four maximum problems*, Ontario Math. Gazette, **32**, (1994) 21–23.

[19] Lévy P., *Sur une généralisation du théorème the Rolle*, Comptes Rendus de l'Académie des Sciences Paris, **198**, (1934) 424–425.

[20] Oxtoby J.C., *Horizontal chord theorems*, Am. Math. Monthly, **79**, (1972) 468–475.

CLAREMONT MCKENNA COLLEGE, CLAREMONT, CA 91711
E-mail address: mmartelli@cmc.edu

The University of Chicago's VIGRE REU and DRP

J.P. May

The University of Chicago has been called "the place where fun comes to die," but it is more accurately called a place where students come to learn. This article will describe programs that are intended to entice students into wanting to teach and to do mathematics. We like them to create mathematics now, but we are far more interested in their wanting to do mathematics always. Our programs are leading significant numbers of students to make such lifelong commitments.

For background, Chicago is now entering the seventh year of an NSF VIGRE Program. (VIGRE stands for Vertical InteGration of Research and Education). This has large graduate student and postdoctoral components, but the most significant effects have been on the undergraduate level. We have run an eight week summer VIGRE REU for the past seven years. It is restricted to University of Chicago undergraduates who have just completed their first, second, or third year of undergraduate studies. We originally budgeted for 18 participants. In the seven years, we have had, in order, 22, 36, 45, 42, 59, 65, and 70 such participants. The 70 this year were chosen from an applicant pool of over 90, and some of the 70 participated without funding since we could not afford to support that many students. In addition, for the past four years our graduate students have run an academic year Directed Reading Program (DRP) for undergraduates, and this has had around 15 undergraduate participants each quarter. In combination, these programs offer year round mathematical enrichment.

This large number of participants comes from a small undergraduate college, now approaching 5000 students after a period of expansion, ensconced within a major graduate university. The number and percentage of graduating mathematics majors has increased substantially and for the past five years has numbered between 75 and 81 BA's, accounting for over 7% of the total graduating class. That puts Mathematics in a statistical dead heat with English as the fourth most popular major at Chicago, bested only by Economics, Biological Sciences, and Political Science. Moreover, an average of 12 students per year have gone on to graduate study in mathematics over the past five years, with more going into other mathematical sciences, especially computer science and physics. These current mathematics graduate students (or in a few cases new PhD's) include 4 at Harvard, 7 at MIT, 7 at Berkeley, 5 at Stanford, 4 at Michigan, 4 at UCLA, 3 at Rutgers, 3 at Texas, and 1 or 2 at each of Chicago, Yale, Columbia, CalTech, Wisconsin, Cornell, Brown,

Received by the editor October 8, 2006.

Indiana, Northwestern, Duke, Rice, Virginia, Ohio State, Maryland, San Diego, Florida, Missouri, Tufts, and a few schools in other countries. A quarter of these mathematics graduate students from Chicago are women.

I wish that other schools had similar records. As director of graduate admissions, I have to report a static or somewhat diminishing pool of very highly qualified applicants over these same years. A striking feature of our first year graduate classes is that the few University of Chicago undergraduates permitted to take them are among the very strongest and best prepared of all the students.

What does this have to do with undergraduate research? The truth, perhaps, is not very much. The NSF has been persuaded to allow the "R" in "REU" to be interpreted broadly. Chicago is a center of advanced mathematical research. Most of our faculty, including many but not all who teach in our REU, have little idea of how to give genuine research problems to undergraduates. For example, I teach in the REU, and nothing related to my own research could possibly be farmed out to undergraduates: far too much background is required. As a teacher of graduate students (40 have obtained PhD's under my supervision), I like students to find their way to their own problems. As a teacher of undergraduates, at least in the REU, I like students to find things to learn on their own, with the cooperation and help of a graduate student or faculty mentor. It does not much matter to me whether or not they do anything genuinely new, as long as it is new to them. Fortunately, others who teach in our REU know better how to give research problems. Lazlo Babai, in particular, has written joint papers with several participants and has even featured some joint work on "the Abelian sandpile model" in a Distinguished Lecture Series at Emory University. While we can point to other original research, that is not the central focus of our program.

Our REU is deeply intertwined with our outreach teaching programs, which are organized and run by Paul Sally, Diane Herrmann, and John Boller. The regular REU participants serve as counsellors for part of the time that they are in the REU, working with either our YSP program for middle school and high school students or our SESAME program for grade school teachers. It came as a welcome surprise to us how much the participants enjoy the dual experience of learning and teaching that our program provides. In the first few years of the REU, our evaluation forms had leading questions aimed at soliciting complaints of too much teaching. There were none. In our dual program, students develop a simultaneous commitment to teaching and research. This is crucially important to the effectiveness of our program in attracting people into graduate mathematics as a career.

Our REU also includes an apprentice program tailored for all but the most advanced of those participants who have completed only one year of college. This is officially a four week program, but some of its participants spend the entire eight weeks. The apprentice program has only been in operation for three summers. It was created in large part to meet the growing popularity of the REU. There is a limit to how many participants can productively be used to teach in the outreach programs, since a ratio of fewer than five to one of outreach participants to student counsellors seems to become counterproductive. This past summer there were 48 participants in the regular REU and 22 apprentices. Interestingly, applicants were asked to select which program they preferred, and most of those who attended as apprentices had expressed a preference to participate in that program. There were also 3 exceptional participants, all women, who were not part of the prescribed pool (a high school student, a student who had just graduated, and a graduate student in another field), making a total of 73 active participants.

The reader interested in a detailed description of the various components of our VIGRE program and of the REU and DRP can find them on our web site

http://www.math.uchicago.edu/ may/VIGRE/index.html.

Links to the successive yearly operation of the REU can be found there. The direct link to the full description of the 2006 REU is

http://www.math.uchicago.edu/ may/VIGRE/VIGREREU2006.html.

Detailed course descriptions, hour by hour schedules, and the organization of the mentorship pairings can be found there. These give an idea of how intensive this eight week program is. Nobody, or at least nobody in his right mind, would start out with the idea of organizing and running a program with over 70 participants, mentored by 28 graduate students, and taught by 11 faculty members. The program has evolved over time, and the best piece of advice to anyone starting out as an organizer of any such program is to listen to the participants at all levels, and learn from them. Evaluations and requests for suggestions have been made every year, and the program has been changing in response to suggestions. This year, for the first time, there was a requirement of either an end of program presentation or a written report for all regular participants. Next year, a written report will be required of everyone. The presentation must of course be optional, since there simply isn't time to listen to 70 presentations.

Most of the 2006 undergraduate papers are also posted on the web site. They range over a wide spectrum of mathematics at varied levels. Some of them are seriously sophisticated. Others are playful. Some are elementary learning experiences. The students are urged to work at their own level of comfort, always mentored, but free to explore any direction. Many of the papers are tied to one or another of the course offerings, others are not. I urge the reader to read a few of these papers, and at least to look at their titles. They run the gamut, including serious algebra, geometry, topology, analysis, probability theory, and logic. Several of them contain interesting original research; many of them include interesting surveys and perspectives on fields. The amount of mathematics the students have absorbed is prodigious. The reader is also urged to look at the course listings. The faculty teaching them include Dickson instructors, assistant professors, and tenured faculty. The subjects include topics in discrete mathematics, algebraic geometry, algebraic topology, geometry, analysis, and even a segment on mathematics and music.

This year's evaluations asked even more explicitly than usual for suggestions for future improvement. The answers seem to indicate that we are getting it right. A few of the responses nicely sum up the concensus and make clear that the respondents know that their answers are taken seriously.

"Overall I thought the program was better this year for two reasons: the mentoring program was much more coordinated, and the requirement of a paper/presentation was enjoyable and worthwhile. These two additions made the program more organized and should definitely be retained for the future."

"Having done the program for three years, I feel that I am in a good position to comment on how effective the current structure is. I felt that in this year, more than in any other, everyone participated in the class/research aspect of the REU. I feel that the paper was largely responsible for this. All in all I really would not change anything about the REU."

"I have done the REU three years and enjoyed it greatly each time, but this year even more so. This year I felt that much more of the material presented was accessible, a lot of which I think has to do with the two additional years of math I've had since my first year in the program. In that vein, I really like how there are classes offered specially for apprentices and more advanced ones as well.

I also liked the addition of a final project requirement. Not only did it force me to learn TeX (a definite plus), but it gave me a goal to work for, which helped give my work more direction. Also, it's nice to have something tangible to present or have at the end of the REU that is representative of the work I put in over the summer.

Another aspect I like is the teaching aspect (I did SESAME this year), as I really enjoy that and feel it's an important part of mathematics.

Those are the things that stood out this year, though of course other aspects are excellent and I think should stay the way they are: the mentorship program, a wide variety of classes on many subjects not usually covered in "regular" math classes, and the positive environment and freedom to learn whatever type and as much mathematics as you like.

My grad student mentor was [name], and I thought he did an excellent job. He was enthusiastic and helpful whenever I had questions.

Overall my experience in the REU has been very positive and has solidified and confirmed my desire to further pursue mathematics (and teaching it) as a career. Thanks for the opportunity and for providing such an excellent program."

Our REU works, but on this scale it is very expensive. I would be delighted to see it emulated at other major research universities. On a smaller scale, aspects of it, at least, could be emulated in many other places. The DRP, in contrast, is very inexpensive, and all it requires in the way of organization is willing and able graduate students. It can easily be emulated elsewhere. In fact, one past participant has organized a replica of the program at Rutgers. The program at Chicago is entirely organized and run by a self-perpetuating committee of graduate students, under very minimal supervision. A detailed description may be found at

http://www.math.uchicago.edu/drp.

The student and mentor agree on a topic of mutual interest and the student learns and discusses all that he or she can. Undergraduates apply to and are screened by the committee. Participants are required to meet with their mentor once a week during the quarter and to put in at least four hours of work on their topic per week. They are required to give a presentation at the end of the quarter. Projects must be approved by the committee. The program description advertises the following benefits of the program. "Participating undergraduates will learn to work independently through studying a topic of their choice, well-suited to their interests. They will develop relationships with graduate student mentors and receive a good deal of personal attention focused on their mathematical studies. Finally, they will gain valuable experience in mathematical communication by giving a presentation on their work to an audience of their peers."

It is to be emphasized that undergraduates who participate obtain no course credit and are not paid, and they do the work on top of their usually heavy course loads. Many of them are also working to help defray their college expenses. The fact that more and more are participating makes a powerful argument that the advertised benefits are being delivered. The graduate student mentors are also volunteering their time, and they receive just token payment (currently $300 per quarter, plus $100 for books for the participants, with a little more for the organizing committee). This is a true labor of love on both sides.

The REU and the DRP both depend on close mentoring, and often joint learning, between undergraduates and graduate students. Students learn an enormous amount of genuinely deep and interesting mathematics that way. The students may or may not do research, but they get a taste of it, and they get a serious feeling for the world of graduate mathematics and beyond, serious enough to make many of them aspire to that as the way to spend their entire lives.

In both the REU and DRP, rigorous mathematics and a sense of community are combined. The presentations are coupled with dinners and are social occasions. Students bring their friends. Mathematics is many Chicago students' view of fun. Once critical mass has been achieved, it is amazing to see just how contagious a love of mathematics can be. There is one prerequisite. The mathematics must be real. It must not be dumbed down. Rigor is essential. Our students, at all levels, learn what a proof is and how to appreciate a beautiful argument or an interesting conclusion. Here are some quotes from this year's applications to the REU, in answers to the request to "Explain briefly why you want this position". Incidentally, the answers show why we no longer have to make any real effort to recruit applicants. Word of mouth is enough.

"I am in love with and absolutely addicted to mathematics and plan to continue its study my entire life. This program is an excellent opportunity both to further my knowledge and to begin exploring this new* area — research. I want to play, I want to learn, and I want to do this with others who are just as excited and passionate about mathematics as I am.
*new to me, that is."

"... the YSP counselor work I've done this year has been some of the most fun I've had. Taking such young, malleable, minds that are unaware of what they "cannot" do, setting them in front of esoteric mathematics, and watching them go forth has been an utter joy this year so far; doing this for an extended period of time over the summer would, I believe, be amazing."

"I have heard from friends that it is an excellent program. I have heard glowing reviews of the participating faculty. In short, I believe that this is as close to an ideal math-learning situation as I can ever hope to encounter."

"I have heard wonderful things about the REU at Chicago ... and have wanted to do it since my first year. Finally, I have obtained my parents' permission to not come home this summer (for the first time), and so I am jumping on a chance to take an intensive math program."

"I love mathematics." [That is a leit motif running through the applications.]

"I could say 'I love mathematics,' but that is cliché. Math is for me not a simple 'love' but more of an all-consuming fascination, something that I just need to do."

"I really truly enjoy mathematics. It is what I plan on doing the rest of my life."

"Well, *briefly*, because I had a blast in the REU last year."

"Because it's a summer full of math."

"Because it means spending my summer studying, researching, and teaching math, which sounds like an amazing way to spend the summer."

"This program brings together two of my favorite things: teaching mathematics and learning mathematics."

"Of all of the classes that I have taken in my life, Honors calculus has probably been the most stimulating. This course made math not only interesting, but also made it challenging for me for the first time in my life. Seeing the proofs that π and and the square root of two are irrational was one of, I have to say, the most thrilling moments of my life. I think that this program offers me a unique opportunity to pursue this recently discovered passion."

DEPARTMENT OF MATHEMATICS, THE UNIVERSITY OF CHICAGO, CHICAGO, IL, 60637
E-mail address: may@math.uchicago.edu

Twenty Years of MAA Student Paper Sessions

Betty Mayfield

The printed program of the Summer 1987 meeting of the Mathematical Association of America at the University of Utah contained a small note: "Experimental Student Paper Session." Four undergraduate students gave talks there, to an audience of about seventy meeting participants. Everyone agreed that the students had performed admirably, and that the sessions should continue at future meetings. And so the MAA Undergraduate Student Paper Sessions were born.

From a vantage point twenty years later, it is remarkable that the notion of undergraduate research was so unusual, and that only four students could be found as pioneers for this inaugural session. Undergraduate research, and the conference talks associated with it, have become an accepted part of many conferences, but especially of MathFest, the summer meeting of the MAA. In fact, even in years when the MAA did not hold an official summer meeting, undergraduate students gathered to talk about mathematics: Robert Smith recalls the first-ever joint MAA-Pi Mu Epsilon student conference in the Summer of 1992. Over one hundred students and their advisors met at Miami University to listen to lectures, attend workshops, and give talks on their own research. At that meeting, 20 MAA students (and 20 PME students) presented papers, and the winners of the Mathematical Contest in Modeling spoke about their solutions.

Since that first experimental session, the number of student talks has grown almost exponentially. Students report on research they have done as projects in mathematics classes and as independent studies, at their home institutions and in summer REUs, alone and as part of a collaborative research group. They come from small liberal arts colleges and large state universities. Their topics range from pure to applied mathematics, including its history and pedagogy. There are usually eight to ten different MAA student paper sessions over the course of two afternoons at MathFest, in addition to separate Pi Mu Epsilon paper sessions and modeling contest presentations. Prizes are given for the most outstanding talks in each session; in addition, special prizes are awarded by SIAM, by the Environmental Mathematics SIGMAA, and by the Council on Undergraduate Research.

To illustrate the breadth and depth of the mathematics undertaken by these young people, we list the titles of some of last year's prize-winners:

Received by the editor December 27, 2006.

Many thanks to Richard Neal of the MAA Committee on Undergraduate Student Activities and Chapters, and to Robert S. Smith and Jennifer Galovich of Pi Mu Epsilon, for their excellent memories, detailed records, and general helpfulness.

Magic Squares and Elliptic Curves;
Centers and Eccentricities of Finite Simple Graphs;
Shaping Things Up: The Smallest Enclosing Ellipsoid of Random Knots;
Optimal Resource Allocation to Deter a Terrorist;
Peopling of America with Logistic-Diffusion Simulations.

The impact of REUs on the summer paper sessions cannot be over-emphasized. As more and more students participate in these summer programs, they are eager to communicate their results to their peers and to the mathematical community at large – and the summer meeting of the MAA is a natural place for them to do that. In Boulder in 2003, for example, half of the MAA student sessions in the program contained a note such as, "The last five speakers in this session are students of Prof. X in an REU at Y College." There were 55 MAA student talks at that meeting; two years later, in Albuquerque, there were 67. Space and time for student talks was becoming scarce; we had become victims of our own success.

And so a current issue facing the MAA is how to accommodate the ever-increasing number of students engaged in undergraduate research and to find ways for them to share their results. An undergraduate poster session at the Joint Mathematics Meetings in January is also bursting at the seams. One REU sponsors its own paper session at MathFest; others are investigating ways to include all of their young investigators in some way at the summer meetings.

We have come a long way from our 1987 experiment; our only problem now is dealing with too much success.

Hood College, 401 Rosemont Ave., Frederick, MD 21701
E-mail address: `mayfield@hood.edu`

Mathematics In Multi-disciplinary Research-focused Learning Communities

Jason E. Miller

1. Introduction

Undergraduate research in mathematics is booming at Truman State University. For many years, it existed quietly in the offices of the mathematics faculty disguised as a graduation capstone requirement for students. Outside the mathematics department, the rest of the campus openly celebrated the fact that Truman was heavily invested in providing open-ended research experiences for students. Truman had been making a name for itself as a powerhouse in this area by annually sending strong students with excellent work to the National Conference on Undergraduate Research (NCUR) and professional conferences, and by having many faculty serve as Counselors in the Council of Undergraduate Research (CUR). For nearly 20 years, Truman has held a conference that celebrates student research and scholarly activity, and since 1998 it has closed down campus for one day to do so. In 2006, this conference had 319 presentations by 476 students that represented the mentoring of 162 faculty members.

Until the early 2000s, mathematics student participation in undergraduate research activities on campus was marginal. We sent students to REUs, and we sent students to regional MAA meetings to give presentations on their capstone projects, but we seldom called what they were doing "research." This changed when a group of faculty from the mathematical sciences teamed up with a group of faculty from the life sciences to use undergraduate research as a vehicle for teaching and learning.

The following describes how a capstone requirement created an environment that allowed a group of mathematics and computer science faculty to work with colleagues from science to create a common, coordinated summer research experience program and academic-year follow-up activities that enhance both student and faculty scholarship at Truman. Involvement in these activities has provided students with a unique way to better understand the nature of mathematics and what it means to do research in the mathematical sciences.

Received by the editor December 1, 2006.

2. Undergraduate Research in the Mathematics Curriculum

Like many American mathematics programs, Truman's major degree program requires its advanced students to complete a Senior Integrating Capstone Experience before graduation. The requirement has a student work with a faculty mentor on an independent exploration of a mathematical topic that synthesizes information from multiple sources. The effort results in a locally peer-reviewed paper and a public oral presentation to peers and faculty.

The requirement began to be instituted campus-wide around 1986 as part of Truman's "Degrees with Integrity" initiative [3]. It was intended to increase the coherence of major programs at Truman and to allow disciplines to enhance important collegiate skills such as higher order thinking, writing, speaking, and reflection in the major. Each discipline approached the requirement differently, with some requiring their majors to take a special capstone course and others having their students do more self-directed work. While the structure of the mathematics capstone is suggestive of a research-like experience, the mathematics faculty did not view it as such. In practice, however, Truman's capstone requirement in mathematics was and is an instantiation of undergraduate research in the major curriculum.

Faculty assess the educational outcomes of the capstone through senior exit interviews. These interviews show that, although initially students find the mathematics capstone formidable and overwhelming, after a short time of reflection upon the experience once it's complete, they view it as one of the most positive experiences in the program. This attitude persists in alumni.

By early 2001, Truman's capstone requirement had allowed several mathematics faculty to cultivate skills in mentoring students in independent, research activity. In 2002, the author was working with a mathematics major on a bat echolocation question posed by a biology colleague. The student's results were interesting, and he submitted an abstract to the 2003 NCUR program to get more experience giving public presentations. The abstract was accepted along with dozens of others from Truman students, and when the author traveled to the conference with all the Truman students, he was astounded at the quality of their work. The Truman biology students were particularly impressive. Their projects were thorough and advanced, and their presentations were very professional. It was then that a group mathematics faculty reached out to the biology faculty to learn to more effectively mentor undergraduates in high-quality research.

3. An Experiment in Interdisciplinary Training

At that time, professional societies were abuzz with talk about the growing importance of the mathematical sciences to the life sciences. A month didn't go by without the MAA, the AMS, or SIAM publishing an editorial or review article on the topic, and the National Academies of Science published *Bio2010* [1]. While a group of Truman mathematics and computer science faculty were weighing the implications of this national buzz, we learned that some of our students were out in front of us on this; they were already actively pursuing double majors in biology and the mathematical sciences to take advantage of opportunities in, for example, bioinformatics. It didn't take much for a group of mathematicians and biologists to convince ourselves that together we needed to catch up to our students. In 2003, we submitted a proposal to a pilot NSF supplement program called "Interdisciplinary Training for Undergraduates in Biology and Mathematics" (UBM) that allowed us to establish Truman's Mathematical Biology Initiative. Through this initiative, we started to develop new courses for students and opportunities for faculty to immerse themselves in mathematical biology.

Truman's initiative's main goal was to cultivate long-term interdisciplinary collaborations between faculty in biology and faculty in the mathematical sciences through the use of high quality, faculty mentored undergraduate research experiences. These experiences would also serve as a tool for training students to work at the intersection of the life and mathematical sciences. The grant allowed us to fund two research teams consisting of a cross-disciplinary pair of faculty (one from biology, one from the mathematical sciences) and a cross-disciplinary pair of undergraduates to work over the summer. The work of each team focused on a question in mathematical biology, and the grant allowed the researchers to travel to professional conferences to present the work and learn about the growing importance of interdisciplinary collaboration between professionals in these fields. In this way, we aimed to use undergraduate research experiences to train students to interact and collaborate with experts in fields outside of their own.

During the first summer, two teams worked independently on very different projects. Students were recruited by the individual faculty mentors. Two students worked with a statistician and an ecologist on building and evaluating habitat suitability models for the Missouri Bladderpod, a federally endangered plant native to Missouri. This plant is a research focus of the ecologist, and he had built a gold-standard data set through many years of field work. The second team consisted of more students and a group of five faculty members, two from a local medical school and three from the mathematical sciences department. They worked on an image analytic approach to understanding a phenomena in anatomy and another in physiology.

Both teams made reasonable progress toward their scientific goals, with the image analysis team producing two senior capstone projects and the bladderpod team adding to an already impressive data set and doing some predictive and explanatory modeling that impressed the National Park Service. From a program perspective, we learned two important lessons. First, mentors from different disciplines must be in continual communication about their project and the students they are mentoring; opportunities and expectations for this communication must be built into the program. Second, putting students from the mathematical sciences into the field or at the laboratory bench with their biology collaborators is a priceless learning experience that helps them respect the work it takes to create a good data set.

The following semester, the Mathematical Biology Initiative ran a public bi-weekly seminar on the topic of mathematical biology. For each meeting, the seminar invited a research-active biology faculty member to put on a "research fashion show" to talk about his or her research program. The mathematical scientists in the audience would listen and share thoughts on mathematical or computational approaches that might assist them in their work. Each meeting had 20-25 attendees (faculty and undergraduate). The topics spanned the range of biological scales, and many mathematical and computational ideas were brought to the attention of presenters. The organizers of the seminar quickly recognized the potential to expand the Mathematical Biology Initiative and submitted a grant proposal for the 2004 UBM solicitation. The proposed program, titled "Research-focused Learning Communities in Mathematical Biology" (RLC), built upon the faculty and student interest generated by the seminar, and the knowledge we gained through the first summer of interdisciplinary research with undergraduates.

4. Research-focused Learning Community in Mathematical Biology

The new program differs from the previous in four ways. First, each year it allows Truman's mathematical biology program to support four interdisciplinary research teams of faculty and undergraduates. The NSF does not allow the grant to fund faculty or students from computer science, which is a great disappointment. Fortunately, the Truman administration recognizes the importance of computer science to our efforts and funds the participation of one faculty member and two students each year for the duration of the grant. Second, the grant supports long-term research experiences for each team. Work continues through a calendar year or more. Third, students are selected for the RLC program through a competitive application process that's widely publicized around campus. They are assigned to teams by the mentors through a consensus process. Fourth, our program mandates regular team and community meetings. During the academic year, teams are expected to attend the biweekly Mathematical Biology Seminar, and during the summer, teams participate in community events and a weekly Mathematical Biology meeting. In this way, the program creates and maintains a sense of community based on shared research interests. We further enhance the sense of community during the summer program by requiring all participating students to live in the same wing of a residence hall so they interact with one another daily.

Everyone in the RLC program participates in Truman's 10-week Summer Undergraduate Research Experience program, which is described in the next section. During this time, the students dedicate all their time and energy to research activity and participation in summer program events. Teams are often in daily contact with one another, and all teams gather for a weekly RLC meeting during which everyone articulates short-term and long-term goals for themselves, shares updates on progress toward their team goals, and asks questions or asks for assistance. These meetings create program community and identity, and they allow students to practice communicating with a multi-disciplinary audience.

We evaluate the program by looking at program products and where its student go after graduation. At last count, our students have given over 28 presentations or posters at regional, national, and international professional conferences. We have 12 manuscripts in preparation for submission to peer reviewed journals, 3 accepted in peer reviewed journals, and several others submitted. Four grants have been submitted by teams to support their work, and two of those grants have been funded. Of the 10 RLC participants who have graduated from Truman, 8 have gone on to interdisciplinary graduate programs. Insofar as the intended program outcomes include the training of students for interdisciplinary work and an elevated level of interdisciplinary faculty scholarship, that above data suggests that the program is working.

The RLC program has taught us that an extended research experience dramatically improves team productivity. Team members can spend the spring semester reading, perfecting lab and computer techniques, and planning experiments. They can also plan to spend the fall reflecting and writing about what was learned during the summer. This leaves summer as a time that can be entirely and without apology devoted to carrying out planned research activities. In addition, the extended involvement of undergraduates allows them to be peer-mentors for and help train the next cohort of students that will be involved in the team's work.

5. Talent Expansion Through Summer STEM Learning Community

At about the time the mathematical biology group received funding for its RLC program, Truman's Division of Science and Division of Mathematics and Computer Science received funding from the NSF's *Science, Technology, Engineering, and Mathematics (STEM) Talent Expansion Program* (STEP) to use high quality, faculty mentored undergraduate research experiences to improve student success in STEM courses and increase the number of students who graduate with a baccalaureate degree in a STEM area. Students from backgrounds underrepresented in STEM areas, from community colleges, and from other at-risk groups would be targeted for the program.

The STEP grant established Truman's "The Next STEP" office which coordinates grant related activities, including a bridge program for transfers students in STEM areas and some interdisciplinary curriculum development activities. Its centerpiece activity is the 10-week Summer Undergraduate Research Experience (SURE) program. Through activities intended to foster student personal, professional, and academic growth, the SURE program creates and nurtures a multi-disciplinary community of mentors and students united by their shared interest in STEM research. Prior to the STEP grant, many Truman STEM faculty and students were involved in summer research, but activity in each STEM discipline occurred in isolation from activity in the others. The new "The Next STEP" office and the SURE program brought the disciplines, including mathematics, together for mutual benefit.

To bring faculty in all STEM areas together, each summer the STEP grant provides research stipends for 20 students and their faculty mentors. Stipends are awarded to students through a competitive application process that includes an on-line application and letters of reference. As part of the application, student rank order four projects from among approximately fifty listed by Truman faculty. Student applications are ranked by an anonymous, multi-disciplinary panel of faculty, and the most meritorious students are chosen to be assigned to projects that appealed to them. The review criteria emphasizes the importance of choosing students who are in their first- or second-year, students from backgrounds that are underrepresented in STEM fields, or students who are otherwise good candidates for expanding the talent pool in STEM fields. The final cohort of participating students is selected to insure that all STEM fields are represented and that underrepresented students are included.

Undergraduates who receive a stipend live on-campus with other program participants and students in the RLC program. They attend all SURE events such as a weekend field trip to visit industry partners in St. Louis or Kansas City, weekly community lunches in the dining hall, and weekly workshops that deliver information and training to help students get the most out of their research experience and to understand the value of a STEM degree. Examples of some workshop and discussion topics are: Using LaTeX, Writing a Research Proposal, Giving an Oral Presentation, Reading the Primary Literature, Image Analysis using ImageJ, Ethics and Responsible Research Conduct, The Importance of Writing and Submitting Papers for Peer Review, Data Analysis with R, and Women and Underrepresented

Groups in STEM. Thanks to funding from the academic divisions, all STEM students involved in on-campus research over the summer are allowed to participate in SURE program activities. This has allowed these NSF programs to affect a much larger group of students than those directly funded by the programs

Participating in this program has been particularly good for mathematics students. By living and working with students from the sciences, they grow to understand both the role mathematics plays in the sciences and the merits of pursuing questions in "pure" mathematics. What's more, they can articulate these ideas to non-mathematicians. The faculty who participate in the program by attending the workshops and discussions also grow to have a better understanding of the sciences and how the academic and political cultures differ. Involvement has also mathematics faculty to make professional connections with faculty in science that are having mutual scholarly benefits.

We have a comprehensive assessment plan for the program. The program aims to increase the number of students who graduate with a baccalaureate degree in a STEM area, and because this program targets younger students, we have not yet seen a change in the numbers. Our end-of-summer evaluations are very positive, and data from our participation in the HHMI Survey of Undergraduate Research Experiences II (SURE II) [2] suggests our program compares well with others nationally. We also assess the growth of students through analysis of electronic journals they keep over the summer. The results of those analyses have provided us with important information about the student experience that improve mentoring.

6. Conclusion

Truman's thoughtful approach to creating multi-disciplinary research-focused learning communities has benefited mathematics faculty and students in many ways. Student enthusiasm for science and mathematics has increased, as has faculty understanding of disciplinary differences and similarities. Cross-disciplinary interaction has opened new lines of communication that are inspiring curricular reform in interdisciplinary directions. The opportunities created by these programs, and the long-term and short-term benefits to faculty and students make Truman an especially exciting place to be a teacher-scholar in mathematics.

References

[1] National Research Council. *BIO 2010: Transforming Undergraduate Education for Future Research Biologists*. Committee on Undergraduate Biology Education to Prepare Research Scientists for the 21 st Century. National Academy Press, Washington, DC, 2003.
[2] David Lopatto. Survey of undergraduate research experiences (SURE): First findings. *Cell Biology Education*, 3:270–277, 2004.
[3] Northeast Missouri State University and American Association of State Colleges and Universities. *In pursuit of degrees with integrity : a value added approach to undergraduate assessment*. American Association of State Colleges and Universities, 1984.

DIVISION OF MATHEMATICS AND COMPUTER SCIENCE, TRUMAN STATE UNIVERSITY, KIRKSVILLE, MO 63501
    E-mail address: millerj@truman.edu

Life on Both Sides of the Fence: Mentoring Versus Being Mentored

Mason A. Porter

My mentoring style is an outgrowth of both my rigorous undergraduate career at Caltech and my interdisciplinary scientific training. Caltech undergraduates are expected to drink from the firehose of knowledge, and I try to give the same opportunities to my students. Additionally, to convey an appreciation for science, it is extremely important to impart not only knowledge that is directly germane to a project and how to attack it but also to illustrate just how much wonderful stuff there is to study and some of the places where a student's particular research problem fits into the big picture.

In this article, I will indicate how undergraduate research is organized at Caltech and review my own experiences there. I will subsequently discuss my mentoring history and summarize with a few pithy pieces of advice. I examine several of these points in further detail in an expanded version of this essay [1].

1. Undergraduate Research at Caltech

Caltech's primary research program for undergraduate students is the Summer Undergraduate Research Fellowship (SURF) program, which has been around since 1979 and currently offers research opportunities—both on campus and at the Jet Propulsion Laboratory (JPL), which is run by Caltech—for about 250 Caltech students and 150 students from other institutions each year. The SURF program, a campus-wide research initiative, permeates Caltech's culture. It encompasses all academic fields and is a fundamental part of Caltech's undergraduate education, as roughly three quarters of all Caltech undergraduates students participate in SURF at least once before they leave.

As described at *http://www.surf.caltech.edu/*, the SURF program is modeled on the grant-seeking process. Each student starts by collaborating with a potential mentor to develop a project. Applicants then write research proposals, which are reviewed by a faculty committee that recommends awards. Students work during a 10-week period from the middle of June until late August, earning a salary of $500 per week. During this time, they submit two progress reports that detail unexpected challenges, how their goals have changed from those in their initial proposal, and other similar items. At the conclusion of the program, students submit a technical paper and give an oral presentation at SURF Seminar Day, a symposium modeled

Received by the editor February 7, 2007.

on a professional technical meeting.[1] A draft of the final report is due right before Caltech's fall quarter starts at the end of September and a final version approved by the mentor is due on November 1st.

Every December, the entire Caltech faculty is solicited to advertise SURF projects that they are offering the following summer. As they are received, research opportunities are compiled on a website, which is organized according to Caltech "Divisions" such as "Physics, Math, and Astronomy"). Each opportunity includes a terse description of the problem(s) available and other germane information such as allowed majors; coursework requirements; whether the work is theoretical, computational, or experimental; whether non-Caltech students will be considered, and so on. Students then contact the appropriate professor (or faculty who have not actually advertised projects, which is also often successful) and arrange a meeting. Potential mentors typically have informal interviews with several students for a given project before they decide which one(s) they want to advise.

Caltech's SURF program also includes a variety of extracurricular activities— including weekly seminars by Caltech faculty and JPL technical staff; a participatory discussion series on developing a research career, graduate school admissions, and other topics of interest to future researchers; and social and cultural endeavors. This facilitates interactions between students from other universities and those who attend Caltech, which tends to have a fairly insular student body.

2. My Undergraduate Research Experiences

As a Caltech undergraduate, I undertook three research projects. My work did not yield any publications, but my experiences played strong roles in shaping my mentoring style. I also learned a lot of technical material, how to use LaTeX, how to write scientific reports, and developed strong opinions about what to do (and not to do) as an advisor. Here I will briefly describe my two SURF projects, which both resulted from unsolicited e-mails I sent to professors.

Seeking a career in dynamical systems, I worked in summer 1996 with Jerry Marsden on a project entailing the writing of an expository article on the Hopf fibration and its applications in mechanics. I learned a lot not only about geometric mechanics but also about mathematical exposition and LaTeX. Some of my other lessons were more surprising. For example, I attempted unsuccessfully to get my expository article published on my own. This was a mistake, as I should have instead asked Marsden to be a coauthor, worked with him further to revise the paper, and sought more extensive advice about appropriate journals for the paper. While such misadventures are probably rare for undergraduates, this experience has compelled me to always discuss the publication and dissemination of work with all my research students. I have published papers with several of them, but I let them know very early that this is something we do together.

My summer 1997 SURF project helped me refine my research interests and learn some things one should *not* do as a mentor. I decided to undertake a pure mathematics project in which I was to prove a result motivated by a physical process known as diffusion limited aggregation. Unfortunately, I spent the summer in a constant state of frustration. I did learn a lot and had the chance to read some seminal physics papers, but my advisor was absent from campus virtually

---

[1]There is a seminar day predominantly for non-Caltech students in the middle of August and one for Caltech students in the middle of October.

the entire time and I was horribly stuck with nobody to ask for help. I did not know how to find journal articles on my own, which journal articles I should get, what techniques I should try to learn, etc. However, the physics papers I read were extremely interesting, and I did learn that pure mathematics was not for me. I also learned, by counterexample, additional ways advisors can help. Because of this experience, I always teach my students how to use the research literature.

3. The Transition from Mentee to Mentor

3.1. The Mathematical and Theoretical Biology Institute (MTBI).
My first significant mentoring experience came during the summers of 2000 to 2002 as part of Carlos Castillo-Chavez's Mathematical and Theoretical Biological Institute (MTBI). My role included assisting students with homework problems and acting as an advisor and critic for research projects.

I strongly encouraged the MTBI students to attend research conferences. In 2002, the Society for Industrial and Applied Mathematics's annual summer meeting was held in Philadelphia. We thus arranged a road trip so that the current MTBI students could attend the meeting, where Stephen Wirkus and I co-organized two sessions on mathematical biology whose speakers were all MTBI alums [3]. This "minisymposium" allowed the students to see what people who used to be in their shoes had accomplished, meet with them and discuss both academic and social issues, and experience the usual benefits of attending conferences.

At MTBI, I was known for my tendencies to ask tough questions during presentations and return manuscript drafts containing numerous suggestions written lovingly in red. I know from my own education that academic rigor should be stressed even at the birth of a scientific career. It's also important to ask tough questions in a relatively friendly environment to prepare students for future situations, such as doctoral thesis defenses and conference talks, in which the stakes are higher. On occasion, students would try to finagle answers to my questions, which compelled me to ask even tougher follow-up questions to teach them that that is simply not permissible. I find that many students don't appreciate that it is acceptable, and in fact preferable, for them to give an honest answer of 'I don't know.' and (ideally) to ask the questioner to discuss the matter further offline.

My first experience as the primary advisor on an MTBI research project was a bit rough around the edges. My regular meetings with my students were reasonably productive, but I had a personality conflict with one of them. My initial expectations for her were too high, and she did not appreciate my attempts to push her. Thus, while my belief that one should have high expectations for every student was untempered, I realized that what actually constitutes this varies greatly from student to student. Accordingly, I revised my advising goal as one of making sure that students left my charge in a more advanced state than when they entered it. For some students, this means a peer-reviewed publication, but for others it may be as simple as understanding (and duplicating) some calculations, proofs, or numerical computations from a book or article. I also learned that it is essential that I show my frustration only when I think a kick in the butt will help the student. I still experience considerable frustration when advising many of my students, but I am purposely better at showing it only when I have judged that it's what the student needs to see. Finally, I learned that for me to be a good mentor, it is imperative not only that the students be interested in the project but that I am as well.

3.2. Georgia Tech's VIGRE Program. I began advising projects at Georgia Tech in summer 2003 through the math department's REU, which was funded through the department's National Science Foundation VIGRE grant. When a student first contacts me about research, we meet in person so I can ascertain his/her background and interests. Many students are ready to undertake research without extensive coursework, and I like to be able to accommodate that. (Some of my best students initially approached me during their freshman years.) Alternatively, one can adjust the scope of projects to accommodate students who haven't had as many courses. Progress might be slower, but the first term of a project typically entails more time learning background material than producing original results anyway and student projects that lead to publications almost always last longer than one term. During our first conversation, I also encourage my prospective advisees to contact my past students. Most students seem to be ready to sign up on the spot, but I want them to get into the habit of seeking advice.

My ideas for student projects come from several sources. One of the first I advised, on modeling bipolar disorder, was an extension of an MTBI project. Another, entailing the construction of a graphical user interface to simulate billiard systems, was motivated by a question that arose during a seminar at the Mathematical Sciences Research Institute. Other projects have developed through traditional means such as exploration of the research literature. Many of these have been in network theory, a subject that is particularly suited for undergraduate research.

To facilitate group meetings, establish collaborations, and delve into new research areas, I occasionally co-advised student projects with Georgia Tech faculty. Starting in summer 2003, for example, I co-advised a pair of students who were studying two different networks but needed to learn similar concepts. I found that students typically advance much faster when they are working on similar projects and can bounce ideas off each other. Assigning two students to the same project can be beneficial as well, but I think it's much better if their projects overlap rather than duplicate each other. Additionally, the presence of two advisors with complementary skills is wonderful for both students and mentors. In terms of my own research program, I had learned some network theory in graduate school but had never conducted research in it and was very keen to do so. These student projects were my first forays into network theory (in which I have remained active). My subsequent research groups have included students from multiple majors and faculty from other departments, allowing my students to be introduced not only to research but also to interdisciplinary collaboration. For example, one group included a biology professor, a math professor, an electrical engineering major, a math major, and me.

Motivated by Caltech's framework, I insisted that my students write final reports that we would then polish together by working through several drafts. (I also insisted that they use LaTeX to write these reports and provided them with a tutorial for first-time users that I had written while at MTBI [2].) This not only improved my students' communication skills but also allowed us to better appreciate the gaps that still needed to be filled when it came to possible publication. I also arranged for my students to give short talks in group meetings of Georgia Tech's Center for Nonlinear Science (with which I was affiliated) and to present posters and talks at conferences.

3.3. On the Other Side at Caltech. I have continued my student mentoring as a Caltech postdoc. The 2006 version of my research-opportunity announcement included the following information:

```
Projects in Nonlinear Dynamics and Complex Systems

Michael Cross, Professor, Department of Physics

Mason Porter, Postdoctoral Scholar, Department of Physics
and Center for the Physics of Information

Majors: Any major is good, but students in applied math, physics,
and math are likely to be especially interested in these projects.

Prerequisites: It depends on the specific project. Some require more
than others. I will work with the student's background to design something
appropriate. This should not be considered any sort of obstacle.

Type: Theoretical and/or Computational

Note: This is an on-campus SURF. Caltech students only.

Nonlinearity and complexity abound throughout science, nature, and
technology, as their understanding helps to provide explanations of
myriad phenomena---including synchronization of flashing fireflies and
lasers, chaotic motion in double pendula, the formation of patterns in
chemical reactions, species co-existence in plankton populations,
correlations between political ideology and congressional committee
structure, and chaotic dynamics in both classical and quantum systems.
In this project, the student(s) who work with me will work on some
mathematical modeling of some phenomenon (to be discussed in private
communication) using computational and/or analytical techniques. To get
an idea of the types of things I like to study, please see
www.its.caltech.edu/~mason/research or drop me a line. Possible projects
include ones involving Bose-Einstein condensates, complex networks (such
as congressional networks), quantum chaos, billiard systems, pattern
formation, synchronization, and others.
```

Several of these items require some explanation. First, a member of the Caltech faculty (in this case, my postdoctoral supervisor Michael Cross) must be listed as the official mentor. Caltech asks mentors to list "allowed" majors, although I instead indicated who was more likely to enjoy my projects. I purposely included several project ideas with the intention that the particular projects students chose would be functions of not only my interests but also theirs.[2]

One change I have made in my advising since returning to Caltech has been to establish group meetings along the lines of what is perhaps more familiar in physics departments than in math departments. Hence, in addition to meeting

---

[2]One of my past students was sufficiently advanced that I asked him to come up with his own project, which proved to be very successful.

individually with each student twice a week (for roughly 30 minutes each time) and communicating over e-mail and instant messaging, my students and I meet collectively for about an hour every week. In a typical meeting, two of my students will present material in front of the whole group, the others are expected to ask questions, and I will occasionally speak up when I want to highlight a particular point (which may be related to science, exposition, or both). In some meetings, students practice more formal oral presentations or read and discuss each others' written reports. I specifically instruct my students to ask tough questions and make critical comments, and I expect my them to be similarly critical when reading their peers' papers. While my purpose with these exercises is to improve their oral and written presentation skills, I think the students also might learn a bit about peer review as a byproduct. Once my students have given comments, I add a few of my own, and I accentuate the previously-raised points with which I particularly agree.

The final SURF reports are supposed to be written in the format of an appropriate scientific journal. Caltech's SURF office recommends *Nature* as a default but leaves the final decision to the research mentors. I discuss publishing issues with all my students (even the weaker ones) and select a journal style that is appropriate for their particular project and situation. My past journal models have included both broadly-oriented venues such as *Nature* and the *Proceedings of the National Academy of Sciences* and archival journals such as *Physical Review E* and *Chaos*. As with the initial proposal, I go through four–five drafts of the final report with each student.

4. Conclusions

As I have discussed at length, my undergraduate research experiences at Caltech and my interdisciplinary training have fundamentally shaped my mentoring style. Students obtain an optimal research experience when expectations are high (but not too high). It is extremely valuable for students and faculty from multiple backgrounds to interact regularly, and weekly group meetings provide an excellent supplement to one-on-one meetings by reinforcing this and other boons. From a selfish perspective, I have found advising student projects to be an ideal means to enter new research areas. Although it can be very frustrating at times, my close involvement with talented students is perhaps my favorite academic pursuit.

Acknowledgements

I gratefully acknowledge Stephen Wirkus and Ed Mosteig for reading an early version of this manuscript and giving me numerous helpful suggestions.

References

[1] M. A. PORTER, *Life on Both Sides of the Fence: Mentoring Versus Being Mentored (Extended Version)*. physics/0611046, 2006.

[2] M. A. PORTER, *A Hitchhiker's Guide to LaTeX (or how I learned to stop worrying and love writing my dissertation)*. http://www.math.gatech.edu/~mason/papers/draft/lala.pdf, 2002.

[3] S. WIRKUS AND M. A. PORTER, *SIAM hears from next-generation mathematical biologists at Philadelphia meeting*, SIAM News, 35 (2002).

DEPARTMENT OF PHYSICS AND CENTER FOR THE PHYSICS OF INFORMATION, CALIFORNIA INSTITUTE OF TECHNOLOGY, PASADENA, CA 91125, USA

E-mail address: mason@caltech.edu

Math 485: Seminar in Mathematical Problem Solving

Steven J. Winters

Mathematics majors at the University of Wisconsin Oshkosh are required to take a problem solving class. Math 485: Seminar in Mathematical Problem Solving is a course that meets this requirement and is offered annually during the spring term. It is considered a capstone course for mathematics majors. I have taught this two-credit course twice in the last three years. It meets once per week for 14 weeks. The first class had 11 students enrolled and the second class had 17 students enrolled.

This course is also taught by other faculty. Faculty have the option to teach this course alone or to team teach the course. Additionally, faculty have the freedom to design the seminar using whatever problem solving methods they choose. The topics vary by the instructor. I have chosen to focus the design of the class with a research basis to achieve the problem solving outcome competency.

My goal for the students is that each will discover within themselves their abilities, creativities, and desires to engage in mathematical research. The primary objective of the course is to introduce students to an area of mathematics that they may not be familiar with and lead them to discover how problem solving can lead to research. In this case that area is graph theory.

The first few class sessions are spent learning general graph theory concepts. Once the students have an understanding of graph theory I begin to introduce my own research topics to the students as a way of funneling them toward the actual research project they will work on as the final class project.

Generally, three areas of my research interests are chosen as the direction for the course. The students work individually and in groups during class and on homework assignments. The topic areas are introduced one at a time and designed to build students' skills in both solving and designing research problems. As they progress through the homework assignments they learn how to introduce variations to the problems and attempt to solve them as a means to learn how to come up with new research questions.

Received by the editor December 18, 2006.

> **For example:** One topic covered is on distance related sub-
> graphs of graphs such as the center, median, and periphery. A
> well known problem in this area is the fire station problem. In
> class we talk about recent results of the topic and think of new
> questions. I give them a variation such as; changing the defini-
> tion of distance. The students try to solve the problem with this
> variation. They also try to introduce their own variations.

These exercises build student confidence throughout to be able to introduce their
own variations and solve them or attempt to solve increasingly difficult variations
that I create.

> **For example:** One student's variation was based on a pizza
> delivery restaurant where she worked. The restaurant manage-
> ment defined distance as delivering two pizzas and returning to
> the restaurant. They call this the two-stop and return distance.

The final research project is a group project based on one of the three topics that
they have the most interest. Class time is used as a time for the groups to discuss
what they have been working on. I meet with the students individually or in groups
at least once per week outside of class to give individual attention on their research
project. I use this time to check their work and guide them in the research. I may
present them with additional questions based on the direction and progress they
are making.

To date this course has resulted in one published paper, two papers submitted for
publication, and five papers in progress.

Other research projects:
In addition to this course I have had the opportunity to work with undergraduate
students on research projects through various grant opportunities. The University
of Wisconsin Oshkosh has a Student and Faculty Collaboration Research Program
that is funded through faculty development dollars. The purpose of the program is
to develop student researchers. The grant provides a monetary stipend of $2500 to
the student for participating in the project. The faculty receives a $500 stipend to
be used for supplies needed as part of the research.

The University of Wisconsin System also has a research program called WiscAMP
(Wisconsin Alliance for Minority Participation) which has also given me the opportunity to work with undergraduate students on research. The program was
established in 2004 and one of the goals of the program is to increase the number of baccalaureates awarded to underrepresented minority students in STEM
disciplines (science, technology, engineering, mathematics). The funding source is
different than the Student and Faculty Collaboration Research Program; however,
the elements of the program for myself and the undergraduate student are very
similar. The student receives the grant award of $3000 and we work together on
a research project during the summer. I receive a $500 stipend for this project as
well, to be used for the purchase of supplies needed for the research.

MATHEMATICS DEPARTMENT, UNIVERSITY OF WISCONSIN OSHKOSH, OSHKOSH WI, 54901
E-mail address: `winters@uwosh.edu`

Part IV

Summaries of Conference Sessions

Helping Undergraduate Students into Graduate School

Murli M. Gupta – Facilitator
Joshua D. Laison – Scribe

We discussed several elements of a successful graduate school application.

Mathematics Subject GRE Scores

Many graduate schools do not consider a student's mathematics subject GRE scores to be a big component of their application; in fact, several graduate schools do not require them. Most schools value other elements of an application above GRE scores, and the more information the school has about a student, the less important the student's GRE scores are. It is generally true that the more mathematics courses a student takes, the better the student will do on the mathematics subject GRE. For this reason, students should wait as long as possible to take the exam. On the other hand, it looks bad on a graduate school application when a student who has taken many mathematics courses still receives a low score. In general, it is difficult for a graduate program to weigh mathematics subject GRE scores very highly. For example, since the GRE emphasizes pure mathematics, students who are more interested in applied mathematics do not perform as well. One member of the discussion felt that American students perform worse than non-American students on the mathematics subject GRE. At least one graduate school represented in our discussion values the general GRE score, and more specifically the verbal GRE score, more highly than the mathematics subject score. For schools that do require mathematics subject GRE scores, there is a wide range of opinion about what constitutes an acceptable score, from below the 30th percentile for some schools to above the 70th percentile for others. Of course not all graduate programs were represented in our discussion, and probably the range is even larger.

There was some debate about whether an applicant should avoid sending their mathematics subject GRE scores to schools that don't require them if they determine that their scores are below par. Some members of the discussion group felt that mediocre mathematics subject GRE scores can only hurt an application, and students should not send them unless they're required. Others felt that students should not attempt to manipulate the system, and they should send them along if they have taken the exam.

Received by the editor December 1, 2006.

Recommendation Letters

There was uniform agreement that recommendation letters are the most important part of an application to graduate school in mathematics. Students should get to know their professors well so that their professor will have a lot of positive things to say in a recommendation letter. If a student has participated in a summer research program at a different institution, getting a letter from the director of the program, or the student's advisor in the program, is essential for a strong application; if readers of an application notice that the letter is missing, they may infer that the student's performance in the program was unsatisfactory.

A recommendation letter should be no more than 2 pages long, and should talk about the student specifically. It is generally bad form for the director of a summer research program to include more than a brief description of the program and research the student has done. If a longer description is required, members of the discussion group suggest including it as an appendix to the letter, or as a link to a description on a website.

A recommender who writes many recommendation letters and rates every student very highly becomes remembered as such, and their letters are discounted. Recommendation letters should always be positive, but also provide specific and honest information about the particular student under consideration. The opinion of a faculty member who is known for writing honest letters is weighed very highly. Having said that, recommenders must be very careful not to make statements in a recommendation letter that can be construed as a negative comment about the student. For example, if a letter describes a student as "hard-working" without also specifically mentioning mathematical talent, readers of the application might infer that the student is weak. Recommenders should be careful not to accidentally discredit their students in this way.

Cover Letter

The group spent some time discussing whether a student should specifically mention faculty members they wish to work with in their letter of application. We agreed that students should not list faculty members in their letter based solely on their field. A professor might not be accepting graduate students, or turn out not to be a good match for the student, or be close to retirement. At the least, the student should spend time reading the faculty member's research to make sure it's something that she's interested in. It is very useful for a student to know as much as possible about the school and the department, and helpful if they know members of the department, but students should be careful not to ask a professor to commit to advising before they have spent time at the institution and the professor has gotten to know their work.

Best Versus Right

When choosing the school, it's easy for a student to think that he should attend the highest ranked school he's accepted at. However, members of the discussion group agreed that this is generally a bad idea. Students should seek out a school that fits them, a school that is "right" for them, and should resist pressure to attend the "best" school in terms of rankings. Students should talk to a large number of faculty and graduate students in the department to find out if graduate students are happy there, and talk to them one-on-one, so that graduate students can speak honestly about their experiences. In a worst-case scenario, when the student determines that a school is a bad choice only after they begin as a graduate student, they can always transfer to a different school later on.

Support, Managing, Philosophical

Cindy Wyels

Group 3 discussed challenges inherent in promoting undergraduate research that fell into three broad categories: support (both funds and people), managing REU programs, and philosophical or attitudinal.

Support (funding)

Jim reports that the typical NSF grant to run an REU does not stretch sufficiently far to cover salaries for graduate assistants or expenses and stipends for local students. His solution is to seek outside funding to supplement NSF funding; he mentions that he has developed a relationship with a helpful person in his university's development office. The group discussed other potential sources of funding, as follows:

- the local community
- the alumni of the institution
- local industry
- the Howard Hughes Medical Foundation (focus on supporting students, in particular 1st and 2nd-year students from minorities underrepresented in mathematics and the sciences)
- writing other grants (e.g. research grants) so as to allow 1-2 undergraduate students to "piggyback" on these other grants to pursue undergraduate research
- NSA (funds REUs directly under the right conditions)

Cautionary tales of approaching outside entities without the blessing of the appropriate entities within the university were told, with the moral of working within the system to avoid causing headaches (for project directors and for others).

Support (people)

Some group members indicate that they had trouble convincing colleagues and/ or graduate students to make the time commitment necessary to participate in summer undergraduate research programs. This difficulty reflected both the perception that the funds available to compensate faculty for their time were insufficient as well as that faculty - particularly at RI institutions - prefer to pursue their own research and plans during the summers. A second difficulty in recruiting student mentors

Received by the editor December 5, 2006.

seems to be that some potential mentors are intimidated by the thought of guiding undergraduate students through research projects.

Managing Undergraduate Research Programs
The workload - particularly the administrative and even clerical details - can be overwhelming. When one group member referred to the work of merely managing an REU as " a full-time job" most of the group nodded knowingly. Facets of this job include administrative work, finding funding, recruiting students and mentors, maintaining websites, carrying out program assessment, tracking program graduates, etc. With the reminder that the goal was to provide opportunities for undergraduates to do research, the question was raised as to whether REUs might not be the best model for reaching this goal. How do we systematically change the way that undergraduate mathematics is taught and learned in the U.S.? "Out-of-the-box" thinking is needed. Examples such as that of BYU's providing small grants to make it possible for 1 - 2 students to work closely with a faculty mentor over the course of an academic year were mentioned. One caveat to this particular alternative is the need to have department chairs and/or administrators recognize the significant time commitment that quality undergraduate research demands of faculty, and to recognize such activities as part of the faculty member's work load.

Philosophical/ Attitudinal
In spite of all the progress achieved in the last twenty years, many mathematics faculty simply do not believe that undergraduates are capable of carrying on research in mathematics. A cultural change is required, but not yet accomplished. This change may be evident at some non-RI institutions.

Moderator: Jim Morrow, University of Washington
Scribe: Cindy Wyels, California State University Channel Islands

CALIFORNIA STATE UNIVERSITY CHANNEL ISLANDS, CAMARILLO CA 93012
E-mail address: cynthia.wyels@csuci.edu

Undergraduate Biology and Mathematics Programs

Anant Godbole, Suzanne Lenhart and Margaret Robinson

Introduction

At the PURM conference, we had a breakout session discussing undergraduate biology and mathematics programs. We wanted to inform the mathematics community about interdisciplinary programs involving research experiences for undergraduates. This includes several REU programs, in particular the programs at Arizona State University and Loyola Marymount University were represented in our group. We also discussed new programs for interdisciplinary research work and curriculum development involving students in mathematics and biology. The particular programs that we discussed were the NSF Interdisciplinary Training for Undergraduates in Biological and Mathematical Sciences (UBM) and the NSF STEM Talent Expansion Program (STEP).

Below we briefly describe the goals of the UBM program and describe a particular program at Murray State University. We also describe the NSF-STEP Talent Expansion in Quantitative Biology program at East Tennessee State University (ETSU).

UBMs

The goal of the NSF Interdisciplinary Training on Undergraduate Biology and Mathematics (UBM) program is to enhance undergraduate education and training at the intersection of mathematics and biology and to better prepare these students for careers in interdisplinary work in these areas. The main activity involves long-term research experiences for balanced teams of students, representing both mathematics and biology. Faculty members from mathematics and biology are involved in these programs as mentors. To influence the academic programs for a broad range of students, curriculum development may be a significant part of the effort. Through a mentored experience, undergraduates will gain insight into the connections between biology and mathematics.

The first of these UBM programs were funded as supplements to existing NSF research grants in mathematical biology and most of the work consisted of the research experiences part of this program for two years (about 2002). Full UBM programs with research experiences and curriculum and program development were

Received by the editor December 1, 2006.

funded under this NSF program starting in 2004. Currently there are 24 UBM programs, and these efforts represent a large amount of current work in undergraduate research in mathematics. See the articles about the programs at University of Tennessee and at Truman State University elsewhere in this proceedings.

UBM at Murray State University

The UBM program at Murray State University in Murray, Kentucky, is titled "BioMaPS," Biology and Mathematics in Populations Studies, and is directed by Renee Fister. The goal is to create a research environment for mathematics and biology students to study the designs and patterns that exist in populations at the organismal and cellular levels. The objective is to equip 25 students over a five-year period with the skills to understand these patterns and to develop accurate models of biomedical and ecological phenomena. The students and faculty work together as research teams on integrative projects, asking innovative questions from both biological and mathematical viewpoints. The projects include modeling of ecological and evolutionary processes relating to fecundity, parameter identification, developmental stability, biodiversity, anthropogenic disturbances, and population fluctuations.

A defining aspect of the project is the collaboration of four female and three male mentors across two disciplines, who have the ability to attract a diverse group of students based on their different but complementary backgrounds. These mentors will also be team teaching some courses, including restructuring the Math Modeling course to have a biological emphasis. The BioMaPS program builds on existing partnerships with Rocky Mountain Biological Laboratory and St. Jude Children's Research Hospital. A minor in mathematical biology is being developed.

East Tennessee State University STEP and Symbiosis Program

During the summers of 2004 and 2005, ETSU ran a UBM program as a supplement to an NSF-CCLI grant. The PIs were Istvan Karsai (Biology) and Jeff Knisley (Mathematics). Sixteen students conducted interdisciplinary REU-style research in a team environment where each project was co-directed by one person from each department, and featured both dry- and wet-lab experiences. This project was overseen by the umbrella organization *The Institute for Quantitative Biology*. UBM activity ceased after the summer of 2005, even though, paradoxically, the projects described in the next two paragraphs *were very appropriate for and could very well have been funded by UBM*.

The overarching goal of the NSF-STEP program is to increase the number of US students pursuing and completing degrees in STEM fields. The majority of STEP programs across the nation have multiple disciplines represented. While many of them do feature undergraduate research, it can be conducted in any of several disciplines. At ETSU, however, our focus is on quantitative biology and the five year (2005–2010) $1M project is accordingly titled *Talent Expansion in Quantitative Biology* (TEQB). The PI is Godbole, while co-PIs are Lev Yampolsky and Hugh Miller (Biology), Jack Rhoton (Curriculum and Instruction) and Jay Boland (Math and Honors Programs Director). Three cohorts of 20 students will start with a bridge program between high school and college. The five week program features a course in computational biology and hands-on projects in computational biology. Students earn a $500 stipend over the summer, and tuition is paid. Students declare a major in either mathematics or biology (the quantitative biology track in each case) at the end of the summer. Over the next four years, participants receive *four distinct* undergraduate research experiences: They participate in a year round freshman lab rotation (the labs may be in math or biology) for which a $1500 stipend is earned. During the summer following their sophomore year, they participate in a UBM-style two month research program ($2500) and we will place them in an external REU or UBM program at the end of their junior year. Last but not least, students will write a senior thesis in quantitative biology.

Reviewers of the STEP proposal had several queries. Most significantly, they wanted to know what the curriculum for the students was going to be. Other members of the Institute for Quantitative Biology are working on this aspect. They were Istvan Karsai (PI), Karl Joplin, and Darrell Moore (Biology); and Jeff Knisley, Edith Seier and Michel Helfgott (Mathematics). Godbole and Moore from the STEP project are senior personnel on the corresponding curricular Howard Hughes Medical Institute proposal, which was funded for $1.7M for the period 2006–2010. The HHMI project, titled *Symbiosis*, features the development of a totally integrated first year and sophomore curriculum in math and biology which will be taken by the second and third cohorts in the STEP program. The curriculum will be redesigned as follows: Biology 1,2, and 3; Calculus 1; Probability and Statistics (Calculus-based); and parts of Calculus 2 and Linear Algebra will be deleted from the curriculum and replaced with an integrated team-taught series of three courses (6 credits each). These will be titled Symbiosis 1, 2, and 3, and the ultimate goal is to blur the lines between our disciplines. More curriculum reform at the upper level is also under way, but is not part of the Symbiosis project.

STEP and Symbiosis are major undertakings and have indeed transformed our two departments and the interactions between them. We hope to report on project outcomes in the next PURM proceedings.

Concluding Comments

A measurement of success of these programs includes the number of students that go onto graduate programs, not just the students who go onto mathematics graduate programs. Training students for an interdisciplinary graduate program is different from traditional training for a pure mathematics program.

We also call attention to a new NSF program in interdisplinary training. The goal of Computational Science Training for Undergraduates in the Mathematical Sciences (CSUMS) is to enhance computational aspects of the education and training of undergraduate students in the mathematical sciences – mathematics and statistics – and to better prepare these students to pursue careers and graduate study in fields that require integrated strengths in computation and the mathematical sciences.

When considering involvement in a variety of interdisciplinary activities with students, we recommend consulting the following two publications:

Math & Bio 2010 Linking Undergraduate Disciplines, editor L. A. Steen, Mathematical Association of America, 2005.

Bio 2010: Transforming Undergraduate Education for Future Research Biologists, National Research Council, 2002.

GODBOLE, EAST TENNESSEE STATE UNIVERSITY, DEPARTMENT OF MATH, JOHNSON CITY, TN 37614

E-mail address: godbolea@etsu.edu

LENHART, UNIVERSITY OF TENNESSEE, DEPARTMENT OF MATHEMATICS, KNOXVILLE, TN 37996-1300

E-mail address: lenhart@math.utk.edu

ROBINSON, MOUNT HOLYOKE COLLEGE, DEPARTMENT OF MATH. AND STAT., SOUTH HADLEY, MA 01075

E-mail address: robinson@mtholyoke.edu

Using Undergraduates Effectively to Further One's Research and Dealing with Faculty Burn Out

Facilitator: Kurt Bryan
Scribe: Sarah Spence Adams

Our discussion group centered on two main themes: first, how to use undergraduates effectively to further one's research program, and second, how to deal with faculty "burn out." In each case we tried to address these issue in the contexts of a traditional summer REU and an ongoing research program using students from one's home institution.

With regard to the first issue, some points raised were

- Properly utilized, undergraduates can really further a faculty member's research program, which is especially helpful at smaller schools with heavier teaching loads. The involvement of undergraduates "kills two birds with one stone," by simultaneously educating students while furthering faculty research.

- It can be helpful to start students off by having them work specific examples, especially via computer experiments.

- Perhaps we should make more use of the group research model from the sciences, in which students commit to joining a research group for a period of time (a year, maybe longer). Such students can make more substantial contributions to the program, and then act as mentors to newer students.

- Students themselves are often capable of posing good research questions that can assist a faculty research program.

On the second issue, that of "burn out," most discussion focused on problem generation. Some points raised were

- Problem generation:
  - The new column in FOCUS, in which research problems suitable for undergraduates will be published, could be a source, but may have some difficulties of its own. Specifically, what's to prevent 10 students from attacking, solving, and trying to publish the same result? It was suggested that perhaps some kind of "wiki" approach, in which various research groups update a central site to keep other researchers posted could be helpful. There was no clear consensus on whether undergraduates should care about being "scooped."

Received by the editor November 25, 2006.

- It can be helpful for a faculty member to attend a professional conference which is not in his/her area, and look for research questions which are more easily accessible to a "beginner" in the area. Sometimes one can ask good questions that even the experts had not considered.
  - Students from one year can (and should) pose questions for future research, which can form the basis for the work of students in a following year.
  - In general it is very difficult for students to pick their own original research problems, but one can try to find problems which have many facets or variations and let students choose which variation to work on.
- Other issues raised were
  - Cycling faculty through an REU program can help reduced burnout and give those involved a break.
  - The NSF doesn't really provide enough funds for REU faculty. The amount of work involved isn't commensurate with the salary one can ask for (within the current budgetary guidelines). This contributes to a sense of burnout, and prevents the program from involving too many mentors.

Getting Support from Colleagues; Graduate Student Retention

Facilitator: Judy Walker
Scribe: Joel Foisy

Group Leader: Judy Walker
Group Scribe: Joel Foisy

We discussed several challenges faced by proponents of undergraduate research. The first challenge brought up was the difficulty some mathematics faculty have had with poor judgments by non-mathematicians on internal grant funding committees. At one school in particular, a good mathematics proposal was not funded, whereas in a different year, a much weaker proposal was funded. Several viable solutions emerged from the resulting discussion. One is to make sure that a mathematician is on the school-wide committee that determines grant funding. Another is to stress applications in the grant proposal. For example, if the research is on algebraic number theory, then the proposal should mention that the National Security Agency uses number theory for encryption and code breaking. Finally, one could include an explanation of what mathematics research is in the proposal in order to education the non-mathematicians (see the MAA CUPM report on Mathematics Research by Undergraduates for ideas, www.maa.org/cupm).

We also discussed the challenge of getting more faculty on board with undergraduate research. Most agreed that in order to get more faculty on board, it is important to have institutional buy-in. To this end, participating in undergraduate research should be linked to rewards and promotion. At doctoral granting institutions, graduate students should be brought in on undergraduate research projects, so that they will begin their faculty careers primed to lead undergraduate projects.

Our third challenge involved helping students have better success in persisting in their graduate studies. One participant pointed out that a large percentage of mathematics graduate students drop out before receiving a doctoral degree. Some participants felt that this was because there is such a different level of challenge in the undergraduate and graduate levels. Many felt it would be good to have more summer programs for students between their senior year in undergraduate studies and their first year in graduate studies. Currently, there are few opportunities available for such students; for example, all NSF-funded REU programs explicitly

Received by the editor October 27, 2006.

disqualify this group of students. Some programs, such as PCMI, the IAS Mentoring Program for Women, and the Director.s Summer Program at the NSA, are targeted to undergraduates but are also open to students between the undergraduate and graduate level. Other national programs, such as EDGE and Nebraska IMMERSE, are specifically designed to help this group make a successful transition from undergraduate study to graduate study. There was not agreement on whether or not there should be more NSF-funded programs like these available for graduate school bound students. Many faculty members at the meeting expressed that they felt that students who were already committed to attending graduate school really did not need NSF support for their summer. They felt that the institutions themselves should provide a transitional program, as they deemed it necessary. Some schools, including the University of Virginia and the University of Iowa, already offer summer courses for their incoming graduate students. Finally, we also discussed the difficulty of recruiting minority students into the mathematical sciences, but by the time we got to this important point, time had run out.

Lack of Respect; Targeting Women; Personality Problems

Francine Blanchet-Sadri

Participants of this group: Francine Blanchet-Sadri (scribe), Dennis Davenport, David Farmer, Joe Gallian, Stephen Hartke, Charles Johnson, Thomas Mattman, Sivaram Narayan, Javier Rojo, Zsuzsanna Szaniszlo, David Uminsky, Michelle Wagner, and Stephen Wirkus (facilitator) discussed challenges they have faced in their undergraduate research activities and how they have overcome these challenges. After introducing themselves and their programs, they discussed mainly the following topics:

- *Lack of Respect* It was mentioned that there is lack of respect for level of research of the program, that some administrative personnel at some institutions even view it as a remedial mathematics program.

  It was suggested to disseminate information such as putting on the web some technical reports and doing poster presentations to change the mentality of those administrators.
- *Females into Mathematics* The question was raised: "Is it a good idea to set aside programs only for females?" Although it may strengthen the number of women participating in REU programs, it may not be a nice choice for females to be only with women.

  Some participants answered that these programs give a chance to a woman even if not as well prepared. Others mentioned that there was a problem to get females in programs that are not targeted only to women. Yet, others said that some of the female programs may not be as strong as some other programs.

  It was mentioned that some institutions have put considerable effort to set workshops where there is a good percentage of women attending. An environment is thus provided where females can hear other females. Women do participate in these workshops, feel less isolated, and feel less pressure from men. Maybe women would be more willing to participate in some programs if the ratio of participation between males and females was closer to 50% men and 50% women. Topics may also influence women to choose a particular program. It was recommended that strong females should be involved in strong programs.
- *Personality Problems* Another topic of discussion was personality issues related to students and research assistants. Collaborative learning is a

---

Received by the editor November 18, 2006.

focus so problems occur with some anti-social students, etc. For instance, teams may end up getting broken by conflicting personalities.

The application process includes mathematical background, personal statements, letters of reference, etc. It was mentioned that the recommenders may indicate personality problems but that the letters cannot always be believed. It was suggested that REU directors should consider contacting recommenders for more information. It was also suggested that phone interviews may help to see if potential participants are "normal" individuals. However, some felt that these interviews could eliminate some very strong applicants. Program directors should talk with these anti-social students, should try to incorporate them, and should try to get other students to connect with them. As for research assistants, it was recommended to try to get some alumni of the program.

- *Application Form* Faculty who write recommendation letters usually attach their letter to the application form (if there is a form). It is too time consuming for them to fill out a form for a particular applicant. So the same letter is sent to every REU program!

UNIVERSITY OF NORTH CAROLINA, GREENSBORO, NC 27402-6170
E-mail address: blanchet@uncg.edu

Uniform Acceptance Date

Gary Gordon and J.D. Phillips

Individual REU sites currently set their own application deadlines, ranging from mid-February to mid-March. (Application deadlines of the REU programs represented by the participants in this session ranged from mid-February to early March. Most of these programs make offers to students within one week of the application deadline.) Some students may need to respond to an acceptance by one program before another program's application deadline, leaving the student in an uncomfortable position.

The majority of the participants in our discussion came into this session thinking that a uniform acceptance date was a good idea. But after a spirited and salutary conversation, we left the session unanimously *opposed* to a uniform acceptance date. In a nutshell, we agreed that the potential benefits of a such a deadline are relatively minor, while the potential drawbacks are less so, as outlined below.

Here were some of the arguments we discussed *in favor* of a uniform acceptance date:

(1) A uniform acceptance date would eliminate the initial problems faced by students who receive multiple offers. This would especially benefit the best students.

(2) More programs would see more of the applicant pool.

(3) Informal surveys indicate that students want a uniform acceptance date.

(4) It appears to have worked well for the NSF post-doc program, so why not here?

The following scenario exposes a potential problem with uniformity that the group could not satisfactorily resolve. If all offers are made a certain date, then any student *not* receiving an offer will know they were not in the first round of students selected. Having students come to a program with this kind of baggage seems to be counterproductive. Currently, it is possible to 'hide' some of the ordering of the applicants (since the applicants don't know precisely when offers are made).

In summary form, here are the arguments *against* a uniform acceptance date:

(1) It would only help the best students. As noted above, students would know exactly where they ranked in the selection process, viz, first choice, second choice, etc., based on when they receive their selection letters.

(2) The chaos it's designed to prevent will necessarily unfold anyway after those first students admitted to multiple programs select only one, and programs then must scramble to make subsequent offers.

(3) It would be difficult (some believe impossible) to enforce uniformity, and there are program directors who do not support the idea.

In spite of our reluctance to recommend a fixed uniform date, we do believe steps can be taken to alleviate some of the chaos we see now. We recommend that programs do not accept students prior to some fixed date. We list some possibilities below:

(1) As mentioned above, set an "earliest possible" date (i.e., programs should set no deadlines *before* this date).

(2) Set a uniform "decision date" (i.e., students must decide by this date which program's offer they intend to accept).

(3) Use "windows" or "bracket" dates. That is, instead of a one-date deadline, perhaps set a *range* of dates for programs to accept the first round of students.

(4) Have students sign a contract or a letter of intent (in response to concerns about students backing out of REU's whose invitation they've accepted).

Finally, we need more data to make any recommendation with confidence. In particular, we don't know how many students who *apply* to at least one REU are actually *selected* by at least one REU. Obviously, this would be pertinent information in deciding whether or not there should be a uniform acceptance date. Therefore, we agreed that it would be good to collect this information, if legal. Perhaps the AMS could arrange this.

Here is a list of the session participants:

(1) Scott Chapman, Trinity University

(2) Dennis Davenport, Miami University

(3) Joe Gallian, University of Minnesota, Duluth

(4) Gary Gordon, Lafayette College

(5) Charlie Johnson, College of William and Mary

(6) Ellen Maycock, American Mathematical Society

(7) J.D. Phillips, Wabash College

(8) Michelle Wagner, National Security Agency

(9) Stephen Wirkus, California State Polytechnic University

DEPARTMENT OF MATHEMATICS, LAFAYETTE COLLEGE, EASTON, PA 18042
E-mail address: `gordong@lafayette.edu`

DEPARTMENT OF MATHEMATICS AND COMPUTER SCIENCE, WABASH COLLEGE, P.O. BOX 352, CRAWFORDSVILLE, IN 47933
E-mail address: `phillipj@wabash.edu`

Working with Local Students

Ermelinda Delavina and Daniel Schaal

This session focused on working with local students. All participants in this session felt that engaging the students at one's own college or university in mathematics research is important and worthwhile. However, this activity has its own unique set of challenges, as well as its own unique rewards. The discussion that this articles summarizes focused on those unique challenges and rewards.

When many people think of undergraduate research in mathematics they immediately think of the well-known REU (Research Experiences for Undergraduates) programs. These programs are funded by the National Science Foundation and are advertised nationally. Highly talented undergraduate students from around the country apply to these programs and the organizers of the REU programs usually have a large pool of very qualified students to choose from. The programs generally run for eight to ten weeks in the summer, so the participants have a considerable block of time to devote solely to research without other demands on their time. The participants in these programs often achieve results publishable in mainstream mathematics journals.

A faculty member that is working with undergraduate students from his or her own institution may face difficulties not faced by the director of an REU. The most obvious of these is funding. Finding funds to compensate students and faculty for their time can be a major challenge at many institutions. Recruiting qualified students can be a problem. A particular institution may not have many students with the academic background needed to investigate some research topics and it may be difficult to motivate the students to try original research. The research may occur during the academic year when both the students and the advisor have many other duties. If the research occurs in the summer the students might need to have summer jobs, so total devotion to the research might not be possible. Under these circumstances, it is often the case that the research does not lead to results publishable in a mainstream mathematics journal. The participants in the session had suggestions for each of these potential problems.

Much of the discussion involved funding issues. The participants all agree that if a student is devoting a considerable amount of time to a project, it is nice to be able to compensate the student financially, especially if the student is working on research in the summer instead of taking on a summer job. Likewise, a faculty member that spends the summer advising students instead of teaching summer

---

Received by the editor December 27, 2006.

courses is also deserving of financial compensation. It was mentioned that there are several possible sources of funding for research with local students. If the advisor already has a grant or is applying for a grant to fund personal research, it is often possible to request additional funds to support one or more undergraduate students to participate in the research. If the grant is from the National Science Foundation, it is possible to request funds for a Supplemental REU to provide support for local students. At many institutions there are several possible sources of internal funding. A research office often has money dedicated to supporting faculty research and research involving students is usually encouraged. Department Heads and Deans might have discretionary money that can be used to support student research. It is sometimes even possible to utilize work-study money where working on research with the faculty member is the student's job. Finally, many local undergraduate research programs function successfully with no funding at all. As long as the students and faculty members are willing to devote their time to the project no funding is needed. This is especially common for research that occurs during the academic year.

It was noted in the discussion that a faculty member might not be able to find local students capable of solving the type of the problems investigated at the REU programs. A general consensus was that independent investigation was beneficial to students at all levels of ability and that faculty interested in advising students should work with the students that they have. If no local students are qualified to work on a particular topic of interest to the faculty member, then the faculty member should choose a topic appropriate for the students. This might mean investigating problems that are not deep enough to be published in a mathematics journal if they are solved. This may mean investigating solved problems where the solutions are unknown to the students or trying to apply a known result in a new way. This may also involve reading some recent journal articles. It was mentioned that the ability to read higher mathematics is a valuable skill not often learned in a traditional curriculum and a skill very beneficial to students considering graduate school. Several participants in the discussion noted that there is a considerable gap between the mathematics taught in an undergraduate classroom and the research conducted at an REU. These participants stressed the importance of projects designed to bridge this gap.

There was a consensus that the types of investigations mentioned above are very valuable for the students involved and that the advisors should choose the most ambitious projects that still give the students a reasonable chance of success. There was however a difference of opinion about what these projects should be called. Some participants in the session felt that the term undergraduate research should be reserved for investigations of original problems of publishable quality. It was felt that when lesser projects were called undergraduate research is could give undergraduate research a bad reputation. Other participants disagreed with this idea. They noted that other disciplines use the term "student research" very loosely. A chemistry student helping to set up a research experiment, a biology student gathering data to be analyzed and a psychology student researching a topic in the library were given as examples of activities often referred to as research in other disciplines. Regardless of what we call it, it was agreed that projects that go beyond what is taught in a typical classroom and require the students to do some independent investigation are of great value to the students and should be encouraged.

A goal at most REU programs is for the students to publish research articles in mathematics journals. While this often occurs when faculty work with local students, this may not be a reasonable goal depending on the students involved and the topics of investigation. It was mentioned that many institutions publish a collection of research articles by the local students and there are also a number of national journals devoted exclusively to undergraduate research. These journals do not have as high of a standard as mainstream research journals and give the students a chance to learn the art of writing mathematics. Another excellent goal is for the students to prepare a short talk based on their investigation and present this talk at a mathematics conference.

Finally, it was mentioned in the discussion that there are several distinct advantages to working with local students. The most obvious is the length of time that a research project can last. While REU programs are usually limited to one summer, a research project with local students could last for a few years. REU directors choose participants based on an application, letters of recommendation and possibly a phone interview, but rarely know the students on a personal level. The potential for a personality conflict definitely exists. Faculty member working with local students can select students they know well and with whom they have already established a good working relationship. This also allows faculty members to tailor the projects to fit a particular students strengths and interests.

Notes on PURM Conference

Philip Matchett Wood

1. Welcome

1.1. Jim Schatz, NSA.

- The NSA is the largest employer of mathematicians in the country. Mathematics has long been useful to the intelligence community and has been a huge success story for it. The NSA needs a healthy mathematics community and thus is interested in the state of math in the US and not just short term recruiting to the NSA. Part of a strong community requires encouraging diversity.
- We should be open to all kinds of mathematics, and not look down upon applied mathematics, which can also be very elegant. We shouldn't limit ourselves.

1.2. John Ewing, AMS.

- Some advice graduate students give to undergraduates: math research is hard so make sure you really want to do it. REUs are important to help students figure out what they want to do.
- Goal of conference is to understand different types of undergraduate research programs.

2. Opening Address

2.1. Joe Gallian: Undergraduate Research 1987-2006.

1986 MAA president said, "Undergraduates can neither participate in nor understand the research of math professors." While this may be true, we can take a better attitude.

1987 NSF starts funding REUs, with expectations of students doing examples, generating data, or doing a literature search. This is now what happens in the first week of an REU! There were 8 REUs, mostly 4-6 weeks, and there was nothing at the Joint Meetings related to undergrad research.

1988 MAA formed an Undergraduate Research Committee (which has been a big influence on undergraduate research). There were 14 REUs.

1989 16 REUs.

Received by the editor October 7, 2006.

1990 The NSA starts the Director's Summer Program. AWM starts the Schafer Prize. 15 REUs.

1991 MAA committee gets poster session at Joint Meetings for undergrads, and also a panel on undergrad research, and a three part special session for research papers by undergraduates (22 talks representing 54 students).

1994 At the Joint Meetings: First MAA minicourse on undergraduate research, and the second MAA poster session for undergrad research (and made an annual event–though starting with small numbers, recently the poster sessions involves hundreds of students and is limited by the room size), and the AMS had another special session for undergrad research (38 talks, 70 students).

1995 The AMS has another special session at the Joint Meetings for undergrad research. Aparna Higgins gives the first Project NExT course on undergrad research. The Morgan Prize is established.

1996 Undergrads speaking at the Joint Meetings are identified in the program. There are 6.

1997 MAA minicourse on undergrad research at Joint Meetings.

1998 Now 171 undergraduates at the Joint Meetings, and 36 doing posters. At the Joint Meetings, there was an MAA session for faculty about undergrad research.

1999 First PURM conference, predecessor to this conference.

2001 MAA adds "undergraduate research" to their mission statement.

2002-6 At the Joint Meetings, things like special sessions for undergrad research and minicourse become annual. 49 REUs, 377 undergrads at JM, 192 students in poster session at JM.

- People with NSF grants can get supplements for undergrads to do research.
- Michael Dorff got an NSF grant to fund academic year undergrad research at various schools.
- FOCUS will have a column with good undergraduate research problems.
- In the right circumstances undergraduates can do mathematics research.
- Reasons for changes: NSF and NSA funding, popularity of REUs, DSP program at NSA, minicourses, VIGRE, CUR (Council for Undergraduate Research) played a big role behind the scenes, people who were in REUs now running their own, more pressure on faculty to do research.

2.2. Aparna Higgins: Undergraduate Research at your own institution.

- REU research is great, but there are benefits of non-REU research done during the year.
- Consider: is a measure of success of these programs how many students go onto graduate school? While REUs may help in this regard, so could in-house research. Maybe REUs just confirm people's interest in grad school?
- Ways to do undergrad research: REUs, honors or scholars programs, senior projects, and one-on-one work with faculty
- Reasons to work with undergrads: introduce them to profession, advance mathematics. These goals are not mutually exclusive.

- Advantages of in-house research: introduces students to profession, it is fine if results are not publishable "The only person that an undergraduate research problem needs to be important to is the student." The experience is valuable, whether the student goes to grad school or into the workforce. More benefits: more time than REU, students can learn more broadly, easier to find a doable challenge, time to tailor the project to the student, no funding required.
- Research experience for undergrads is a pedagogical exercise. Discovery learning is a trend in education. The skills that come out of this are good for people who go into math, people who become teachers, and people who go into industry where they need problem solving skills.
- Let's find out lots of ways to do undergrad research!

3. Panel: The Spectrum of Undergraduate Research Programs

Panelists described current forms of various undergraduate research programs: NSF-funded REUs, other summer undergraduate research programs, minority-serving programs, and academic year undergraduate research programs.

3.1. Erika Camacho, Loyola Marymount University. Runs the Applied Math and Science Summer Institute for minorities. It is important to recruit minorities to have a diverse workforce and give research opportunities to students who wouldn't otherwise have them (as per NSF guidelines). In 2000, 14% of Bachelor's degrees are from minorities; from 1997-2004, 5.7% of PhDs were minorities, and 31.3% were women. We need to not only introduce students to the academic environment of mathematics, but also provide them a good personal and emotional environment to make them comfortable, build their self esteem, show them role models not just in academia (also PhDs in industry). AMSSI unique features: held at two campuses (one and then the other) in a summer–faculty are resident and move also, they recruit from schools without strong math programs, they set dates to accommodate quarter systems, look for potential in students not just good past work. They have weekly group meetings where everyone (including faculty) share their worries, goals, family background, etc. They have a first week math boot camp, 7 days from 8AM to midnight. To encourage self esteem: pick students randomly from groups to present and then give a lot of positive feedback, make them work hard so a lot is accomplished.

3.2. Zsuzsanna Szaniszlo, Valparaiso University. Valparaiso does academic year research as well as having an REU. Math majors are required to take a colloquium every year (no credit first three years, 1 credit hour one senior semester) which meets once a week for an hour and each year has different goals. In the year long projects, they put together younger and older students, and find seniors are often too busy to participate. The goal is a good research experience, not necessarily publication. The project work is divided up some by experience and age: younger students produce examples and counterexamples, middle students produce conjectures, and the older students produce proofs. Every week each student has their own task for the week ahead. Recently, faculty get some teaching credit for participating. Students usually don't participate for four years, maybe freshman year and then again junior year. They recruit by personal contacts.

3.3. Suzanne Weekes, Worcester Polytechnic Institute. At WPI, research is integrated into the curriculum and there is a collaborative atmosphere. They also have an REU, which has involved 200 students, 60 industrial projects, and 30 companies. The motivation is applications to real world and to society, and the problems come from industry partners. They work on communication and working with a diverse team. They have good percentages of women and students not at PhD institutions, but have trouble with minority recruitment. One goal is showing the role of math and industry and working with constraints. Students' interest particularly in their REU is important in the application. They do a phone interview/information exchange. Students are in teams of 2-5 students, meeting with faculty at least once a day. The final reports are to the companies and thus usually not allowed to be published.

3.4. Peter May, University of Chicago. At Chicago 7-8%, of students are undergraduate math majors. Seventy students participate in their REU, closed to seniors because of lack of space (28 grad mentors, 11 faculty mentors). The students take courses and do projects, and also teach high school students or grade school teachers. During the 8 week program, there is a 4 week program for high school students, and a 2 week program for local grade school teachers. The program is very popular and they get more applications than they can take. The goal is not to publish but to get students excited about math. The math only needs to be new to the student. Chicago also has a directed reading program run entirely by graduate students, where they meet one hour a week. Grad students are paid a nominal $300.

4. Small Group Reports: Overcoming Challenges[1]

Participants broke up into 6 small groups, each with a facilitator. In each group, the participants discussed challenges they have faced in their undergraduate research activities, and how they have overcome these challenges.

- How to get undergrads to help your research instead of hindering it? Get them started earlier so they are around for longer and can mentor younger students. They can do examples on computers, with examples of success both in PDEs and in number theory.
- How to deal with director burnout? It is hard to keep finding problems! FOCUS is going to print a column with suggested problems, but it would be bad if lots of groups started working on the same one and competing. There should be a wiki webpage, clearly given in the column, where people can say what they are working on and what results they have on these problems. Try getting problems by going to specialized conferences outside of your area and asking simple questions.
- NSF REU mentors are not paid enough compared the pay for standard NSF research grants. People on NSF review panels shouldn't look down on applications asking for faculty support. You can have support staff or undergrads do some of the administrative work to take the burden off faculty.
- How to keep a program going? Get more professors involved. Have the head of the REU doing minimal mentoring. Pick a faculty successor to

---

[1]Jeffrey Adler contributed to this section.

the director. To get more funding, you could ask alums of the university (sometimes you need to go through the Development office). Another idea–Howard Hughes Medical Institute supports summer research for undergraduates. NSF research grants in the department can help a student or two. If you can't get teaching credit, at least get recognition from the deans. The NSA has some start-up funds, but not continuing funds.

- Get the CUPM document on costs and benefits of undergraduate research programs. It is written to convince administrators. It is available at: `www.maa.org/cupm/CUPM-UG-research.pdf`. It is very important to make sure that the administration knows what we're doing. This came through loud and clear.

- To run undergrad research at a PhD institution, you need external funding, smaller schools don't expect much funding (faculty are more expected to do it as part of their job). Make the university aware of the work you are doing.

- How large should groups be? A lot of people think two students is good. Maybe with three, someone gets left out.

- What happens when students fail to show up at the REU? Or accept and then later turn it down?

- What do you do with unfinished projects? Should you reuse them?

- How to improve communication ability of students? Practicing their talks helps a lot, and instills a value that it is important to give good talks.

- How to get support for academic year research with students? Don't be afraid to use buzzwords. Get on university committees. Toot your own horn.

- What to do the summer between undergrad and grad school? (There is a bridge program at University of Nebraska-Lincoln.) The focus would be different. Instead of promoting graduate study, we are preparing them for graduate study.

- Are women only programs good? They have positives: strength in numbers, good role models.

- Do phone interviews help? Maybe you can weed out students with bad attitudes but some good students don't interview well on the phone.

- Jim Schatz from the NSA says *the NSA can provide a letter explaining the importance of undergraduate research from their perspective.*

- How do we prevent fighting between REU directors over students?

- NSF has a program to fund REUs for U.S. citizens to do research in other countries. Some people hope for funding for non-U.S. citizens in NSF REUs.

5. Panel: Diversity Issues in Undergraduate Research

The panelists in this session discussed the importance of involving minority communities in undergraduate research in mathematics in order to increase their representation in the area. They also discussed effective recruiting techniques and successful strategies.

5.1. Herbert Medina, Loyola Marymount University. 54.9% of PhDs in math in the US are to US citizens or permanent residents (46.2% just to US citizens). These have been decreasing over the last 10 years. 28% of math PhDs are women, 6.2% are black, Hispanic or Native American. US citizens are more likely than non-citizens to take initial jobs in the US. There is a downward trend in math BAs, both among all college degrees, and the percentage that are US citizens. One way to combat these issues is to tap into the resources of the underrepresented groups that may not have the tradition of grad school in their culture and have pressure to get a job immediately, and the mathematics programs aren't working well to develop this talent. Undergrad math degrees to blacks, Hispanics, and Native Americans are on the rise (despite the decline overall), so this is a good resource to focus on.

5.2. Darren Narayan, Rochester Institute of Technology. Narayan ran an NREUP program. His goal is to get more underrepresented groups into PhD programs. NREUPs run 6 weeks typically, with 4 students and 1 mentor ($5000 to program director and $3000 to each student), the students must be *named in the proposal*. The problem used was good because it was easy to describe to others and had applications. Students cooperated to built up examples and the whole team counted down on finitely many cases left to resolve, which was exciting and motivating.

5.3. Ricardo Cortez, Tulane University. Challenges are students with low income and no family or friends who went to grad school or even college. The strategy is to aim for students after their sophomore or junior year, do a research program, and then do a lot of work following-up with them. Talk to them about graduate school and fellowships, make them part of a larger network of mentors and get them tied into the community. For this, groups of three are best. Set the right tone and atmosphere for cooperative learning. Also, it has been good to start with a lot of structure to the program and then reduce it slowly giving students more freedom and responsibility. It is great if students end up working 15 hour days, by their choice. Lead by example and really be there for the students. The program avoids accepting students already on a fast-track to grad school.

5.4. Dennis Davenport, Miami University. To generate applications: go to conferences with minority students, advertise at national and regional conferences, do mass emailings but pick specific people who care about what you do and not just department chairs, contact institutions that graduate lots of minorities (see www.diverseeducation.com/index.asp), and have a good website. Students who apply without a proof based course aren't ready for the REU but they need to be recruited early into mathematics, so we need some kind of pre-REU program to reach out to these students. The top minority students get a lot of offers, but there are still minority students with a lot of potential we can reach out to. Calling people with many offers has been successful to recruit them.

5.5. More comments.
- To recruit minority students, it will take time and effort over years. Use the contacts from this conference to help!
- To try to recruit more low income students, we could ask for more information about this on applications; for example asking students to talk about their background, or asking specifically if they are first generation college.

- Is it better to have minorities in all minority programs or ones with a critical mass, or is it better to spread them out? We need to be finding new students, not competing over the ones we already know about. Look for talented non-math majors.
- When "taking a risk" on a student with less background but perhaps potential, you can call people who wrote letters of recommendation to learn more about the student before deciding to make an offer.
- Pairing weak students with weak students is better for the students than weak with strong.

6. Open Forum

6.1. Faculty Compensation for REUs.

- Almost everyone present has directed an REU, but almost none had local compensation. Someone wrote *local* support into their grant (e.g., 15% locally and 85% from NSF for compensation, in trade they took $\frac{1}{3}$ of their students from the local university). Some schools already have programs in place for local students to do summer research, and the REU can be tied in with this.
- At one place, students were enrolled in "summer classes" as extension students, the grant paid the tuition, and then the mentors were paid by the school as teachers.
- Should we ask for more from the NSF? (Yes, is the consensus.) It is more work than doing your own research, but yet you would get paid more to do your own research. Some people have successfully written in two months of support into their REU grant. Someone on a grant panel pointed out that review panels will notice if you ask for more support, so you should explain why.

6.2. Have REUs caused the quality of entering grad students to improve?

- Some people have noticed the quality of entering grad students improving.
- More students are now going to REUs since there are more REUs. REU applications are getting better, at least anecdotally. This should be documented.
- Someone says grad students who had been to REUs had an easier time in grad school. Can this be documented? Someone else says REU alums have a harder time because PhD problems take longer and more background.
- Lacking a recommendation from an REU director (whose REU you attended) is a red flag on a student's grad school application.
- When an REU director is writing a recommendation, it should concentrate on the student, not the REU. It shouldn't say too much about the REU, and should speak first about the student, mentioning the REU only at the end.

6.3. What is the right size group for students at an REU?.

- Students can work alone on related problems and then discuss them a lot. It is important to still develop communication skills even if students are working alone.

- It is good we have many models that work for different students, e.g. group and individual work.

6.4. Adding undergraduate research to the curriculum.

- If we add undergraduate research to the curriculum, how can we justify it (and what will be cut out) to other faculty? You could just add research without cutting anything, making the program harder for students. You could try to incorporate research projects into the regular classes, but that will necessarily reduce course content. Maybe if students are going into industry, a research project would be more useful than some of the standard curriculum.
- Some state schools have a revenue generation model for deciding what gets taught. This is bad for undergraduate research.

6.5. Are the numbers of Master's degrees in math going down, as the PhDs are? Is this important? Should we consider the mathematics community we would like our students to join to only include PhDs, or Master's also? Yes, we care about the students with Master's. The NSA likes REU students even if they don't have PhDs.

6.6. How do we define undergraduate research? A student has to produce something credible *to the student*. Students get "beat up" in math classes, in research they come up with new answers together. The MAA Committee on undergraduate research has a document defining this, which is now available online.

6.7. If an REU student becomes a high school teacher... it is great! They can teach with that perspective.

6.8. What can you do with half finished projects? One student later used them while he was tenure track (i.e. students can finish them later on their own). The mentor can finish them. The REU mentor could find a professor at the student's school to help the student finish the project.

6.9. Where do we get models for undergraduate research? We should try looking at other sciences and adapting their models.

6.10. Announcements.

- Michael Dorff at Brigham Young has a $1.28 million NSF grant to give out mini-grants to do in-house research. They buy out a faculty course for $5000 and give $3000 to the student. email: `mdorff@math.byu.edu`
- July 16-19 there will be a NSA sponsored conference in Duluth Minnesota with colloquium style talks.

7. Panel: Assessment of Programs

The speakers summarized information gleaned from assessments now available to determine the impact that research programs have had on students. The assessment techniques that some REUs now use were presented.

7.1. Michelle Wagner, NSA. NSA funds 14 REUs, with a strong minority focus (8 of the 14 have significant minority participation). To assess, the NSA does site visits and performance reports. They want to see objectives achieved and summaries of student projects. The site visits also help develop relationships with students and faculty.

The NSA did a survey asking REU directors for information on previous students from 1988-2005: where are they now, what degrees did they get, what jobs are they in. (Note: this does not include DSP.) The data has a lot of missing points, and some students appear twice. The data show 59% female and 41% male, though if you take out the two women only programs it is 49% female and 51% male. They are very balanced racially, including 26% Hispanic/Latino, which a few programs are responsible for. The graduate school data is 45% unknown and 35% have gone to grad school in math. Of the 527 in math or statistics graduate school, 183 are minorities and 187 are women (a drop from the REU participation). In the future, the NSA would like to get more and better data, and get more underrepresented groups in REUs.

Question: How can REU directors keep track of students? Answers: email, Google, forced exit surveys, yearly questionnaires, ask other students. You can ask the NSA for administrative assistant funding to do this and other administrative work in your grant.

Comment: one reason it is harder to track female REU students is some of them change their names.

7.2. Frank Connolly, University of Notre Dame. The AMS did a survey of 411 REU students from 1997-2001, with 262 responses. The respondents were 55% male and 44% female, 9% Latino/Hispanic and 2% African-American. Of the 262 respondents, 75% did an REU not at their home institution; 36% did an REU with a strong instruction component and 61% went to an REU solely focused on research. Student opinion: 84% said it was valuable, 16% said somewhat or not valuable (100% of Latinos said it was valuable); 83% said it did not shorten their time to a PhD. Did it accelerate their development as mathematicians?: 22% said definitely, 68% of Latinos said definitely, 44% said slightly, 23% said no. 53% of students said it had no influence on their choice of thesis topic, 36% said it had a slight or definite effect. Was the REU a factor in going to graduate school (for those who did)?: 78% said definitely or somewhat, 82% of women and 90% of Latinos said definitely or somewhat. Did an REU get you into a better graduate program?: 76% said definitely or somewhat. Did it help you win a fellowship? (the question didn't make it clear what to do if you didn't get a fellowship, or if a teaching assistantship was a fellowship): 48% said yes, (79% of Latinos), 24% said no, 28% no fellowship/can't answer. Of the 262 replies, 72% went to PhD programs and 13% went to Master's program, with 66% in math and 30% in other mathematical sciences.

A question to ask next time: did the REU increase your commitment to mathematics?

There were lots of positive comments on the survey by the students, and 5 negative comments.

7.3. Dan Isaksen, Wayne State University. When assessing our programs, we all look for future math achievement. What else should we look for?

Different programs have different goals. Here are some examples to consider. Example 1: After attending an REU, a student decides she doesn't want to do math and goes to electrical engineering graduate school. Example 2: Because of a good recommendation from an REU director, a student gets into a difficult PhD program and leaves with a Master's. Example 3: A woman is having good success in a PhD program with a few publications, but she puts her career on hold to start a family. Example 4: A student does not have a mathematically successful REU experience but has fun, and goes on to become a tenured math professor. Example 5: A student is an REU purely to pad his resume and goes to law school as previously planned. Example 6: A student's REU problem turns out to be previously solved, but the student later becomes a math professor. Success or failure? Some points of success for REUs are better networking, helping students make better career decisions (even possibly to leave mathematics) and developing professional communication skills.

Audience comments and questions:

- There may be a disconnect between goals that are good and the goals that the NSF has for REUs. One challenge is how to measure our goals.
- Should the quality of research produced by a student be a measure of REU success?
- One way to assess (according to the NSF) is whether the program transformed a dependent student into an independent student.
- Howard Hughes Medical Institute invites programs to participate in their survey. Contact David Lopatto at Grinnell College about SURE II survey.
- Do grad schools like REU experience? Many schools see it as a positive. When students work in teams, it is hard to tell what individual students contributed. Students should write about their REU research, explaining how it fits into other mathematics (e.g., what other mathematics is related to the work that they did). The recommendation letter from the REU director is very important, and its absence is a problem.
- One concern is that REUs give students a "too easy" impression of what math research is like.

7.4. Websites to look at. Beyond Bias and Barriers: Fulfilling the Potential of Women in Academic Science and Engineering
`http://www.nap.edu/catalog/11741.html`

National Academies Committee on Women in Academic Science and Engineering `http://www7.nationalacademies.org/womeninacademe/`

8. Panel: Perspectives from Students

Current graduate students described their undergraduate research experiences and the impact these have had on their mathematical development. They also discussed the way their time in an undergraduate research program affected their graduate experiences.

8.1. David Uminsky, Boston University. David's first REU was at the SIMU program at Puerto Rico in 2000.

- How did he choose SIMU? He saw a poster by chance in the hallway. It was the only REU he applied to.

- Did the REU affect his graduate school decision? Yes. He switched to being a math major after the REU because he saw a glimpse of life as a mathematician.
- The mathematical content of the research did not matter as much as the experience, but he is still working on a project that grew out of what he did at the REU.
- Did an REU better prepare him for graduate school? Yes. The tangible benefits were how to write in LaTeX, Maple and Mathematica skills, mathematical speaking skills. He got a chance to go through the refereeing process and it built confidence. It helped him get into graduate school.

8.2. Melanie Matchett Wood, Princeton University. Melanie did the Duluth REU in 2000 and also did academic year research at her undergrad institution, Duke. Her research experience was a big part of her decision to go to math graduate school. It showed her the most exciting part of mathematics and gave her confidence she could do math research. She applied to seven graduate schools, and ended up going to the school she least expected to attend, in a large part due to two contacts she had made through her research experiences. The REUs helped her prepare for graduate school by helping her develop research skills. She loves communicating about mathematics, and the lack of it she finds problematic. In both of her research experiences she worked alone. One big positive effect of REUs is students getting in touch with experts in a field, and it is good to have feedback from a professional in the exact area. At Duke, her adviser was in a related field, but she was put in touch with experts in the field and that helped a lot. Her experience is unique because she only likes certain parts of mathematics, so it was very helpful that she was able to choose problems she really liked in her research experience.

8.3. Paul Gibson, Northwestern University. Paul was an undergraduate at Notre Dame. Notre Dame had an honors math track, which he felt was very important in his development. His first summer of college he was a lifeguard. In the following year he did more math and the honors track math adviser suggested that he participate in the Notre Dame REU. The Notre Dame REU was a lot of fun working in a group. He met an adviser every day for two hours. His junior year, while abroad, he stayed in contact with the math department, keeping him on track for graduate school. That summer he went to an REU at Indiana University, with a mini-course (not tied to the research) and a project he worked on alone. He met with an adviser once a day. He liked both the group work and working alone. In his senior year, he participated in a seminar on undergraduate research at Notre Dame, which was a reading course in Riemannian geometry, his current field in graduate school. He went to Northwestern because it was a good school and he got lucky there were a lot of professors there doing what he likes.

8.4. Rana Mikkelson, Iowa State University. She went to Kalamazoo for undergrad, and she didn't like linear algebra, but it is now her research area. She attended the 2002 Grand Valley State REU, where she studied dynamical systems, and is now a mentor at the Iowa State REU. At her REU there were a lot of good social activities with other students, including a daily group breakfast, and official and unofficial activities, for example Xbox in the math building. It was a good way to meet professors in a nonacademic context, which makes it easier to talk to professors about academic things. Her research partner at the REU was older and more experienced so it was important she felt comfortable talking to a professor. The group work was successful because the pair complemented each other: her partner had lots of big ideas and she worked on the rigorous details. The students and mentors found problems to work on together by looking at examples for patterns. It was a great way to spend the summer instead of just taking more courses. It confirmed she shouldn't be afraid of doing math research. She learned LaTeX and math writing skills, had practice giving talks and heard a lot of good talks. She went to an REU at the suggestion of a professor and from looking at a poster. Her other research experience was her senior thesis, which was a bad experience because the question she was working on had a false assumption. This was exacerbated by the fact that the problem was outside her adviser's specialty, but even so she learned from the bad experience. She attended the Nebraska Undergraduate Conference for Women, which was a great experience.

8.5. Questions from the Audience.

- How do you keep students happy at an REU? Do students miss the freedom of a car? It was great to have fun weekend activities. Living and cooking with other students was a good experience. Having rented cars sometimes available helps. Students can find fun things to do on their own. Students can join university activities, like intramural sports.
- Given that math is different from experimental sciences, how did an REU help with your courses, and could REUs be different to help you more with coursework? REUs help develop good work habits. They might make students want to know more how courses are relevant to other mathematics. The REUs show you the light of research at the end of the tunnel of courses. REUs address different topics from coursework.
- Could an REU experience help borderline students to graduate school? Most students in a grad program had done research as undergrads. People who are double majors could benefit from an REU experience, but most REUs focus on people just doing math. REUs would be good for non-honors track students or students who wouldn't otherwise have research experiences. It is good to check up on students after the summer.
- How could we measure value added of REU programs? You could give before and after surveys, but make surveys anonymous.
- What is your family background? DU: Mother Spanish teacher, father eye doctor. His high school was not so great, and his Dad wanted him to major in physics and be a doctor. MMW: Mom is a public school teacher, and was 100% supportive of school and graduate school. PG: Father is an engineer and mother a lab assistant. His high school wasn't good. RM: Father a farmer and switched to construction, mother in corrections, and neither went to college.
- Poll to the audience: How many of your programs could help borderline students (i.e. students who aren't sure of their interest in mathematics)? Lots of hands went up. How many of your programs advertise this? Few hands.
- What to tell students who don't get into REUs? Help them connect with local professors to do research.

- Was money a factor in your decision to go to an REU? Not at all. A little bit. It helped decide between REUs. It was important that the REU didn't cost anything to the student.
- Do students expect to get a paper published from an REU? Not for me. In some competitive grad schools, most students have undergraduate research papers. A submission is not so important but writing something up was good. I didn't expect a paper but I did ended up submitting one. The students' impression is that grad schools look favorably on papers.
- There is a concern about the feeling that research papers help students get into graduate school. The NSF graduate fellowship application supports this by asking students for a research plan.

9. Reports from Parallel Sessions

9.1. Uniform Acceptance Date. It is good for transparency but there are numerous problems. Mainly, students will know if they are second or third choices for an REU. This might affect their interactions at the REU. A big question we need to know the answer to: How many students apply to REUs and do not get in any program?

Moderator asks: Could we agree to a date before which students would not be required to accept an REU offer? We need to make sure we can still be competitive with offers from industry internships. Some people want to make their acceptance dates later. There is a lot of confusion between uniform offer dates and uniform acceptance dates, and also between these and a date *before which* no student could be asked to respond. The graduate school acceptance model seems to work well. No one objects to agreeing that students should not have to respond to offers before February 20.

9.2. Generating Research Problems. Good things in a research problem: in the adviser's area, a careful literature search is part of the research, the adviser should have ideas for the first few steps, the problem should be easy to get into, students should learn some new math, the results should "in theory" be publishable. It is good to help students connect with a field of mathematics. The problem should be open ended. One concern is that the student makes a small advance, but then the problem is solved entirely by a professional. One solution is to tell the students it is okay if they can't publish their results. Being able to do computations is great. It is good if advisers at least guide students in selecting problems.

9.3. Undergraduate biology and mathematics programs. Some students want to see the relevance of their research, and working in mathematical biology is a good way to do this. Graduate school requirements for mathematical biology are different from pure mathematics and so the definition of success of these programs must be different: students going to any graduate school is good, not just math graduate school.

9.4. Working with Local Students. Usually it is not part of an REU, but when it is part of an REU it is important to make sure students are full participants in the REU and live with the rest of the group. It is important to give students problems that are good for *them*. If students are working on a non-publishable problem, should it be called research? Know your audience: yes to the deans, tell colleagues it is an independent investigation. Can we bridge the gap between an REU and the classroom? We could have a project or investigation in class.

9.5. Using graduate students as advisers. They are likely to be committed to the program and they can do lots of work. How to pay them? Maybe the department can, or you can put it in the grant. What can they do? They can teach LaTeX, they can make mathematical suggestions, they can design the t-shirt, and they can plan social activities. The grad students are closer to the students and so they can find issues with students and group work and report it sooner to the director. How to make sure the graduate students are working? It is not an issue if you pick students you know are hard working.

9.6. Helping undergraduate students into graduate school. What if students have low subject GREs? Look to see if the schools they are applying to require the subject test and if they have a strict cutoff. A poor verbal GRE is viewed as a negative. A letter of recommendation from past REU directors is critical. Letters should be no more than two pages, and should talk about the student at the beginning, with information on the program or details of the research as an appendix. Instead of getting students into the best school, you should get them into the right school.

10. Closing Panel

Joe Gallian "Good things happen when people get together." He learned a lot about REUs during this conference. Mark your calendars for the next PURM conference in 2012!

Frank Connolly said the value added to REUs is developing a student's career, showing that math can be fun, developing their confidence in research, and increasing their commitment to mathematics. He feels that $6000-$6500 per student is not enough funding for REUs. We should look into levels of funding of REUs in other disciplines. While there are lots of definitions of success, the NSF and NSA definitions matter, since they are funding the programs. Colleagues don't always recognize the importance of undergraduate research.

Ivelisse Rubio said the expectation of this conference was to exchange ideas and that has been a success. It is great that many people are increasingly aware of the issue of getting underrepresented groups into research and REUs. Summer research is great, but we can also work on more academic year research.

Aparna Higgins said the highlights of the conference were meeting research directors, and learning about the diversity of research experiences. We can't just have one model. She is delighted to hear people trading suggestions about issues with research programs, and to hear people talking about year-round research programs. She would like to see more programs for people who haven't had as much preparation. She was disappointed by the focus on REUs since not all students can go to REUs. Contact Michael Dorff (`mdorff@math.byu.edu`) for more information about mini-grants for year-round research.

Acknowledgment: The editor wishes to thank Aparna Higgins for assisting with this article.

DEPARTMENT OF MATHEMATICS, RUTGERS, THE STATE UNIVERSITY OF NEW JERSEY, PISCATAWAY, NJ 08854-8019

E-mail address: matchett@math.rutgers.edu

Part V

Surveys of Undergraduate Research Programs

e-mail questions to: dxc@ams.org

Please note: This survey is being conducted in conjunction with the Promoting Undergraduate Research in Mathematics (PURM) conference that was held on September 28-30, 2006. If this form is inappropriate for your program and if you are interested in submitting an article for publication in the conference proceedings, then please contact the proceedings editor, Joe Gallian (jgallian@d.umn.edu).

Name of respondent: Jeffrey Adler

Name of program: Algebra, Number Theory, and Applications

Program held at what institution? The University of Akron

Check here if this program is no longer active: [ ]
If so, last year of operation:

1. Briefly describe the kind of program you direct:
[✓] REU
[ ] REU program for certain group:
[ ] Other:
Describe:

2. How long has your program existed (including other directors)?
2 years

3. What is your adviser/student ratio? 1/2.7

4. In all the years of program operation, what is the total number of students who have participated in the program? 17

5. Of the students who have attended your program in the past five years, what percent were women? 47 %
If this figure is an estimate, check here: [ ]

6. Of the students who have attended your program in the past five years, what percent were Hispanic/Latino, Native American or African American? 0 %
If this figure is an estimate, check here: [ ]

7. Is your program:
[ ] instructional in nature
[✓] research oriented
[ ] both
Comments:

8. How many weeks does your program run?
[ ]1 [ ]2 [ ]3 [ ]4 [ ]5 [ ]6 [ ]7 [✓]8 [ ]9 [ ]10+

9. If your program has an instructional component, are students asked to work (check both if necessary):
[ ] in groups   [ ] individually

If your program has a research component, do students work on their research projects (check both if necessary):
[✓] in groups   [✓] individually
Comments:

For questions 10-16, please give an estimate if exact figures are not available (indicate when the figures are estimates).

10. Of the students whose academic career you are knowledgeable since they participated in your program, give the number who fit into the following categories:
[11] are still undergraduates
[2] are currently in a graduate program in the mathematical sciences
[0] have received a Masters in the mathematical sciences (and are no longer in graduate school)
[0] have received a Ph.D. in the mathematical sciences
[0] are currently in a graduate program in the sciences (excluding the mathematical sciences)
[0] have received a Masters in the sciences (excluding the mathematical sciences)
[0] have received a Ph.D. in the sciences (excluding the mathematical sciences)
[4] other (e.g., did not obtain an advanced degree, or are pursuing graduate studies outside of the sciences)

Comments:

One former participant is an actuary, one works for NSA, one is taking time off and will apply to grad school. Most or all seniors are applying to math grad school.

11. In the last five years, what percentage of your program participants have received national graduate fellowships (NSF, Hertz, NDSEG, Ford, NPSC, GEM, etc.)?

☑ 0-10% ☐ 11-20% ☐ 31-40% ☐ 41-50% ☐ 51-60%
☐ 61-70% ☐ 71-80% ☐ 81-90% ☐ 91-100%
This figure is an estimate ☐

12. In the past five years, approximately how many oral presentations at conferences have resulted from your program per year? 4

13. In the past five years, approximately how many poster presentations at conferences have resulted from your program per year? 1

14. Approximately, how many publications in refereed journals have resulted from your program per year? 0

15. What is the total annual budget for your program?
61273

16. Typically, how much direct financial support does your institution provide? 6966

17. Typically, what kind of non-financial support does your institution provide to your program?
☑ waiver of overhead/reduction of normal overhead
☐ faculty salaries
☑ release time
☐ summer support for director
☑ clerical/administrative support (full or part-time)
☐ graduate tutoring
☑ school facilities (including office space)
☑ telephone/photocopying/postage
☑ computer use/computer support
☐ free food/entertainment and/or transportation for students
☐ free housing for students
☑ other: waiver of out-of-state tuition fees

18. Which agency is the principal source of your support (NSF, NSA, others)?
NSF

19. What measures do you use to assess the success of the program?
For now, an anonymous exit survey. In the future, we'll do followup surveys.

20. Does your program target a particular audience? ☐ Yes ☑ No
Please describe:

21. Please describe any unusual features of your program.

22. Briefly describe the goals of your program.
We want students to learn what mathematical research is like so that they can make better decisions about (and be better prepared for) mathematical careers. To this end, we do not offer any courses within this program. Rather, we have the students work on problems whose answers are, as far as we can tell, unknown.

2006 Survey of Undergraduate Programs in Mathematics

Please note: This survey is being conducted in conjunction with the Promoting Undergraduate Research in Mathematics (PURM) conference that was held on September 28-30, 2006. If this form is inappropriate for your program and if you are interested in submitting an article for publication in the conference proceedings, then please contact the proceedings editor, Joe Gallian (jgallian@d.umn.edu).

Name of respondent: Tuncay Aktosun

Name of program: NREUP

Program held at what institution? University of Texas at Arlington

Check here if this program is no longer active: ☐
If so, last year of operation:

1. Briefly describe the kind of program you direct:
☐ REU
☑ REU program for certain group: minority students
☐ Other:

Describe:

involving 4 undergraduate minority students in summer research

2. How long has your program existed (including other directors)?
Since the summer of 2006

3. What is your adviser/student ratio? 2/4

4. In all the years of program operation, what is the total number of students who have participated in the program? 4

5. Of the students who have attended your program in the past five years, what percent were women? 50 %
If this figure is an estimate, check here: ☐

6. Of the students who have attended your program in the past five years, what percent were Hispanic/Latino, Native American or African American? 100 %
If this figure is an estimate, check here: ☐

7. Is your program:
☐ instructional in nature
☑ research oriented
☐ both
Comments:

8. How many weeks does your program run?
☐ 1 ☑ 2 ☐ 3 ☐ 4 ☐ 5 ☑ 6 ☐ 7 ☐ 8 ☐ 9 ☐ 10+

9. If your program has an instructional component, are students asked to work (check both if necessary):
☐ in groups ☐ individually

If your program has a research component, do students work on their research projects (check both if necessary):
☑ in groups ☐ individually
Comments:

For questions 10-16, please give an estimate if exact figures are not available (indicate when the figures are estimates).

10. Of the students whose academic career you are knowledgeable since they participated in your program, give the number who fit into the following categories:

4 are still undergraduates

0 are currently in a graduate program in the mathematical sciences

0 have received a Masters in the mathematical sciences (and are no longer in graduate school)

0 have received a Ph.D. in the mathematical sciences

0 are currently in a graduate program in the sciences (excluding the mathematical sciences)

0 have received a Masters in the sciences (excluding the mathematical sciences)

0 have received a Ph.D. in the sciences (excluding the mathematical sciences)

0 other (e.g., did not obtain an advanced degree, or are pursuing graduate studies outside of the sciences)

Comments:

11. In the last five years, what percentage of your program participants have received national graduate fellowships (NSF, Hertz, NDSEG, Ford, NPSC, GEM, etc.)?
[✓] 0-10% [ ] 11-20% [ ] 31-40% [ ] 41-50% [ ] 51-60%
[ ] 61-70% [ ] 71-80% [ ] 81-90% [ ] 91-100%
This figure is an estimate [ ]

12. In the past five years, approximately how many oral presentations at conferences have resulted from your program per year? [2]

13. In the past five years, approximately how many poster presentations at conferences have resulted from your program per year? [1]

14. Approximately, how many publications in refereed journals have resulted from your program per year? [0]

15. What is the total annual budget for your program?
$25,000.00

16. Typically, how much direct financial support does your institution provide? $0.00

17. Typically, what kind of non-financial support does your institution provide to your program?
[ ] waiver of overhead/reduction of normal overhead
[ ] faculty salaries
[ ] release time
[ ] summer support for director
[ ] clerical/administrative support (full or part-time)
[ ] graduate tutoring
[✓] school facilities (including office space)
[✓] telephone/photocopying/postage
[✓] computer use/computer support
[ ] free food/entertainment and/or transportation for students
[ ] free housing for students
[ ] other:

18. Which agency is the principal source of your support (NSF, NSA, others)?
MAA through funds from NSF, NSA, and Moody's Foundation

19. What measures do you use to assess the success of the program?
e-mail survey after the program; independent evaluation of the program by a group at University of Oregon

20. Does your program target a particular audience? [✓] Yes [ ] No
Please describe:
4 undergraduate minority students

21. Please describe any unusual features of your program.
intensive summer research, small program, close interaction between mentors and participants, minority participants only

22. Briefly describe the goals of your program.
Active research under supervision of mathematicians who are dedicated researchers and mentors. The objectives are to provide the participants with meaningful research experience, to show them the enjoyment of doing research, to encourage them to pursue advanced degrees in mathematical sciences, and to increase research participation by minority groups. At the end of the program the participants prepare a written report and give a presentation of their research. The program is enhanced with some scientific, cultural, and social activities.

For the details of our program, see:

http://omega.uta.edu/~aktosun/nreup

2006 Survey of Undergraduate Programs in Mathematics

Please note: This survey is being conducted in conjunction with the Promoting Undergraduate Research in Mathematics (PURM) conference that was held on September 28-30, 2006. If this form is inappropriate for your program and if you are interested in submitting an article for publication in the conference proceedings, then please contact the proceedings editor, Joe Gallian (jgallian@d.umn.edu).

Name of respondent: David Auckly

Name of program: Brain Storming and Barn Storming

Program held at what institution? Kansas State University

Check here if this program is no longer active: ☐
If so, last year of operation:

1. Briefly describe the kind of program you direct:
☐ REU
☑ REU program for certain group: sophomores & juniors
☐ Other:

Describe:

We run an exciting REU site, combining mathematics related to flight, flight lessons and research projects in pure mathematics.

2. How long has your program existed (including other directors)?

Since June 2005

3. What is your adviser/student ratio? 1 advisor for every 2 students; and equal numbers of faculty and students.

4. In all the years of program operation, what is the total number of students who have participated in the program? 23

5. Of the students who have attended your program in the past five years, what percent were women? 17 %

If this figure is an estimate, check here: ☐

6. Of the students who have attended your program in the past five years, what percent were Hispanic/Latino, Native American or African American? 12 %

If this figure is an estimate, check here: ☐

7. Is your program:
☐ instructional in nature
☐ research oriented
☑ both

Comments:

We have many seminars/classes that expose our students to as much mathematics as possible. We also have each student pursue some topic in depth as a research problem in pure mathematics.

8. How many weeks does your program run?
☐ 1 ☐ 2 ☐ 3 ☐ 4 ☐ 5 ☐ 6 ☐ 7 ☑ 8 ☐ 9 ☐ 10+

9. If your program has an instructional component, are students asked to work (check both if necessary)?

☑ in groups   ☑ individually

If your program has a research component, do students work on their research projects (check both if necessary):

☑ in groups   ☑ individually

Comments:

Students may choose to work together or in groups; most choose to work in groups.

For questions 10-16, please give an estimate if exact figures are not available (indicate when the figures are estimates).

10. Of the students whose academic career you are knowledgeable since they participated in your program, give the number who fit into the following categories:

16 are still undergraduates

7 are currently in a graduate program in the mathematical sciences

☐ have received a Masters in the mathematical sciences (and are no longer in graduate school)

☐ have received a Ph.D. in the mathematical sciences

☐ are currently in a graduate program in the sciences (excluding the mathematical sciences)

☐ have received a Masters in the sciences (excluding the mathematical sciences)

☐ have received a Ph.D. in the sciences (excluding the mathematical sciences)

☐ other (e.g., did not obtain an advanced degree, or are pursuing graduate studies outside of the sciences)

Comments:
We expect that almost all of our students will obtain an advanced degree in Math/Science.

11. In the last five years, what percentage of your program participants have received national graduate fellowships (NSF, Hertz, NDSEG, Ford, NPSC, GEM, etc.)?
☐0-10% ☐11-20% ☐31-40% ☐41-50% ☑51-60%
☐61-70% ☐71-80% ☐81-90% ☐91-100%
This figure is an estimate ☑

12. In the past five years, approximately how many oral presentations at conferences have resulted from your program per year? 30%

13. In the past five years, approximately how many poster presentations at conferences have resulted from your program per year? 10%

14. Approximately, how many publications in refereed journals have resulted from your program per year? 2

15. What is the total annual budget for your program?
$102,000

16. Typically, how much direct financial support does your institution provide? $26,000

17. Typically, what kind of non-financial support does your institution provide to your program?
☐ waiver of overhead/reduction of normal overhead
☑ faculty salaries
☑ release time
☑ summer support for director
☑ clerical/administrative support (full or part-time)
☑ graduate tutoring
☑ school facilities (including office space)
☑ telephone/photocopying/postage
☑ computer use/computer support
☑ free food/entertainment and/or transportation for students
☑ free housing for students
☐ other:

18. Which agency is the principal source of your support (NSF, NSA, others)?
NSF

19. What measures do you use to assess the success of the program?
Student evaluations, publications.

20. Does your program target a particular audience? ☑ Yes ☐ No
Please describe:
Juniors in college. We want our students to share their experiences with other students at their undergraduate schools and want to help them get into graduate school.

21. Please describe any unusual features of your program.
We discuss mathematics related to flight and give the students flight lessons.

22. Briefly describe the goals of your program.
We want to attract more students into math/science, and help those students who are interested in preparing for graduate school and a career in math/science.

2006 Survey of Undergraduate Programs in Mathematics

Please note: This survey is being conducted in conjunction with the Promoting Undergraduate Research in Mathematics (PURM) conference that was held on September 28-30, 2006. If this form is inappropriate for your program and if you are interested in submitting an article for publication in the conference proceedings, then please contact the proceedings editor, Joe Gallian (jgallian@d.umn.edu).

Name of respondent: **Bela Bajnok**

Name of program:

Program held at what institution? **Gettysburg College**

Check here if this program is no longer active: ☐
If so, last year of operation:

1. Briefly describe the kind of program you direct:

   ☐ REU

   ☐ REU program for certain group:

   ☑ Other: **Research classes at Gettysburg College**

   Describe: **Three levels of research classes every semester**

2. How long has your program existed (including other directors)?

   Three years in this format.

3. What is your adviser/student ratio? **About 15 to 1.**

4. In all the years of program operation, what is the total number of students who have participated in the program? **Maybe 100.**

5. Of the students who have attended your program in the past five years, what percent were women? **50**%

   If this figure is an estimate, check here: ☑

6. Of the students who have attended your program in the past five years, what percent were Hispanic/Latino, Native American or African American? **2**%

   If this figure is an estimate, check here: ☑

---

7. Is your program:

   ☐ instructional in nature

   ☐ research oriented

   ☑ both

   Comments:

8. How many weeks does your program run?

   ☐ 1 ☐ 2 ☐ 3 ☐ 4 ☐ 5 ☐ 6 ☐ 7 ☐ 8 ☐ 9 ☑ 10+

9. If your program has an instructional component, are students asked to work (check both if necessary):

   ☑ in groups   ☑ individually

   If your program has a research component, do students work on their research projects (check both if necessary):

   ☑ in groups   ☑ individually

   Comments:

   Usually, students start working in small groups of 2-3, then most choose a sub-project on which they work alone.

For questions 10-16, please give an estimate if exact figures are not available (indicate when the figures are estimates).

10. Of the students whose academic career you are knowledgeable since they participated in your program, give the number who fit into the following categories:

   20 are still undergraduates

   10 are currently in a graduate program in the mathematical sciences

   2 have received a Masters in the mathematical sciences (and are no longer in graduate school)

   3 have received a Ph.D. in the mathematical sciences

   20 are currently in a graduate program in the sciences (excluding the mathematical sciences)

   10 have received a Masters in the sciences (excluding the mathematical sciences)

   5 have received a Ph.D. in the sciences (excluding the mathematical sciences)

   30 other (e.g., did not obtain an advanced degree, or are pursuing graduate studies outside of the sciences)

Comments: [ ]

11. In the last five years, what percentage of your program participants have received national graduate fellowships (NSF, Hertz, NDSEG, Ford, NPSC, GEM, etc.)?
☒ 0-10%   ☐ 11-20%   ☐ 31-40%   ☐ 41-50%   ☐ 51-60%
☐ 61-70%   ☐ 71-80%   ☐ 81-90%   ☐ 91-100%
This figure is an estimate [ ]

12. In the past five years, approximately how many oral presentations at conferences have resulted from your program per year? 15

13. In the past five years, approximately how many poster presentations at conferences have resulted from your program per year? 3

14. Approximately, how many publications in refereed journals have resulted from your program per year? 0

15. What is the total annual budget for your program?
0

16. Typically, how much direct financial support does your institution provide? 0

17. Typically, what kind of non-financial support does your institution provide to your program?
☐ waiver of overhead/reduction of normal overhead
☐ faculty salaries
☐ release time
☐ summer support for director
☐ clerical/administrative support (full or part-time)
☐ graduate tutoring
☐ school facilities (including office space)
☐ telephone/photocopying/postage
☐ computer use/computer support
☐ free food/entertainment and/or transportation for students
☐ free housing for students
☐ other: Classes count as one course towards my regular teaching load.

18. Which agency is the principal source of your support (NSF, NSA, others)? [ ]

19. What measures do you use to assess the success of the program?
class participation, presentations, formal paper

20. Does your program target a particular audience? ☒ Yes   ☐ No
Please describe:
current Gettysburg students

21. Please describe any unusual features of your program. [ ]

22. Briefly describe the goals of your program.
1. Develop a thorough understanding of a topic in advanced mathematics which is of current research interest; in particular, learn about its concepts, past achievements, and open questions.
2. Develop some understanding of several topics in advanced mathematics which are of current research interest.
3. Learn some of the general methods of mathematics research.
4. Continue to develop an ability to think and reason at an advanced level; in particular, learn how to read, write, and present proofs, explore problems which require substantial effort, present extensive arguments, and evaluate such arguments provided by others, think, write, and talk abstractly, succinctly, and convincingly.
5. Carry out an investigation and give weekly reports on the findings.
6. Write a formal research paper.

2006 Survey of Undergraduate Programs in Mathematics

Please note: This survey is being conducted in conjunction with the Promoting Undergraduate Research in Mathematics (PURM) conference that was held on September 28-30, 2006. If this form is inappropriate for your program and if you are interested in submitting an article for publication in the conference proceedings, then please contact the proceedings editor, Joe Gallian (jgallian@d.umn.edu).

Name of respondent: Francine Blanchet-Sadri

Name of program: Algorithmic Combinatorics on Words

Program held at what institution? University of North Carolina at Greensboro

Check here if this program is no longer active: ☐
If so, last year of operation:

1. Briefly describe the kind of program you direct:
   ☑ REU
   ☐ REU program for certain group:
   ☐ Other:

   Describe:

2. How long has your program existed (including other directors)?
   Program started in Summer 2005

3. What is your adviser/student ratio? 1/8 + 2 assistants

4. In all the years of program operation, what is the total number of students who have participated in the program? 16

5. Of the students who have attended your program in the past five years, what percent were women? 30 %

   If this figure is an estimate, check here: ☐

6. Of the students who have attended your program in the past five years, what percent were Hispanic/Latino, Native American or African American? 5 %

   If this figure is an estimate, check here: ☐

7. Is your program:
   ☐ instructional in nature
   ☑ research oriented
   ☐ both
   Comments:

8. How many weeks does your program run?
   ☐ 1 ☐ 2 ☐ 3 ☐ 4 ☐ 5 ☐ 6 ☐ 7 ☑ 8 ☐ 9 ☐ 10+

9. If your program has an instructional component, are students asked to work (check both if necessary):
   ☐ in groups   ☐ individually

   If your program has a research component, do students work on their research projects (check both if necessary):
   ☑ in groups   ☑ individually

   Comments:

For questions 10-16, please give an estimate if exact figures are not available (indicate when the figures are estimates).

10. Of the students whose academic career you are knowledgeable since they participated in your program, give the number who fit into the following categories:

   [11] are still undergraduates

   [5] are currently in a graduate program in the mathematical sciences

   [1] have received a Masters in the mathematical sciences (and are no longer in graduate school)

   [ ] have received a Ph.D. in the mathematical sciences

   [ ] are currently in a graduate program in the sciences (excluding the mathematical sciences)

   [ ] have received a Masters in the sciences (excluding the mathematical sciences)

   [ ] have received a Ph.D. in the sciences (excluding the mathematical sciences)

   [2] other (e.g., did not obtain an advanced degree, or are pursuing graduate studies outside of the sciences)

18. Which agency is the principal source of your support (NSF, NSA, others)?

NSF

19. What measures do you use to assess the success of the program?

Publications in leading journals and presentations at international conferences. Other research products include software that is made available through the web.

20. Does your program target a particular audience? ☐ Yes ☑ No
Please describe:

21. Please describe any unusual features of your program.
Students establish World Wide Web server interfaces for automated use of the programs related to our combinatorial algorithms. This objective involves extensive computer programming and requires some experience using a programming language such as Java.

22. Briefly describe the goals of your program.
A first objective is to introduce students to various challenging algorithmic combinatorial problems on partial words, through lectures and reading. A second objective of the program is for students to develop superior skills in mathematical writing and oral communication. A third objective is to submit the produced original collaborative research on algorithmic combinatorics on words to leading journals. Also, students gain experience in communicating mathematics verbally through presentations at national professional meetings and national/international conferences. A fourth objective is for students to gain experience in the use of computers and their interaction in mathematical research. A fifth objective of the program is to strongly encourage underrepresented groups to participate.

Comments:

11. In the last five years, what percentage of your program participants have received national graduate fellowships (NSF, Hertz, NDSEG, Ford, NPSC, GEM, etc.)?
☑0-10% ☐11-20% ☐31-40% ☐41-50% ☐51-60%
☐61-70% ☐71-80% ☐81-90% ☐91-100%
This figure is an estimate ☐

12. In the past five years, approximately how many oral presentations at conferences have resulted from your program per year? 6

13. In the past five years, approximately how many poster presentations at conferences have resulted from your program per year?

14. Approximately, how many publications in refereed journals have resulted from your program per year? 6

15. What is the total annual budget for your program?
Approximately $93,000

16. Typically, how much direct financial support does your institution provide?

17. Typically, what kind of non-financial support does your institution provide to your program?
☐ waiver of overhead/reduction of normal overhead
☐ faculty salaries
☐ release time
☐ summer support for director
☐ clerical/administrative support (full or part-time)
☐ graduate tutoring
☑ school facilities (including office space)
☑ telephone/photocopying/postage
☑ computer use/computer support
☑ free food/entertainment and/or transportation for students
☐ free housing for students
☐ other:

2006 Survey of Undergraduate Programs in Mathematics

Please note: This survey is being conducted in conjunction with the Promoting Undergraduate Research in Mathematics (PURM) conference that was held on September 28-30, 2006. If this form is inappropriate for your program and if you are interested in submitting an article for publication in the conference proceedings, then please contact the proceedings editor, Joe Gallian (jgallian@d.umn.edu).

Name of respondent: David Brown

Name of program: Virtual Reality, Robotics, and Visualization

Program held at what institution? Ithaca College

Check here if this program is no longer active: ☐
If so, last year of operation:

1. Briefly describe the kind of program you direct:

☑ REU

☐ REU program for certain group: students from colleges

☐ Other:

Describe: Research classes at Gettysburg College

Interdisciplinary between CS and Math, focusing on visualization & analysis

2. How long has your program existed (including other directors)?

3 years

3. What is your adviser/student ratio? 1/4

4. In all the years of program operation, what is the total number of students who have participated in the program? 30

5. Of the students who have attended your program in the past five years, what percent were women? 37 %

If this figure is an estimate, check here: ☐

6. Of the students who have attended your program in the past five years, what percent were Hispanic/Latino, Native American or African American? 6 %

If this figure is an estimate, check here: ☐

---

7. Is your program:
☐ instructional in nature
☐ research oriented
☑ both
Comments:

8. How many weeks does your program run?
☐ 1 ☐ 2 ☐ 3 ☐ 4 ☐ 5 ☐ 6 ☐ 7 ☑ 8 ☐ 9 ☐ 10+

9. If your program has an instructional component, are students asked to work (check both if necessary):

☑ in groups   ☐ individually

If your program has a research component, do students work on their research projects (check both if necessary):

☑ in groups   ☐ individually

Comments:

Usually, students start working in small groups of 2-3, then most choose a sub-project on which they work alone.

For questions 10-16, please give an estimate if exact figures are not available (indicate when the figures are estimates).

10. Of the students whose academic career you are knowledgeable since they participated in your program, give the number who fit into the following categories:

13 are still undergraduates

4 are currently in a graduate program in the mathematical sciences

0 have received a Masters in the mathematical sciences (and are no longer in graduate school)

0 have received a Ph.D. in the mathematical sciences

3 are currently in a graduate program in the sciences (excluding the mathematical sciences)

0 have received a Masters in the sciences (excluding the mathematical sciences)

0 have received a Ph.D. in the sciences (excluding the mathematical sciences)

0 other (e.g., did not obtain an advanced degree, or are pursuing graduate studies outside of the sciences)

Comments:

11. In the last five years, what percentage of your program participants have received national graduate fellowships (NSF, Hertz, NDSEG, Ford, NPSC, GEM, etc.)?
☑0-10% ☐11-20% ☐31-40% ☐41-50% ☐51-60%
☐61-70% ☐71-80% ☐81-90% ☐91-100%
This figure is an estimate ☐

12. In the past five years, approximately how many oral presentations at conferences have resulted from your program per year? 5

13. In the past five years, approximately how many poster presentations at conferences have resulted from your program per year? 1

14. Approximately, how many publications in refereed journals have resulted from your program per year? 1

15. What is the total annual budget for your program?
165,603

16. Typically, how much direct financial support does your institution provide?
0

17. Typically, what kind of non-financial support does your institution provide to your program?
☐waiver of overhead/reduction of normal overhead
☐faculty salaries
☐release time
☐summer support for director
☐clerical/administrative support (full or part-time)
☐graduate tutoring
☑school facilities (including office space)
☑telephone/photocopying/postage
☑computer use/computer support
☐free food/entertainment and/or transportation for students
☐free housing for students
☐other: Classes count as one course towards my regular teaching load.

18. Which agency is the principal source of your support (NSF, NSA, others)?
nsf

19. What measures do you use to assess the success of the program?
surveys and tracking of former participants

20. Does your program target a particular audience? ☑ Yes ☐ No
Please describe:
Students from non-research institutions

21. Please describe any unusual features of your program.
intensive summer research, small program, close interaction between mentors and participants, minority participants only

22. Briefly describe the goals of your program.
We focus on the role of experimentation and visualization in creation of mathematical ideas. Students from predominantly non-research institutions are provided with mentorship in the process of research in mathematics and computer science.

2006 Survey of Undergraduate Programs in Mathematics

e-mail questions to: dxc@ams.org

Please note: This survey is being conducted in conjunction with the Promoting Undergraduate Research in Mathematics (PURM) conference that was held on September 28-30, 2006. If this form is inappropriate for your program and if you are interested in submitting an article for publication in the conference proceedings, then please contact the proceedings editor, Joe Gallian (jgallian@d.umn.edu).

Name of respondent: Erika T. Camacho & Stephen A. Wirkus

Name of program: Applied Mathematical SciencesSummer Institute

Program held at what institution? Cal Poly Pomona & Loyola Marymount Univ

Check here if this program is no longer active: ☐
If so, last year of operation:

1. Briefly describe the kind of program you direct:

☑ REU

☐ REU program for certain group:

☐ Other:

Describe: 7 week program in applied mathematics; research groups of 4 students

2. How long has your program existed (including other directors)?

2005-present

3. What is your adviser/student ratio? 4:1

4. In all the years of program operation, what is the total number of students who have participated in the program? 32

5. Of the students who have attended your program in the past five years, what percent were women? 59.3 %

   If this figure is an estimate, check here: ☐

6. Of the students who have attended your program in the past five years, what percent were Hispanic/Latino, Native American or African American? 56.3 %

   If this figure is an estimate, check here: ☐

7. Is your program:
   ☐ instructional in nature
   ☐ research oriented
   ☑ both

Comments:

First week is lecture; 2nd and 3rd weeks are transition from lecture to independent research; 4-7th weeks are research under faculty guidance.

8. How many weeks does your program run?
   ☐ 1 ☐ 2 ☐ 3 ☐ 4 ☐ 5 ☐ 6 ☑ 7 ☐ 8 ☐ 9 ☐ 10+

9. If your program has an instructional component, are students asked to work (check both if necessary):

   ☑ in groups   ☐ individually

If your program has a research component, do students work on their research projects (check both if necessary):

   ☑ in groups   ☐ individually

Comments:

Group work during lecture portion is mixed regularly; from the 2nd week and on, students work in a group of 4 and with the help of a graduate RA and a faculty

For questions 10-16, please give an estimate if exact figures are not available (indicate when the figures are estimates).

10. Of the students whose academic career you are knowledgeable since they participated in your program, give the number who fit into the following categories:

   23 are still undergraduates

   7 are currently in a graduate program in the mathematical sciences

   0 have received a Masters in the mathematical sciences (and are no longer in graduate school)

   0 have received a Ph.D. in the mathematical sciences

   0 are currently in a graduate program in the sciences (excluding the mathematical sciences)

   0 have received a Masters in the sciences (excluding the mathematical sciences)

   0 have received a Ph.D. in the sciences (excluding the mathematical sciences)

   2 other (e.g., did not obtain an advanced degree, or are pursuing graduate studies outside of the sciences)

Comments:

11. In the last five years, what percentage of your program participants have received national graduate fellowships (NSF, Hertz, NDSEG, Ford, NPSC, GEM, etc.)?

☒0-10% ☐11-20% ☐31-40% ☐41-50% ☐51-60%
☐61-70% ☐71-80% ☐81-90% ☐91-100%

☐ This figure is an estimate

12. In the past five years, approximately how many oral presentations at conferences have resulted from your program per year? 1

13. In the past five years, approximately how many poster presentations at conferences have resulted from your program per year? 15

14. Approximately, how many publications in refereed journals have resulted from your program per year? 2

15. What is the total annual budget for your program?

$284,000

16. Typically, how much direct financial support does your institution provide? $4,500

17. Typically, what kind of non-financial support does your institution provide to your program?
☑ waiver of overhead/reduction of normal overhead
☐ faculty salaries
☑ release time
☐ summer support for director
☑ clerical/administrative support (full or part-time)
☐ graduate tutoring
☑ school facilities (including office space)
☑ telephone/photocopying/postage
☑ computer use/computer support
☐ free food/entertainment and/or transportation for students
☐ free housing for students
☐ other:

18. Which agency is the principal source of your support (NSF, NSA, others)?

NSF, NSA, DoD

19. What measures do you use to assess the success of the program?

students pursuing advanced degrees; poster and oral presentations at conference; publications in refereed journals; broadening students' perspectives

20. Does your program target a particular audience? ☑ Yes ☐ No
Please describe:

Women, underrepresented minorities, and those students who do not have adequate opportunities to conduct research in their home institutions

21. Please describe any unusual features of your program.

1. Split campuses, 2. partnership with IGERT programs, 3. Recruitment of Students from Non-research/Non-selective Schools, 4. Exposure to Non-traditional Applications of Mathematics, 5. Open Weekly Research Meetings

22. Briefly describe the goals of your program.

The goal of AMSSI is to develop Ph.D. mathematicians from underrepresented backgrounds (female, underrepresented minority, and those who don't have the research opportunities at their undergraduate institution) who will impact the culture and diversity of the U.S. while strengthening our communities. We provide intensive & in-depth research experience in the mathematical sciences with the ultimate goal of encouraging the pursuit graduate work; we foster a collaborative learning environment & strong bond among the students that will aid them in their summer research & provide them with a network of support in graduate school; we expose the students to new mathematical topics, computational tools, and provide a network of individuals to draw from before, during, and after grad school.

Please note: This survey is being conducted in conjunction with the Promoting Undergraduate Research in Mathematics (PURM) conference that was held on September 28-30, 2006. If this form is inappropriate for your program and if you are interested in submitting an article for publication in the conference proceedings, then please contact the proceedings editor, Joe Gallian (jgallian@d.umn.edu).

Name of respondent: Scott Chapman

Name of program: Trinity University REU

Program held at what institution? Trinity University, San Antonio, TX

Check here if this program is no longer active: [ ]
If so, last year of operation: [ ]

1. Briefly describe the kind of program you direct:
[✓] REU
[ ] REU program for certain group:
[ ] Other:
Describe:
Our Program is involves 12 students working on 4 different projects.

2. How long has your program existed (including other directors)?
We have recently completed our 9th year.

3. What is your adviser/student ratio? 4/12

4. In all the years of program operation, what is the total number of students who have participated in the program? 88

5. Of the students who have attended your program in the past five years, what percent were women? 44 %
If this figure is an estimate, check here: [ ]

6. Of the students who have attended your program in the past five years, what percent were Hispanic/Latino, Native American or African American? 19 %
If this figure is an estimate, check here: [ ]

7. Is your program:
[ ] instructional in nature
[✓] research oriented
[ ] both
Comments:
We have students completely focused on their research projects and do not offer accompanying short courses.

8. How many weeks does your program run?
[ ]1 [ ]2 [ ]3 [ ]4 [ ]5 [ ]6 [ ]7 [✓]8 [ ]9 [ ]10+

9. If your program has an instructional component, are students asked to work (check both if necessary):
[ ] in groups   [ ] individually

If your program has a research component, do students work on their research projects (check both if necessary):
[✓] in groups   [✓] individually

Comments:
Usually our students do group work, but if students wish to work individually, then we attempt to arrange this.

For questions 10-16, please give an estimate if exact figures are not available (indicate when the figures are estimates).

10. Of the students whose academic career you are knowledgeable since they participated in your program, give the number who fit into the following categories:
20 are still undergraduates
35 are currently in a graduate program in the mathematical sciences
11 have received a Masters in the mathematical sciences (and are no longer in graduate school)
3 have received a Ph.D. in the mathematical sciences
1 are currently in a graduate program in the sciences (excluding the mathematical sciences)
2 have received a Masters in the sciences (excluding the mathematical sciences)
0 have received a Ph.D. in the sciences (excluding the mathematical sciences)
2 other (e.g., did not obtain an advanced degree, or are pursuing graduate studies outside of the sciences)

2006 Survey of Undergraduate Programs in Mathematics

Please note: This survey is being conducted in conjunction with the Promoting Undergraduate Research in Mathematics (PURM) conference that was held on September 28-30, 2006. If this form is inappropriate for your program and if you are interested in submitting an article for publication in the conference proceedings, then please contact the proceedings editor, Joe Gallian (jgallian@d.umn.edu).

Name of respondent: Michael Dorff

Name of program: BYU Summer REU

Program held at what institution? Brigham Young University

Check here if this program is no longer active: ☐
If so, last year of operation:

1. Briefly describe the kind of program you direct:
   ☐ REU
   ☑ REU program for certain group: students from colleges
   ☐ Other:

Describe:
An 8-week summer REU

2. How long has your program existed (including other directors)?
   2 years

3. What is your adviser/student ratio? 1/4

4. In all the years of program operation, what is the total number of students who have participated in the program? 17

5. Of the students who have attended your program in the past five years, what percent were women? 65 %
   If this figure is an estimate, check here: ☐

6. Of the students who have attended your program in the past five years, what percent were Hispanic/Latino, Native American or African American? 0 %
   If this figure is an estimate, check here: ☐

Comments:

11. In the last five years, what percentage of your program participants have received national graduate fellowships (NSF, Hertz, NDSEG, Ford, NPSC, GEM, etc.)?
    ☑ 0-10%   ☐ 11-20%   ☐ 31-40%   ☐ 41-50%   ☐ 51-60%
    ☐ 61-70% ☐ 71-80%   ☐ 81-90%   ☐ 91-100%
    This figure is an estimate ☑

12. In the past five years, approximately how many oral presentations at conferences have resulted from your program per year? 4

13. In the past five years, approximately how many poster presentations at conferences have resulted from your program per year? 3

14. Approximately, how many publications in refereed journals have resulted from your program per year? 2

15. What is the total annual budget for your program?
    Approximately $70,000

16. Typically, how much direct financial support does your institution provide? $6,000

17. Typically, what kind of non-financial support does your institution provide to your program?
    ☐ waiver of overhead/reduction of normal overhead
    ☐ faculty salaries
    ☐ release time
    ☑ summer support for director
    ☑ clerical/administrative support (full or part-time)
    ☐ graduate tutoring
    ☑ school facilities (including office space)
    ☑ telephone/photocopying/postage
    ☑ computer use/computer support
    ☐ free food/entertainment and/or transportation for students
    ☑ free housing for students
    ☐ other:

2006 Survey of Undergraduate Programs in Mathematics

Please note: This survey is being conducted in conjunction with the Promoting Undergraduate Research in Mathematics (PURM) conference that was held on September 28-30, 2006. If this form is inappropriate for your program and if you are interested in submitting an article for publication in the conference proceedings, then please contact the proceedings editor, Joe Gallian (jgallian@d.umn.edu).

Name of respondent: Michael Dorff

Name of program: BYU Summer REU

Program held at what institution? Brigham Young University

Check here if this program is no longer active: ☐
If so, last year of operation:

1. Briefly describe the kind of program you direct:
   - ☐ REU
   - ☑ REU program for certain group: students from colleges
   - ☐ Other:
   Describe:
   An 8-week summer REU

2. How long has your program existed (including other directors)?
   2 years

3. What is your adviser/student ratio? 1/4

4. In all the years of program operation, what is the total number of students who have participated in the program? 17

5. Of the students who have attended your program in the past five years, what percent were women? 65 %
   If this figure is an estimate, check here: ☐

6. Of the students who have attended your program in the past five years, what percent were Hispanic/Latino, Native American or African American? 0 %
   If this figure is an estimate, check here: ☐

7. Is your program:
   - ☐ instructional in nature
   - ☑ research oriented
   - ☐ both
   Comments:

8. How many weeks does your program run?
   ☐1 ☐2 ☐3 ☐4 ☐5 ☐6 ☐7 ☑8 ☐9 ☐10+

9. If your program has an instructional component, are students asked to work (check both if necessary):
   - ☐ in groups
   - ☐ individually

If your program has a research component, do students work on their research projects (check both if necessary):
   - ☑ in groups
   - ☐ individually
   Comments:

For questions 10-16, please give an estimate if exact figures are not available (indicate when the figures are estimates).

10. Of the students whose academic career you are knowledgeable since they participated in your program, give the number who fit into the following categories:
   - 11 are still undergraduates
   - 6 are currently in a graduate program in the mathematical sciences
   - 0 have received a Masters in the mathematical sciences (and are no longer in graduate school)
   - 0 have received a Ph.D. in the mathematical sciences
   - 0 are currently in a graduate program in the sciences (excluding the mathematical sciences)
   - 0 have received a Masters in the sciences (excluding the mathematical sciences)
   - 0 have received a Ph.D. in the sciences (excluding the mathematical sciences)
   - 0 other (e.g., did not obtain an advanced degree, or are pursuing graduate studies outside of the sciences)

Comments:

11. In the last five years, what percentage of your program participants have received national graduate fellowships (NSF, Hertz, NDSEG, NPSC, GEM, etc.)?
☑ 0-10% ☐ 11-20% ☐ 31-40% ☐ 41-50% ☐ 51-60%
☐ 61-70% ☐ 71-80% ☐ 81-90% ☐ 91-100%
This figure is an estimate ☐

12. In the past five years, approximately how many oral presentations at conferences have resulted from your program per year? `10`

13. In the past five years, approximately how many poster presentations at conferences have resulted from your program per year? `1`

14. Approximately, how many publications in refereed journals have resulted from your program per year? `2`

15. What is the total annual budget for your program?
$65,000

16. Typically, how much direct financial support does your institution provide? `$10,000`

17. Typically, what kind of non-financial support does your institution provide to your program?
☐ waiver of overhead/reduction of normal overhead
☐ faculty salaries
☐ release time
☐ summer support for director
☐ clerical/administrative support (full or part-time)
☐ graduate tutoring
☑ school facilities (including office space)
☑ telephone/photocopying/postage
☑ computer use/computer support
☐ free food/entertainment and/or transportation for students
☐ free housing for students
☐ other:

18. Which agency is the principal source of your support (NSF, NSA, others)?
NSF

19. What measures do you use to assess the success of the program?
Surveys and data

20. Does your program target a particular audience? ☑ Yes ☐ No
Please describe:
Women and students from institutions that do not have a graduate program in mathematics.

21. Please describe any unusual features of your program.

22. Briefly describe the goals of your program.
The objectives of this program are: (1) to provide undergraduate students with the opportunity to experience doing mathematical research; (2) to encourage undergraduate students, especially female students and students from institutions with limited research opportunities, to attend graduate school in mathematics; and (3) to prepare participants to be successful in graduate school.

2006 Survey of Undergraduate Programs in Mathematics

Please note: This survey is being conducted in conjunction with the Promoting Undergraduate Research in Mathematics (PURM) conference that was held on September 28-30, 2006. If this form is inappropriate for your program and if you are interested in submitting an article for publication in the conference proceedings, then please contact the proceedings editor, Joe Gallian (jgallian@d.umn.edu).

Name of respondent: Joel Foisy

Name of program: SUNY Potsdam/Clarkson REU

Program held at what institution? SUNY Potsdam

Check here if this program is no longer active: ☐
If so, last year of operation:

1. Briefly describe the kind of program you direct:

   ☑ REU

   ☐ REU program for certain group:

   ☐ Other:

   Describe:

   Traditional REU;each of 3-5 faculty members works with groups of 3-5 students

2. How long has your program existed (including other directors)?

   Since 1997

3. What is your adviser/student ratio? Usually 4 to 1.

4. In all the years of program operation, what is the total number of students who have participated in the program? 116

5. Of the students who have attended your program in the past five years, what percent were women? 37 %

   If this figure is an estimate, check here: ☐

6. Of the students who have attended your program in the past five years, what percent were Hispanic/Latino, Native American or African American? 4 %

   If this figure is an estimate, check here: ☐

7. Is your program:

   ☐ instructional in nature
   ☑ research oriented
   ☐ both

   Comments:

8. How many weeks does your program run?

   ☐ 1 ☐ 2 ☐ 3 ☐ 4 ☐ 5 ☐ 6 ☐ 7 ☑ 8 ☐ 9 ☐ 10+

9. If your program has an instructional component, are students asked to work (check both if necessary):

   ☐ in groups   ☐ individually

   If your program has a research component, do students work on their research projects (check both if necessary):

   ☑ in groups   ☐ individually

   Comments:

   Students work in groups. We also have a weekly guest speaker who gives an hour-long lecture.

For questions 10-16, please give an estimate if exact figures are not available (indicate when the figures are estimates).

10. Of the students whose academic career you are knowledgeable since they participated in your program, give the number who fit into the following categories:

   26 are still undergraduates

   46 are currently in a graduate program in the mathematical sciences

   11 have received a Masters in the mathematical sciences (and are no longer in graduate school)

   9 have received a Ph.D. in the mathematical sciences

   2 are currently in a graduate program in the sciences (excluding the mathematical sciences)

   1 have received a Masters in the sciences (excluding the mathematical sciences)

   0 have received a Ph.D. in the sciences (excluding the mathematical sciences)

   22 other (e.g., did not obtain an advanced degree, or are pursuing graduate studies outside of the sciences)

Comments:

Nine are teachers.

11. In the last five years, what percentage of your program participants have received national graduate fellowships (NSF, Hertz, NDSEG, Ford, NPSC, GEM, etc.)?

☐ 0-10%   ☐ 11-20%   ☐ 31-40%   ☐ 41-50%   ☑ 51-60%
☐ 61-70%   ☐ 71-80%   ☐ 81-90%   ☐ 91-100%

This figure is an estimate ☑

12. In the past five years, approximately how many oral presentations at conferences have resulted from your program per year? 4

13. In the past five years, approximately how many poster presentations at conferences have resulted from your program per year? 4

14. Approximately, how many publications in refereed journals have resulted from your program per year? 1

15. What is the total annual budget for your program?

Approximately $122,000

16. Typically, how much direct financial support does your institution provide?   Approximately $38,000

17. Typically, what kind of non-financial support does your institution provide to your program?

☐ waiver of overhead/reduction of normal overhead
☐ faculty salaries
☑ release time
☐ summer support for director
☑ clerical/administrative support (full or part-time)
☐ graduate tutoring
☑ school facilities (including office space)
☑ telephone/photocopying/postage
☑ computer use/computer support
☑ free food/entertainment and/or transportation for students
☑ free housing for students
☐ other:

18. Which agency is the principal source of your support (NSF, NSA, others)?

NSF, but NSA contributes too.

19. What measures do you use to assess the success of the program?

Surveys and student tracking.

20. Does your program target a particular audience?   ☐ Yes   ☑ No
Please describe:

No, we do not target a specific group, but we are trying to step up our minority recruitment.

21. Please describe any unusual features of your program.

Two institutions are involved.

22. Briefly describe the goals of your program.

We have ambitious goals for our students. We would like to help them build their confidence in their ability to do research independently, and to help them build their appreciation and understanding of the vast field of mathematics. We also want to help them improve their oral and written communication skills.

2006 Survey of Undergraduate Programs in Mathematics

e-mail questions to: dxc@ams.org

Please note: This survey is being conducted in conjunction with the Promoting Undergraduate Research in Mathematics (PURM) conference that was held on September 28-30, 2006. If this form is inappropriate for your program and if you are interested in submitting an article for publication in the conference proceedings, then please contact the proceedings editor, Joe Gallian (jgallian@d.umn.edu).

Name of respondent: Joseph A. Gallian

Name of program: Duluth REU in Discrete Mathematics

Program held at what institution? University of Minnesota Duluth

Check here if this program is no longer active: ☐
If so, last year of operation:

1. Briefly describe the kind of program you direct:

☑ REU

☐ REU program for certain group:

☐ Other:

Describe:

10 week summer program for 8-10 students

2. How long has your program existed (including other directors)?

29 years

3. What is your adviser/student ratio? 1/3 including two graduate students

4. In all the years of program operation, what is the total number of students who have participated in the program? 143

5. Of the students who have attended your program in the past five years, what percent were women? 38 %

If this figure is an estimate, check here: ☐

6. Of the students who have attended your program in the past five years, what percent were Hispanic/Latino, Native American or African American? 0 %

If this figure is an estimate, check here: ☐

---

7. Is your program:
☐ instructional in nature
☑ research oriented
☐ both
Comments:

8. How many weeks does your program run?
☐ 1 ☐ 2 ☐ 3 ☐ 4 ☐ 5 ☐ 6 ☐ 7 ☐ 8 ☐ 9 ☑ 10+

9. If your program has an instructional component, are students asked to work (check both if necessary):

☐ in groups   ☐ individually

If your program has a research component, do students work on their research projects (check both if necessary):

☐ in groups   ☑ individually

Comments:
Students are encouraged to help each other.

For questions 10-16, please give an estimate if exact figures are not available (Indicate when the figures are estimates).

10. Of the students whose academic career you are knowledgeable since they participated in your program, give the number who fit into the following categories:

18 are still undergraduates

43 are currently in a graduate program in the mathematical sciences

5 have received a Masters in the mathematical sciences (and are no longer in graduate school)

60 have received a Ph.D. in the mathematical sciences

2 are currently in a graduate program in the sciences (excluding the mathematical sciences)

2 have received a Masters in the sciences (excluding the mathematical sciences)

3 have received a Ph.D. in the sciences (excluding the mathematical sciences)

3 other (e.g., did not obtain an advanced degree, or are pursuing graduate studies outside of the sciences)

Comments:

[   ]

11. In the last five years, what percentage of your program participants have received national graduate fellowships (NSF, Hertz, NDSEG, Ford, NPSC, GEM, etc.)?
☐0-10% ☐11-20% ☐31-40% ☑41-50% ☐51-60%
☐61-70% ☐71-80% ☐81-90% ☐91-100%
This figure is an estimate ☐

12. In the past five years, approximately how many oral presentations at conferences have resulted from your program per year? [4]

13. In the past five years, approximately how many poster presentations at conferences have resulted from your program per year? [0]

14. Approximately, how many publications in refereed journals have resulted from your program per year? [8 per year over the last five years]

15. What is the total annual budget for your program?
[$121,000]

16. Typically, how much direct financial support does your institution provide? [$2,000]

17. Typically, what kind of non-financial support does your institution provide to your program?
☐waiver of overhead/reduction of normal overhead
☐faculty salaries
☑release time
☐summer support for director
☐clerical/administrative support (full or part-time)
☐graduate tutoring
☑school facilities (including office space)
☑telephone/photocopying/postage
☑computer use/computer support
☑free food/entertainment and/or transportation for students
☐free housing for students
☐other: [   ]

---

18. Which agency is the principal source of your support (NSF, NSA, others)?
[NSF, NSA]

19. What measures do you use to assess the success of the program?
[Publications; graduate fellowships received; quality of graduate schools attended; presentations at meetings; participants returning as visitors; networking]

20. Does your program target a particular audience? ☑ Yes ☐ No
Please describe:
[I seek participants who intend to become professional mathematicians.]

21. Please describe any unusual features of your program.
[Large number of participants return as visitors to assist. I use former participants who are graduate students as research advisers. I am the only faculty person involved.]

22. Briefly describe the goals of your program.
[Provide a professional level research experience for participants; provide an introduction to the profession; form a network of people who will become professional mathematicians.]

2006 Survey of Undergraduate Programs in Mathematics

Please note: This survey is being conducted in conjunction with the Promoting Undergraduate Research in Mathematics (PURM) conference that was held on September 28-30, 2006. If this form is inappropriate for your program and if you are interested in submitting an article for publication in the conference proceedings, then please contact the proceedings editor, Joe Gallian (jgallian@d.umn.edu).

Name of respondent: Murli M. Gupta

Name of program: Summer Program for Women in Mathematics

Program held at what institution? George Washington University

Check here if this program is no longer active: ☐
If so, last year of operation: [ ]

1. Briefly describe the kind of program you direct:

☐ REU

☐ REU program for certain group: [ ]

☑ Other:

Describe:

Ours is a Summer Enrichment Program for Women

2. How long has your program existed (including other directors)?

Since 1995.

3. What is your adviser/student ratio? Not applicable. This is not an REU.

4. In all the years of program operation, what is the total number of students who have participated in the program? 186

5. Of the students who have attended your program in the past five years, what percent were women? 100 %

If this figure is an estimate, check here: ☐

6. Of the students who have attended your program in the past five years, what percent were Hispanic/Latino, Native American or African American? 8 %

If this figure is an estimate, check here: ☑

---

7. Is your program:

☐ instructional in nature
☐ research oriented
☑ both

Comments:

8. How many weeks does your program run?

☐ 1　☐ 2　☐ 3　☐ 4　☑ 5　☐ 6　☐ 7　☐ 8　☐ 9　☐ 10+

9. If your program has an instructional component, are students asked to work (check both if necessary):

☑ in groups　　☑ individually

If your program has a research component, do students work on their research projects (check both if necessary):

☑ in groups　　☑ individually

Comments:

For questions 10-16, please give an estimate if exact figures are not available (indicate when the figures are estimates).

10. Of the students whose academic career you are knowledgeable since they participated in your program, give the number who fit into the following categories:

20 are still undergraduates

60 are currently in a graduate program in the mathematical sciences

26 have received a Masters in the mathematical sciences (and are no longer in graduate school)

21 have received a Ph.D. in the mathematical sciences

5 are currently in a graduate program in the sciences (excluding the mathematical sciences)

5 have received a Masters in the sciences (excluding the mathematical sciences)

2 have received a Ph.D. in the sciences (excluding the mathematical sciences)

49 other (e.g., did not obtain an advanced degree, or are pursuing graduate studies outside of the sciences)

Comments:

11. In the last five years, what percentage of your program participants have received national graduate fellowships (NSF, Hertz, NDSEG, Ford, NPSC, GEM, etc.)?

☐ 0-10%   ☑ 11-20%   ☐ 31-40%   ☐ 41-50%   ☐ 51-60%
☐ 61-70%   ☐ 71-80%   ☐ 81-90%   ☐ 91-100%
This figure is an estimate ☑

12. In the past five years, approximately how many oral presentations at conferences have resulted from your program per year? 5

13. In the past five years, approximately how many poster presentations at conferences have resulted from your program per year? 5

14. Approximately, how many publications in refereed journals have resulted from your program per year? 0

15. What is the total annual budget for your program?

$220,000

16. Typically, how much direct financial support does your institution provide? $0

17. Typically, what kind of non-financial support does your institution provide to your program?
☑ waiver of overhead/reduction of normal overhead
☐ faculty salaries
☐ release time
☐ summer support for director
☑ clerical/administrative support (full or part-time)
☐ graduate tutoring
☑ school facilities (including office space)
☑ telephone/photocopying/postage
☑ computer use/computer support
☐ free food/entertainment and/or transportation for students
☐ free housing for students
☐ other:

18. Which agency is the principal source of your support (NSF, NSA, others)?

NSA

19. What measures do you use to assess the success of the program?

Surveys during and at the end of program. Subsequent assessment of impacts at SPWM Reunions at JMM. Long term impacts from former participants.

20. Does your program target a particular audience? ☑ Yes ☐ No
Please describe:

Women who are at a point of making decision on their post-graduation careers.

21. Please describe any unusual features of your program.

We have field trips to places in DC area where professional mathematicians are doing important scientific work. We also have guest lecturers who come to talk about what brought them into mathematics, and provide encouragement.

22. Briefly describe the goals of your program.

* provide an immersion program representative of key aspects of graduate school and professional mathematical practice.
* promote active mathematical thinking.
* underscore the beauty and enjoyment of mathematics.
* foster a camaraderie among the participants .
* bring the participants into contact with active mathematical researchers through a program of guest lectures and field trips.
* provide interaction with a wide variety of successful women in mathematical sciences who serve as role models.
* illustrate the role of mathematics as the foundation of the sciences and the wide range of mathematical applications in government, business, and industry.
* provide students with information about graduate schools and careers in mathematics.

2006 Survey of Undergraduate Programs in Mathematics

Please note: This survey is being conducted in conjunction with the Promoting Undergraduate Research in Mathematics (PURM) conference that was held on September 28-30, 2006. If this form is inappropriate for your program and if you are interested in submitting an article for publication in the conference proceedings, then please contact the proceedings editor, Joe Gallian (jgallian@d.umn.edu).

Name of respondent: Deanna Haunsperger

Name of program: Carleton College Summer Mathematics Program

Program held at what institution? Carleton College

Check here if this program is no longer active: ☐
If so, last year of operation:

1. Briefly describe the kind of program you direct:

☐ REU

☐ REU program for certain group:

☑ Other: Summer Enrichment for Women Undergraduates

Describe:

2. How long has your program existed (including other directors)?
Since 1995

3. What is your adviser/student ratio? 2 Instructors, 2 TAs, 18 students

4. In all the years of program operation, what is the total number of students who have participated in the program? 180

5. Of the students who have attended your program in the past five years, what percent were women? 100 %

   If this figure is an estimate, check here: ☐

6. Of the students who have attended your program in the past five years, what percent were Hispanic/Latino, Native American or African American? 2.2 %

   If this figure is an estimate, check here: ☐

7. Is your program:
   ☑ instructional in nature
   ☐ research oriented
   ☐ both
   Comments:

8. How many weeks does your program run?
   ☐ 1 ☐ 2 ☐ 3 ☑ 4 ☐ 5 ☐ 6 ☐ 7 ☐ 8 ☐ 9 ☐ 10+

9. If your program has an instructional component, are students asked to work (check both if necessary):

   ☑ in groups   ☑ individually

   If your program has a research component, do students work on their research projects (check both if necessary):

   ☐ in groups   ☐ individually

   Comments:
   Usually, students start working in small groups of 2-3, then most choose a sub-project on which they work alone.

For questions 10-16, please give an estimate if exact figures are not available (indicate when the figures are estimates).

10. Of the students whose academic career you are knowledgeable since they participated in your program, give the number who fit into the following categories:

   24 are still undergraduates

   53 are currently in a graduate program in the mathematical sciences

   23 have received a Masters in the mathematical sciences (and are no longer in graduate school)

   20 have received a Ph.D. in the mathematical sciences

   3 are currently in a graduate program in the sciences (excluding the mathematical sciences)

   0 have received a Masters in the sciences (excluding the mathematical sciences)

   0 have received a Ph.D. in the sciences (excluding the mathematical sciences)

   57 other (e.g., did not obtain an advanced degree, or are pursuing graduate studies outside of the sciences)

Comments:

11. In the last five years, what percentage of your program participants have received national graduate fellowships (NSF, Hertz, NDSEG, Ford, NPSC, GEM, etc.)?
[✓] 0-10% [ ] 11-20% [ ] 31-40% [ ] 41-50% [ ] 51-60%
[ ] 61-70% [ ] 71-80% [ ] 81-90% [ ] 91-100%
This figure is an estimate [✓]

12. In the past five years, approximately how many oral presentations at conferences have resulted from your program per year? [5-10]

13. In the past five years, approximately how many poster presentations at conferences have resulted from your program per year? [0-5]

14. Approximately, how many publications in refereed journals have resulted from your program per year? [0]

15. What is the total annual budget for your program?
Approximately $140,000

16. Typically, how much direct financial support does your institution provide? [$0]

17. Typically, what kind of non-financial support does your institution provide to your program?
[ ] waiver of overhead/reduction of normal overhead
[ ] faculty salaries
[ ] release time
[ ] summer support for director
[✓] clerical/administrative support (full or part-time)
[ ] graduate tutoring
[✓] school facilities (including office space)
[ ] telephone/photocopying/postage
[✓] computer use/computer support
[ ] free food/entertainment and/or transportation for students
[ ] free housing for students
[ ] other:

18. Which agency is the principal source of your support (NSF, NSA, others)?
NSF

19. What measures do you use to assess the success of the program?
In the past, end-of-program evaluations and the academic career of the participants after leaving the program. Currently, however, we are using survey instruments at various stages of their career to see how our program affects them.

20. Does your program target a particular audience? [✓] Yes [ ] No
Please describe:
Freshmen or sophomore women interested in and talented in mathematics.

21. Please describe any unusual features of your program.
We catch the women at an early age and plug them into a support environment of active, professional mathematicians.

22. Briefly describe the goals of your program.
The objectives are to improve participants' likelihood of persistence to PhD and increase the likelihood of their attaining successful and rewarding careers in mathematics. The means to achieve these ends are: the inspirational experience of the Carleton Summer Mathematics Program; active and supportive personal mentoring from sophomore year in college through and beyond the PhD and early professional career years; and regular community gatherings at the Joint Meetings and at Carleton College each summer. The goal is to nurture, support and advise these young women as they negotiate the manifold transitions facing them on the path from early undergraduate years to established professional mathematician and to have these same young women mentor and support the young women behind them.

2006 Survey of Undergraduate Programs In Mathematics

Please note: This survey is being conducted in conjunction with the Promoting Undergraduate Research in Mathematics (PURM) conference that was held on September 28-30, 2006. If this form is inappropriate for your program and if you are interested in submitting an article for publication in the conference proceedings, then please contact the proceedings editor, Joe Gallian (jgallian@d.umn.edu).

Name of respondent: [Bill Hawkins]

Name of program: [NREUP]

Program held at what institution? [12 sites nationally]

Check here if this program is no longer active: ☐
If so, last year of operation: [　]

1. Briefly describe the kind of program you direct:
   ☐ REU
   ☑ REU program for certain group: [minorities]
   ☐ Other:
   Describe: [　]

2. How long has your program existed (including other directors)?
   [Since Summer 2003]

3. What is your adviser/student ratio? [1/4]

4. In all the years of program operation, what is the total number of students who have participated in the program? [136]

5. Of the students who have attended your program in the past five years, what percent were women? [39]%
   If this figure is an estimate, check here: ☐

6. Of the students who have attended your program in the past five years, what percent were Hispanic/Latino, Native American or African American? [94]%
   If this figure is an estimate, check here: ☐

---

7. Is your program:
   ☐ instructional in nature
   ☑ research oriented
   ☐ both
   Comments: [　]

8. How many weeks does your program run?
   ☐ 1 ☐ 2 ☐ 3 ☐ 4 ☐ 5 ☑ 6 ☐ 7 ☐ 8 ☐ 9 ☐ 10+

9. If your program has an instructional component, are students asked to work (check both if necessary):
   ☐ in groups　　☐ individually

If your program has a research component, do students work on their research projects (check both if necessary):
   ☑ in groups　　☑ individually
   Comments: [　]

For questions 10-16, please give an estimate if exact figures are not available (indicate when the figures are estimates).

10. Of the students whose academic career you are knowledgeable since they participated in your program, give the number who fit into the following categories:
   [100] are still undergraduates
   [7] are currently in a graduate program in the mathematical sciences
   [　] have received a Masters in the mathematical sciences (and are no longer in graduate school)
   [　] have received a Ph.D. in the mathematical sciences
   [5] are currently in a graduate program in the sciences (excluding the mathematical sciences)
   [　] have received a Masters in the sciences (excluding the mathematical sciences)
   [　] have received a Ph.D. in the sciences (excluding the mathematical sciences)
   [　] other (e.g., did not obtain an advanced degree, or are pursuing graduate studies outside of the sciences)

Comments: [ ]

11. In the last five years, what percentage of your program participants have received national graduate fellowships (NSF, Hertz, NDSEG, Ford, NPSC, GEM, etc.)?
☐ 0-10% ☑ 11-20% ☐ 31-40% ☐ 41-50% ☐ 51-60%
☐ 61-70% ☐ 71-80% ☐ 81-90% ☐ 91-100%
This figure is an estimate ☑

12. In the past five years, approximately how many oral presentations at conferences have resulted from your program per year? [100+]

13. In the past five years, approximately how many poster presentations at conferences have resulted from your program per year? [10]

14. Approximately, how many publications in refereed journals have resulted from your program per year? [3]

15. What is the total annual budget for your program?
$300K

16. Typically, how much direct financial support does your institution provide? [NA]

17. Typically, what kind of non-financial support does your institution provide to your program?
☐ waiver of overhead/reduction of normal overhead
☐ faculty salaries
☐ release time
☐ summer support for director
☑ clerical/administrative support (full or part-time)
☐ graduate tutoring
☑ school facilities (including office space)
☑ telephone/photocopying/postage
☑ computer use/computer support
☐ free food/entertainment and/or transportation for students
☐ free housing for students
☐ other: [ ]

18. Which agency is the principal source of your support (NSF, NSA, others)?
NSF, NSA, Moody's Foundation

19. What measures do you use to assess the success of the program?
Outside evaluator collects pre-, post-, and multi-year follow-up data.

20. Does your program target a particular audience? ☑ Yes ☐ No
Please describe:
Underrepresented minorities near the funded sites.

21. Please describe any unusual features of your program.
Currently conducted at 12 sites: Atlanta Metro Coll., CA State Channel Islands, CA State Chico, Clayton State, DE State, DeVry Univ. of NJ, Grambling State, Spelman Coll., St. Mary's Coll. of MD, St. Peter's Coll., UT Arlington, VA State.

22. Briefly describe the goals of your program.
Provide challenging research experiences to minority undergraduates to increase their interest in obtaining advanced degrees in the mathematical sciences.

2006 Survey of Undergraduate Programs in Mathematics

Please note: This survey is being conducted in conjunction with the Promoting Undergraduate Research in Mathematics (PURM) conference that was held on September 28-30, 2006. If this form is inappropriate for your program and if you are interested in submitting an article for publication in the conference proceedings, then please contact the proceedings editor, Joe Gallian (jgallian@d.umn.edu).

Name of respondent: Aloysius Helminck

Name of program: Mathematics

Program held at what institution? North Carolina State University

Check here if this program is no longer active: ☐
If so, last year of operation:

1. Briefly describe the kind of program you direct:

☑ REU

☐ REU program for certain group:

☐ Other:

Describe:

2. How long has your program existed (including other directors)?

2 years

3. What is your adviser/student ratio? 1/3.5 (group size 3 to 4 students)

4. In all the years of program operation, what is the total number of students who have participated in the program? 36

5. Of the students who have attended your program in the past five years, what percent were women? 53 %

If this figure is an estimate, check here: ☐

6. Of the students who have attended your program in the past five years, what percent were Hispanic/Latino, Native American or African American? 8 %

If this figure is an estimate, check here: ☐

7. Is your program:
☐ instructional in nature
☑ research oriented
☐ both
Comments:

8. How many weeks does your program run?
☐ 1 ☐ 2 ☐ 3 ☐ 4 ☐ 5 ☐ 6 ☐ 7 ☐ 8 ☐ 9 ☑ 10+

9. If your program has an instructional component, are students asked to work (check both if necessary):

☑ in groups   ☐ individually

If your program has a research component, do students work on their research projects (check both if necessary):

☑ in groups   ☐ individually

Comments:

For questions 10-16, please give an estimate if exact figures are not available (indicate when the figures are estimates).

10. Of the students whose academic career you are knowledgeable since they participated in your program, give the number who fit into the following categories:

28 are still undergraduates

6 are currently in a graduate program in the mathematical sciences

0 have received a Masters in the mathematical sciences (and are no longer in graduate school)

0 have received a Ph.D. in the mathematical sciences

2 are currently in a graduate program in the sciences (excluding the mathematical sciences)

0 have received a Masters in the sciences (excluding the mathematical sciences)

0 have received a Ph.D. in the sciences (excluding the mathematical sciences)

0 other (e.g., did not obtain an advanced degree, or are pursuing graduate studies outside of the sciences)

Comments:

11. In the last five years, what percentage of your program participants have received national graduate fellowships (NSF, Hertz, NDSEG, Ford, NPSC, GEM, etc.)?

☑ 0-10% ☐ 11-20% ☐ 31-40% ☐ 41-50% ☐ 51-60%
☐ 61-70% ☐ 71-80% ☐ 81-90% ☐ 91-100%

This figure is an estimate ☐

12. In the past five years, approximately how many oral presentations at conferences have resulted from your program per year? 6

13. In the past five years, approximately how many poster presentations at conferences have resulted from your program per year? 5

14. Approximately, how many publications in refereed journals have resulted from your program per year? 3

15. What is the total annual budget for your program?

$75,000 for 2005 and $220,000 for 2006

16. Typically, how much direct financial support does your institution provide? 1000

17. Typically, what kind of non-financial support does your institution provide to your program?

☐ waiver of overhead/reduction of normal overhead
☐ faculty salaries
☑ release time
☐ summer support for director
☑ clerical/administrative support (full or part-time)
☑ graduate tutoring
☑ school facilities (including office space)
☑ telephone/photocopying/postage
☑ computer use/computer support
☐ free food/entertainment and/or transportation for students
☐ free housing for students
☑ other: transportation, food, etc for recreational activities

18. Which agency is the principal source of your support (NSF, NSA, others)?

NSF and NSA

19. What measures do you use to assess the success of the program?

Exit interviews and surveys of REU students. Tracking students after they leave the program. Participation in National and local conferences.

20. Does your program target a particular audience? ☐ Yes ☑ No
Please describe:

21. Please describe any unusual features of your program.

The program incorporates 2 SAMSI workshops and projects from industrial sponsors.

22. Briefly describe the goals of your program.

Objectives of this REU program are to provide a rich applied math research experience to a diverse population of students that will encourage them to continue their academic programs to the graduate level and will help them in making more informed decisions about their academic or nonacademic careers. The programs aims especially to encourage underrepresented groups.

2006 Survey of Undergraduate Programs in Mathematics

Please note: This survey is being conducted in conjunction with the Promoting Undergraduate Research in Mathematics (PURM) conference that was held on September 28-30, 2006. If this form is inappropriate for your program and if you are interested in submitting an article for publication in the conference proceedings, then please contact the proceedings editor, Joe Gallian (jgallian@d.umn.edu).

Name of respondent: Leslie Hogben

Name of program: Mathematics and Computing Research Experience

Program held at what institution? Iowa State University

Check here if this program is no longer active: ☐
If so, last year of operation:

1. Briefly describe the kind of program you direct:

☑ REU

☐ REU program for certain group:

☐ Other:

Describe:

2. How long has your program existed (including other directors)?
1998 (2003)

3. What is your adviser/student ratio? approx 1:2

4. In all the years of program operation, what is the total number of students who have participated in the program? 58

5. Of the students who have attended your program in the past five years, what percent were women? 43 %

If this figure is an estimate, check here: ☐

6. Of the students who have attended your program in the past five years, what percent were Hispanic/Latino, Native American or African American? 19 %

If this figure is an estimate, check here: ☐

---

7. Is your program:
☐ instructional in nature
☑ research oriented
☐ both

Comments:
see report

8. How many weeks does your program run?
☐ 1 ☐ 2 ☐ 3 ☐ 4 ☐ 5 ☐ 6 ☐ 7 ☑ 8 ☐ 9 ☐ 10+

9. If your program has an instructional component, are students asked to work (check both if necessary):

☐ in groups   ☐ individually

If your program has a research component, do students work on their research projects (check both if necessary):

☑ in groups   ☐ individually

Comments:
see report

For questions 10-16, please give an estimate if exact figures are not available (indicate when the figures are estimates).

10. Of the students whose academic career you are knowledgeable since they participated in your program, give the number who fit into the following categories:

21 are still undergraduates

7 are currently in a graduate program in the mathematical sciences

1 have received a Masters in the mathematical sciences (and are no longer in graduate school)

0 have received a Ph.D. in the mathematical sciences

3 are currently in a graduate program in the sciences (excluding the mathematical sciences)

0 have received a Masters in the sciences (excluding the mathematical sciences)

0 have received a Ph.D. in the sciences (excluding the mathematical sciences)

3 other (e.g., did not obtain an advanced degree, or are pursuing graduate studies outside of the sciences)

Comments:

participants supported by several programs
we do not have data for most of the non-NSF funded students

11. In the last five years, what percentage of your program participants
have received national graduate fellowships (NSF, Hertz, NDSEG, Ford, NPSC, GEM, etc.)?

☑ 0-10% ☐ 11-20% ☐ 31-40% ☐ 41-50% ☐ 51-60%
☐ 61-70% ☐ 71-80% ☐ 81-90% ☐ 91-100%

This figure is an estimate ☑

12. In the past five years, approximately how many oral presentations at
conferences have resulted from your program per year? 5

13. In the past five years, approximately how many poster presentations
at conferences have resulted from your program per year? 1

14. Approximately, how many publications in refereed journals have resulted
from your program per year? 3

15. What is the total annual budget for your program?

NSF part about $70,000 per year, also funds from ISU math, Alliance

16. Typically, how much direct financial support does your institution
provide? pays 7-8 grad std RAs at $2000-3000 each

17. Typically, what kind of non-financial support does your institution
provide to your program?
☑ waiver of overhead/reduction of normal overhead
☐ faculty salaries
☐ release time
☐ summer support for director
☑ clerical/administrative support (full or part-time)
☐ graduate tutoring
☑ school facilities (including office space)
☑ telephone/photocopying/postage
☑ computer use/computer support
☐ free food/entertainment and/or transportation for students
☐ free housing for students
☑ other: most of my time is unpaid. ISU pays 9 month salary when a bit is done

18. Which agency is the principal source of your support (NSF, NSA, others)?

NSF

19. What measures do you use to assess the success of the program?

see report

20. Does your program target a particular audience? ☐ Yes ☑ No
Please describe:

part of it does

21. Please describe any unusual features of your program.

see report

22. Briefly describe the goals of your program.

see report

Other comments

e-mail questions to: dxc@ams.org

2006 Survey of Undergraduate Programs in Mathematics

Please note: This survey is being conducted in conjunction with the Promoting Undergraduate Research in Mathematics (PURM) conference that was held on September 28-30, 2006. If this form is inappropriate for your program and if you are interested in submitting an article for publication in the conference proceedings, then please contact the proceedings editor, Joe Gallian (jgallian@d.umn.edu).

Name of respondent: Daniel Isaksen

Name of program:

Program held at what institution? Wayne State University

Check here if this program is no longer active: [ ]
If so, last year of operation: [ ]

1. Briefly describe the kind of program you direct:

[ ] REU

[ ] REU program for certain group:

[✓] Other: Year-round research projects with local students

Describe: Individual undergraduate research projects on an ad hoc basis

2. How long has your program existed (including other directors)? 3 years

3. What is your adviser/student ratio? 1/1 or 1/2

4. In all the years of program operation, what is the total number of students who have participated in the program? 5

5. Of the students who have attended your program in the past five years, what percent were women? 40 %

   If this figure is an estimate, check here: [ ]

6. Of the students who have attended your program in the past five years, what percent were Hispanic/Latino, Native American or African American? 0 %

   If this figure is an estimate, check here: [ ]

7. Is your program:
   [ ] instructional in nature
   [✓] research oriented
   [ ] both
   Comments:

8. How many weeks does your program run?
   [ ]1 [ ]2 [ ]3 [ ]4 [ ]5 [ ]6 [ ]7 [ ]8 [ ]9 [✓]10+

9. If your program has an instructional component, are students asked to work (check both if necessary):
   [ ] in groups [ ] individually

   If your program has a research component, do students work on their research projects (check both if necessary):
   [ ] in groups [ ] individually
   Comments:

For questions 10-16, please give an estimate if exact figures are not available (indicate when the figures are estimates).

10. Of the students whose academic career you are knowledgeable since they participated in your program, give the number who fit into the following categories:
    [1] are still undergraduates
    [3] are currently in a graduate program in the mathematical sciences
    [ ] have received a Masters in the mathematical sciences (and are no longer in graduate school)
    [ ] have received a Ph.D. in the mathematical sciences
    [ ] are currently in a graduate program in the sciences (excluding the mathematical sciences)
    [ ] have received a Masters in the sciences (excluding the mathematical sciences)
    [ ] have received a Ph.D. in the sciences (excluding the mathematical sciences)
    [1] other (e.g., did not obtain an advanced degree, or are pursuing graduate studies outside of the sciences)

Comments:

11. In the last five years, what percentage of your program participants have received national graduate fellowships (NSF, Hertz, NDSEG, Ford, NPSC, GEM, etc.)?
☒0-10% ☐11-20% ☐31-40% ☐41-50% ☐51-60%
☐61-70% ☐71-80% ☐81-90% ☐91-100%
This figure is an estimate ☐

12. In the past five years, approximately how many oral presentations at conferences have resulted from your program per year? 1

13. In the past five years, approximately how many poster presentations at conferences have resulted from your program per year? 0

14. Approximately, how many publications in refereed journals have resulted from your program per year? 0

15. What is the total annual budget for your program?
$2,000

16. Typically, how much direct financial support does your institution provide? $0

17. Typically, what kind of non-financial support does your institution provide to your program?
☐ waiver of overhead/reduction of normal overhead
☐ faculty salaries
☐ release time
☐ summer support for director
☒ clerical/administrative support (full or part-time)
☐ graduate tutoring
☒ school facilities (including office space)
☒ telephone/photocopying/postage
☒ computer use/computer support
☐ free food/entertainment and/or transportation for students
☐ free housing for students
☐ other:

---

18. Which agency is the principal source of your support (NSF, NSA, others)?
NSF

19. What measures do you use to assess the success of the program?

20. Does your program target a particular audience? ☒ Yes ☐ No
Please describe:
Wayne State undergraduates

21. Please describe any unusual features of your program.
It's not really a program. It's just me supervising research projects.

22. Briefly describe the goals of your program.
I want to help students make better career decisions. Students should at least be considering going to graduate school in a mathematical science, but I don't necessarily consider it a failure if the student does not go to graduate school.

Secondary goals include improving the writing and speaking ability of the student.

2006 Survey of Undergraduate Programs In Mathematics

Please note: This survey is being conducted in conjunction with the Promoting Undergraduate Research in Mathematics (PURM) conference that was held on September 28-30, 2006. If this form is inappropriate for your program and if you are interested in submitting an article for publication in the conference proceedings, then please contact the proceedings editor, Joe Gallian (jgallian@d.umn.edu).

Name of respondent: Kevin James

Name of program: REU: Combinatorics & Comp. Number Theory

Program held at what institution? Clemson University

Check here if this program is no longer active: [ ]
If so, last year of operation:

1. Briefly describe the kind of program you direct:
[✓] REU
[ ] REU program for certain group:
[ ] Other:
Describe:
NSF funded research in number theory & combinatorics.

2. How long has your program existed (including other directors)?
Since summer 2002.

3. What is your adviser/student ratio? 2:9

4. In all the years of program operation, what is the total number of students who have participated in the program? 47

5. Of the students who have attended your program in the past five years, what percent were women? 45 %
If this figure is an estimate, check here: [ ]

6. Of the students who have attended your program in the past five years, what percent were Hispanic/Latino, Native American or African American? 13 %
If this figure is an estimate, check here: [ ]

7. Is your program:
[ ] instructional in nature
[✓] research oriented
[ ] both
Comments:
Research teams consisting of 2-3 undergrads, a grad student and at least one faculty member attack unsolved problems.

8. How many weeks does your program run?
[ ]1 [ ]2 [ ]3 [ ]4 [ ]5 [ ]6 [ ]7 [✓]8 [ ]9 [ ]10+

9. If your program has an instructional component, are students asked to work (check both if necessary):
[ ] in groups [ ] individually

If your program has a research component, do students work on their research projects (check both if necessary):
[✓] in groups [ ] individually
Comments:

For questions 10-16, please give an estimate if exact figures are not available (indicate when the figures are estimates).

10. Of the students whose academic career you are knowledgeable since they participated in your program, give the number who fit into the following categories:
[17] are still undergraduates
[16] are currently in a graduate program in the mathematical sciences
[1] have received a Masters in the mathematical sciences (and are no longer in graduate school)
[0] have received a Ph.D. in the mathematical sciences
[3] are currently in a graduate program in the sciences (excluding the mathematical sciences)
[0] have received a Masters in the sciences (excluding the mathematical sciences)
[0] have received a Ph.D. in the sciences (excluding the mathematical sciences)
[3] other (e.g., did not obtain an advanced degree, or are pursuing graduate studies outside of the sciences)

Comments:

11. In the last five years, what percentage of your program participants have received national graduate fellowships (NSF, Hertz, NDSEG, Ford, NPSC, GEM, etc.)?

☒ 0-10% ☐ 11-20% ☐ 31-40% ☐ 41-50% ☐ 51-60%
☐ 61-70% ☐ 71-80% ☐ 81-90% ☐ 91-100%
This figure is an estimate ☑

12. In the past five years, approximately how many oral presentations at conferences have resulted from your program per year? 2-3

13. In the past five years, approximately how many poster presentations at conferences have resulted from your program per year? 1

14. Approximately, how many publications in refereed journals have resulted from your program per year? 2

15. What is the total annual budget for your program?

~$100,000

16. Typically, how much direct financial support does your institution provide? $500

17. Typically, what kind of non-financial support does your institution provide to your program?
☐ waiver of overhead/reduction of normal overhead
☐ faculty salaries
☐ release time
☐ summer support for director
☑ clerical/administrative support (full or part-time)
☐ graduate tutoring
☑ school facilities (including office space)
☑ telephone/photocopying/postage
☑ computer use/computer support
☐ free food/entertainment and/or transportation for students
☐ free housing for students
☐ other:

18. Which agency is the principal source of your support (NSF, NSA, others)?

NSF

19. What measures do you use to assess the success of the program?
1) Grad school attendance, although this is probably not the best measure.
2) Number of publications

20. Does your program target a particular audience? ☑ Yes ☐ No
Please describe:

We reserve 4 spots for participants from the southeastern US.

21. Please describe any unusual features of your program.

We run a weekly colloquium series in order to give participants an opportunity to meet and interact with several professional mathematicians.

22. Briefly describe the goals of your program.

Publication of journal articles is stated as one of our goals, and although such publications are indicators of success in research, this is not our primary measure of success. We are primarily interested in whether our participants can make an informed decision about continuing their education in graduate school and whether they develop a realistic view of mathematical research.

Our participants should learn a good bit of new mathematics while in our program, which they do. They should interact with many professional mathematicians which is achieved through our weekly colloquium series. Another indicator of a successful program is the impact of the program on the research of the undergraduate and graduate students and supervisors.

e-mail questions to: dxc@ams.org

2006 Survey of Undergraduate Programs In Mathematics

Please note: This survey is being conducted in conjunction with the Promoting Undergraduate Research in Mathematics (PURM) conference that was held on September 28-30, 2006. If this form is inappropriate for your program and if you are interested in submitting an article for publication in the conference proceedings, then please contact the proceedings editor, Joe Gallian (jgallian@d.umn.edu).

Name of respondent: Lew Ludwig

Name of program: Summer Scholars at Denison University

Program held at what institution? Denison University

Check here if this program is no longer active: ☐
If so, last year of operation:

1. Briefly describe the kind of program you direct:

☐ REU

☐ REU program for certain group:

☑ Other:

Describe: This program funds Denison students to conduct 8 weeks of summer research with Denison faculty members.

2. How long has your program existed (including other directors)?
Since the early 1990's. I will only comment on the last 5 years - since 2002.

3. What is your adviser/student ratio? 1 to 1, or 1 to 2

4. In all the years of program operation, what is the total number of students who have participated in the program? 41

5. Of the students who have attended your program in the past five years, what percent were women? 29 %

If this figure is an estimate, check here: ☐

6. Of the students who have attended your program in the past five years, what percent were Hispanic/Latino, Native American or African American? 2 %

If this figure is an estimate, check here: ☐

7. Is your program:
☐ instructional in nature
☑ research oriented
☐ both

Comments:

8. How many weeks does your program run?
☐ 1 ☐ 2 ☐ 3 ☐ 4 ☐ 5 ☐ 6 ☐ 7 ☑ 8 ☐ 9 ☐ 10+

9. If your program has an instructional component, are students asked to work (check both if necessary):

☐ in groups   ☐ individually

If your program has a research component, do students work on their research projects (check both if necessary):

☑ in groups   ☑ individually

Comments:
Majority work individually, but some work in pairs.

For questions 10-16, please give an estimate if exact figures are not available (indicate when the figures are estimates).

10. Of the students whose academic career you are knowledgeable since they participated in your program, give the number who fit into the following categories:

9 are still undergraduates

6 are currently in a graduate program in the mathematical sciences

2 have received a Masters in the mathematical sciences (and are no longer in graduate school)

0 have received a Ph.D. in the mathematical sciences

9 are currently in a graduate program in the sciences (excluding the mathematical sciences)

3 have received a Masters in the sciences (excluding the mathematical sciences)

0 have received a Ph.D. in the sciences (excluding the mathematical sciences)

0 other (e.g., did not obtain an advanced degree, or are pursuing graduate studies outside of the sciences)

Comments:

11. In the last five years, what percentage of your program participants have received national graduate fellowships (NSF, Hertz, NDSEG, Ford, NPSC, GEM, etc.)?

☑ 0-10% ☐ 11-20% ☐ 31-40% ☐ 41-50% ☐ 51-60%
☐ 61-70% ☐ 71-80% ☐ 81-90% ☐ 91-100%
This figure is an estimate ☐

12. In the past five years, approximately how many oral presentations at conferences have resulted from your program per year? [5]

13. In the past five years, approximately how many poster presentations at conferences have resulted from your program per year? [1]

14. Approximately, how many publications in refereed journals have resulted from your program per year? [2]

15. What is the total annual budget for your program?

16. Typically, how much direct financial support does your institution provide? [100%]

17. Typically, what kind of non-financial support does your institution provide to your program?
☑ waiver of overhead/reduction of normal overhead
☐ faculty salaries
☐ release time
☐ summer support for director
☑ clerical/administrative support (full or part-time)
☐ graduate tutoring
☑ school facilities (including office space)
☑ telephone/photocopying/postage
☑ computer use/computer support
☐ free food/entertainment and/or transportation for students
☑ free housing for students
☐ other:

18. Which agency is the principal source of your support (NSF, NSA, others)?

Internal endowment (90%) and the Howard Hughes Foundation (10%).

19. What measures do you use to assess the success of the program?

External presentations and publications.

20. Does your program target a particular audience? ☑ Yes ☐ No
Please describe:

Denison students

21. Please describe any unusual features of your program.

Totally self-contained at Denison.

22. Briefly describe the goals of your program.

Provide students with a true research experience and opportunity to work closely with a faculty mentor. Prepare students for experiences beyond Denison.

2006 Survey of Undergraduate Programs in Mathematics

Please note: This survey is being conducted in conjunction with the Promoting Undergraduate Research in Mathematics (PURM) conference that was held on September 28-30, 2006. If this form is inappropriate for your program and if you are interested in submitting an article for publication in the conference proceedings, then please contact the proceedings editor, Joe Gallian (jgallian@d.umn.edu).

Name of respondent: Thomas Mattman

Name of program: Research Experiences in Mathematics for Undergraduates & Teachers

Program held at what institution? California State University, Chico

Check here if this program is no longer active: [ ]
If so, last year of operation:

1. Briefly describe the kind of program you direct:
[✓] REU
[ ] REU program for certain group:
[ ] Other:
Describe: Our program is a joint REU and RET (Research Experience for Teachers). We also focus on women & underrepresented minorities.

2. How long has your program existed (including other directors)?
Summers of 2003 - present.

3. What is your adviser/student ratio? 1 advisor to 6 students (& teachers)

4. In all the years of program operation, what is the total number of students who have participated in the program? 33 students & 6 teachers

5. Of the students who have attended your program in the past five years, what percent were women? 40 %
If this figure is an estimate, check here: [ ]

6. Of the students who have attended your program in the past five years, what percent were Hispanic/Latino, Native American or African American? 46 %
If this figure is an estimate, check here: [ ]

7. Is your program:
[ ] instructional in nature
[✓] research oriented
[ ] both
Comments:

8. How many weeks does your program run?
[ ]1 [ ]2 [ ]3 [ ]4 [ ]5 [✓]6 [ ]7 [ ]8 [ ]9 [ ]10+

9. If your program has an instructional component, are students asked to work (check both if necessary):
[ ] in groups [ ] individually

If your program has a research component, do students work on their research projects (check both if necessary):
[✓] in groups [ ] individually
Comments:

For questions 10-16, please give an estimate if exact figures are not available (indicate when the figures are estimates).

10. Of the students whose academic career you are knowledgeable since they participated in your program, give the number who fit into the following categories:
16 are still undergraduates
5 are currently in a graduate program in the mathematical sciences
[ ] have received a Masters in the mathematical sciences (and are no longer in graduate school)
[ ] have received a Ph.D. in the mathematical sciences
[ ] are currently in a graduate program in the sciences (excluding the mathematical sciences)
[ ] have received a Masters in the sciences (excluding the mathematical sciences)
[ ] have received a Ph.D. in the sciences (excluding the mathematical sciences)
2 other (e.g., did not obtain an advanced degree, or are pursuing graduate studies outside of the sciences)

Comments:

11. In the last five years, what percentage of your program participants have received national graduate fellowships (NSF, Hertz, NDSEG, Ford, NPSC, GEM, etc.)?
☑ 0-10%   ☐ 11-20%   ☐ 31-40%   ☐ 41-50%   ☐ 51-60%
☐ 61-70%   ☐ 71-80%   ☐ 81-90%   ☐ 91-100%
This figure is an estimate ☐

12. In the past five years, approximately how many oral presentations at conferences have resulted from your program per year?   10

13. In the past five years, approximately how many poster presentations at conferences have resulted from your program per year?   5

14. Approximately, how many publications in refereed journals have resulted from your program per year?   2

15. What is the total annual budget for your program?
$100,000

16. Typically, how much direct financial support does your institution provide?   None

17. Typically, what kind of non-financial support does your institution provide to your program?
☐ waiver of overhead/reduction of normal overhead
☐ faculty salaries
☐ release time
☐ summer support for director
☐ clerical/administrative support (full or part-time)
☐ graduate tutoring
☑ school facilities (including office space)
☑ telephone/photocopying/postage
☑ computer use/computer support
☐ free food/entertainment and/or transportation for students
☐ free housing for students
☐ other:

18. Which agency is the principal source of your support (NSF, NSA, others)?
NSF

19. What measures do you use to assess the success of the program?
We give a Leikert scaled questionnaire to measure changes in perception of mathematics. The questionnaire is administered before the program, directly after the program and on an annual basis thereafter. We also track careers of participants and a control group of non-participants.

20. Does your program target a particular audience? ☑ Yes ☐ No
Please describe:
We attempt to recruit a large proportion of women and underrepresented minorities. We also have 2 or 3 teachers participating each year.

21. Please describe any unusual features of your program.
Teachers are part of the research team and work side by side with the undergraduates. To measure the effect of the program, students are selected at random from pools of 3 equally qualified applicants. We then attempt to follow up with the other 2 participants in the pool on an annual basis to see if our program had any effect on their career choices.

22. Briefly describe the goals of your program.
- Encourage undergraduate students, especially those from underrepresented groups, to pursue careers in sciences and engineering, including teaching.
- Help to better prepare students to pursue advanced degrees and careers in the sciences.
- Provide in-service teachers with a research experience that will foster excitement about mathematics, increase content understanding, and inspire pedagogical innovation in their classrooms.
- Promote and enhance mathematical research involving undergraduates at CSU, Chico.

2006 Survey of Undergraduate Programs in Mathematics

Please note: This survey is being conducted in conjunction with the Promoting Undergraduate Research in Mathematics (PURM) conference that was held on September 28-30, 2006. If this form is inappropriate for your program and if you are interested in submitting an article for publication in the conference proceedings, then please contact the proceedings editor, Joe Gallian (jgallian@d.umn.edu).

Name of respondent: Jim Morrow

Name of program: Inverse Problems for Electrical Networks

Program held at what institution? Univ. of Washington

Check here if this program is no longer active: [ ]
If so, last year of operation:

1. Briefly describe the kind of program you direct:
[✓] REU
[ ] REU program for certain group:
[ ] Other:
Describe: Introduce students to research on electrical netoworks

2. How long has your program existed (including other directors)? 19 years

3. What is your adviser/student ratio? (1fac+2TA)/12students

4. In all the years of program operation, what is the total number of students who have participated in the program? 165

5. Of the students who have attended your program in the past five years, what percent were women? 25%

   If this figure is an estimate, check here: [ ]

6. Of the students who have attended your program in the past five years, what percent were Hispanic/Latino, Native American or African American? 2%

   If this figure is an estimate, check here: [ ]

7. Is your program:
[ ] instructional in nature
[✓] research oriented
[ ] both
Comments:

8. How many weeks does your program run?
[ ]1 [ ]2 [ ]3 [ ]4 [ ]5 [ ]6 [ ]7 [✓]8 [ ]9 [ ]10+

9. If your program has an instructional component, are students asked to work (check both if necessary):
[ ] in groups   [ ] individually

If your program has a research component, do students work on their research projects (check both if necessary):
[✓] in groups   [✓] individually

Comments: The instruction is brief and takes place in the first week.

For questions 10-16, please give an estimate if exact figures are not available (indicate when the figures are estimates).

10. Of the students whose academic career you are knowledgeable since they participated in your program, give the number who fit into the following categories:

21 are still undergraduates

26 are currently in a graduate program in the mathematical sciences

5 have received a Masters in the mathematical sciences (and are no longer in graduate school)

23 have received a Ph.D. in the mathematical sciences

2 are currently in a graduate program in the sciences (excluding the mathematical sciences)

2 have received a Masters in the sciences (excluding the mathematical sciences)

1 have received a Ph.D. in the sciences (excluding the mathematical sciences)

10 other (e.g., did not obtain an advanced degree, or are pursuing graduate studies outside of the sciences)

Comments:

11. In the last five years, what percentage of your program participants have received national graduate fellowships (NSF, Hertz, NDSEG, Ford, NPSC, GEM, etc.)?
☐ 0-10% ☐ 11-20% ☑ 31-40% ☐ 41-50% ☐ 51-60%
☐ 61-70% ☐ 71-80% ☐ 81-90% ☐ 91-100%
This figure is an estimate ☑

12. In the past five years, approximately how many oral presentations at conferences have resulted from your program per year? 2

13. In the past five years, approximately how many poster presentations at conferences have resulted from your program per year? 2

14. Approximately, how many publications in refereed journals have resulted from your program per year? .5

15. What is the total annual budget for your program?
65,000

16. Typically, how much direct financial support does your institution provide? 3500

17. Typically, what kind of non-financial support does your institution provide to your program?
☐ waiver of overhead/reduction of normal overhead
☐ faculty salaries
☐ release time
☑ summer support for director
☐ clerical/administrative support (full or part-time)
☐ graduate tutoring
☑ school facilities (including office space)
☑ telephone/photocopying/postage
☑ computer use/computer support
☐ free food/entertainment and/or transportation for students
☐ free housing for students
☐ other:

18. Which agency is the principal source of your support (NSF, NSA, others)?
nsf

19. What measures do you use to assess the success of the program?
fellowships and scholarships, Modeling Contest Winners, graduate degrees

20. Does your program target a particular audience? ☑ Yes ☐ No
Please describe:
Students who want to engage in mathematical research

21. Please describe any unusual features of your program.
Students start work on problems right away. They learn about the latest results in discrete inverse problems.

22. Briefly describe the goals of your program.
1. Introduce students to research
2. Students learn how to write research papers
3. Students give many oral reports
4. Students learn how to interact with others mathematically

2006 Survey of Undergraduate Programs in Mathematics

e-mail questions to: dxc@ams.org

Please note: This survey is being conducted in conjunction with the Promoting Undergraduate Research in Mathematics (PURM) conference that was held on September 28-30, 2006. If this form is inappropriate for your program and if you are interested in submitting an article for publication in the conference proceedings, then please contact the proceedings editor, Joe Gallian (jgallian@d.umn.edu).

Name of respondent: J.D. Phillips

Name of program: Wabash Summer Institute in Algebra

Program held at what institution? Wabash College

Check here if this program is no longer active: [ ]
If so, last year of operation: [ ]

1. Briefly describe the kind of program you direct:
[✓] REU

[ ] REU program for certain group:

[ ] Other:

Describe:
Preference for students from institutions that provide few research opportunities for undergrads.

2. How long has your program existed (including other directors)?
Three years.

3. What is your adviser/student ratio? 1:4

4. In all the years of program operation, what is the total number of students who have participated in the program? 25

5. Of the students who have attended your program in the past five years, what percent were women? 40 %
If this figure is an estimate, check here: [ ]

6. Of the students who have attended your program in the past five years, what percent were Hispanic/Latino, Native American or African American? 12 %
If this figure is an estimate, check here: [ ]

7. Is your program:
[ ] instructional in nature
[ ] research oriented
[✓] both
Comments:

8. How many weeks does your program run?
[ ]1 [ ]2 [ ]3 [ ]4 [ ]5 [ ]6 [ ]7 [✓]8 [ ]9 [ ]10+

9. If your program has an instructional component, are students asked to work (check both if necessary):
[✓] in groups   [ ] individually

If your program has a research component, do students work on their research projects (check both if necessary):
[✓] in groups   [ ] individually

Comments:

For questions 10-16, please give an estimate if exact figures are not available (Indicate when the figures are estimates).

10. Of the students whose academic career you are knowledgeable since they participated in your program, give the number who fit into the following categories:
[ ] are still undergraduates
[ ] are currently in a graduate program in the mathematical sciences
[ ] have received a Masters in the mathematical sciences (and are no longer in graduate school)
[ ] have received a Ph.D. in the mathematical sciences
[ ] are currently in a graduate program in the sciences (excluding the mathematical sciences)
[ ] have received a Masters in the sciences (excluding the mathematical sciences)
[ ] have received a Ph.D. in the sciences (excluding the mathematical sciences)
[ ] other (e.g., did not obtain an advanced degree, or are pursuing graduate studies outside of the sciences)

Comments:

11. In the last five years, what percentage of your program participants have received national graduate fellowships (NSF, Hertz, NDSEG, Ford, NPSC, GEM, etc.)?

☒0-10% ☐11-20% ☐31-40% ☐41-50% ☐51-60%
☐61-70% ☐71-80% ☐81-90% ☐91-100%
This figure is an estimate ☐

12. In the past five years, approximately how many oral presentations at conferences have resulted from your program per year? ☐7

13. In the past five years, approximately how many poster presentations at conferences have resulted from your program per year? ☐4

14. Approximately, how many publications in refereed journals have resulted from your program per year? ☐2

15. What is the total annual budget for your program?

$64,957

16. Typically, how much direct financial support does your institution provide? ☐None

17. Typically, what kind of non-financial support does your institution provide to your program?
☐ waiver of overhead/reduction of normal overhead
☐ faculty salaries
☐ release time
☐ summer support for director
☑ clerical/administrative support (full or part-time)
☐ graduate tutoring
☑ school facilities (including office space)
☑ telephone/photocopying/postage
☑ computer use/computer support
☐ free food/entertainment and/or transportation for students
☐ free housing for students
☐ other: _____

18. Which agency is the principal source of your support (NSF, NSA, others)?

NSF

19. What measures do you use to assess the success of the program?
Participant evaluations.
Contact with participants for 5 years after the program.

20. Does your program target a particular audience? ☐ Yes ☑ No
Please describe:
There is special consideration for women and underrepresented minorities.

21. Please describe any unusual features of your program.
Completely in algebra (3 areas).

22. Briefly describe the goals of your program.

WSIA has several goals:
- to provide a meaningful mathematical research opportunity for undergraduates who are less likely to have a research-oriented experience than other student populations;
- to encourage participants to attend graduate school in the mathematical sciences and to help them develop some of the tools and confidence necessary to succeed there;
- to provide the support and framework for the participants to share their results and experiences with the larger mathematical community;
- to begin to provide the tools and knowledge necessary to become an independent, contributing member of the mathematical community;
- to create a highly diverse and supportive environment where the participants learn to work and live with a wide variety of individuals;
- to give students a significant experience studying the ethical dimensions of science and mathematics; and
- to attempt to accurately measure the success or failure of this program through a wide variety of evaluative processes with the assistance of an experienced agency.

2006 Survey of Undergraduate Programs in Mathematics

Please note: This survey is being conducted in conjunction with the Promoting Undergraduate Research in Mathematics (PURM) conference that was held on September 28-30, 2006. If this form is inappropriate for your program and if you are interested in submitting an article for publication in the conference proceedings, then please contact the proceedings editor, Joe Gallian (jgallian@d.umn.edu).

Name of respondent: Efstratios Prassidis

Name of program: R.E.U.

Program held at what institution? Canisius College

Check here if this program is no longer active: [ ]
If so, last year of operation: [ ]

1. Briefly describe the kind of program you direct:
   - [✓] REU
   - [ ] REU program for certain group: [ ]
   - [ ] Other: [ ]
   - Describe: [ ]

2. How long has your program existed (including other directors)?
   Two Years

3. What is your adviser/student ratio? 1/4

4. In all the years of program operation, what is the total number of students who have participated in the program? 16

5. Of the students who have attended your program in the past five years, what percent were women? 37.5%
   - *If this figure is an estimate, check here:* [ ]

6. Of the students who have attended your program in the past five years, what percent were Hispanic/Latino, Native American or African American? 6.25%
   - *If this figure is an estimate, check here:* [ ]

7. Is your program:
   - [ ] instructional in nature
   - [✓] research oriented
   - [ ] both
   - Comments:

8. How many weeks does your program run?
   [ ]1 [ ]2 [ ]3 [ ]4 [ ]5 [ ]6 [ ]7 [✓]8 [ ]9 [ ]10+

9. If your program has an instructional component, are students asked to work (check both if necessary):
   - [ ] in groups   [ ] individually

If your program has a research component, do students work on their research projects (check both if necessary):
   - [✓] in groups   [✓] individually

Comments: Some projects are done individually and in other projects students work in groups.

For questions 10-16, please give an estimate if exact figures are not available (indicate when the figures are estimates).

10. Of the students whose academic career you are knowledgeable since they participated in your program, give the number who fit into the following categories:
   - 12 are still undergraduates
   - 1 are currently in a graduate program in the mathematical sciences
   - [ ] have received a Masters in the mathematical sciences (and are no longer in graduate school)
   - [ ] have received a Ph.D. in the mathematical sciences
   - 3 are currently in a graduate program in the sciences (excluding the mathematical sciences)
   - [ ] have received a Masters in the sciences (excluding the mathematical sciences)
   - [ ] have received a Ph.D. in the sciences (excluding the mathematical sciences)
   - [ ] other (e.g., did not obtain an advanced degree, or are pursuing graduate studies outside of the sciences)

Comments:

11. In the last five years, what percentage of your program participants have received national graduate fellowships (NSF, Hertz, NDSEG, Ford, NPSC, GEM, etc.)?
☑ 0-10% □ 11-20% □ 31-40% □ 41-50% □ 51-60%
□ 61-70% □ 71-80% □ 81-90% □ 91-100%
This figure is an estimate □

12. In the past five years, approximately how many oral presentations at conferences have resulted from your program per year? 20

13. In the past five years, approximately how many poster presentations at conferences have resulted from your program per year? 4

14. Approximately, how many publications in refereed journals have resulted from your program per year? 2

15. What is the total annual budget for your program?
41,500

16. Typically, how much direct financial support does your institution provide? None

17. Typically, what kind of non-financial support does your institution provide to your program?
□ waiver of overhead/reduction of normal overhead
□ faculty salaries
☑ release time
☑ summer support for director
☑ clerical/administrative support (full or part-time)
□ graduate tutoring
☑ school facilities (including office space)
☑ telephone/photocopying/postage
☑ computer use/computer support
□ free food/entertainment and/or transportation for students
□ free housing for students
□ other:

18. Which agency is the principal source of your support (NSF, NSA, others)?
NSF

19. What measures do you use to assess the success of the program?
- Publications, Presentations (talks and posters)
- Number of students accepted in graduate programs.

20. Does your program target a particular audience? □ Yes ☑ No
Please describe:

21. Please describe any unusual features of your program.

22. Briefly describe the goals of your program.
- Advancing mathematical research in undergraduates.
- Exposing undergraduates to original mathematical research.
- Preparing students for graduate studies in mathematics and/or sciences.
- Finding research collaborators among undergraduate students.

2006 Survey of Undergraduate Programs in Mathematics

Please note: This survey is being conducted in conjunction with the Promoting Undergraduate Research in Mathematics (PURM) conference that was held on September 28-30, 2006. If this form is inappropriate for your program and if you are interested in submitting an article for publication in the conference proceedings, then please contact the proceedings editor, Joe Gallian (jgallian@d.umn.edu).

Name of respondent: Ivelisse Rubio & Herbert A. Medina

Name of program: Sum. Inst. in Math for Undergrad. (SIMU)

Program held at what institution? University of Puerto Rico - Humacao

Check here if this program is no longer active: ☑
If so, last year of operation: 2002

1. Briefly describe the kind of program you direct:
☐ REU
☑ REU program for certain group: Hispanics/Native Amer.
☐ Other:

Describe:

Two weeks of intro to a research topic; then research in groups.

2. How long has your program existed (including other directors)?
It existed for 5 years.

3. What is your adviser/student ratio? 1:12

4. In all the years of program operation, what is the total number of students who have participated in the program? 115

5. Of the students who have attended your program in the past five years, what percent were women? 50 %
If this figure is an estimate, check here: ☐

6. Of the students who have attended your program in the past five years, what percent were Hispanic/Latino, Native American or African American? 93 %
If this figure is an estimate, check here: ☐

7. Is your program:
☐ instructional in nature
☑ research oriented
☐ both

Comments:

There was an instructional two-week component but that was so that students could be ready for the research projects in the last 4 weeks of the program.

8. How many weeks does your program run?
☐1 ☐2 ☐3 ☐4 ☐5 ☑6 ☐7 ☐8 ☐9 ☐10+

9. If your program has an instructional component, are students asked to work (check both if necessary):
☑ in groups   ☑ individually

If your program has a research component, do students work on their research projects (check both if necessary):
☑ in groups   ☐ individually

Comments:

Groups of 3.

For questions 10-16, please give an estimate if exact figures are not available (indicate when the figures are estimates).

10. Of the students whose academic career you are knowledgeable since they participated in your program, give the number who fit into the following categories:

0 are still undergraduates

39 are currently in a graduate program in the mathematical sciences

31 have received a Masters in the mathematical sciences (and are no longer in graduate school)

3 have received a Ph.D. in the mathematical sciences

1 are currently in a graduate program in the sciences (excluding the mathematical sciences)

☐ have received a Masters in the sciences (excluding the mathematical sciences)

2 have received a Ph.D. in the sciences (excluding the mathematical sciences)

☐ other (e.g., did not obtain an advanced degree, or are pursuing graduate studies outside of the sciences)

Comments: [ ]

11. In the last five years, what percentage of your program participants have received national graduate fellowships (NSF, Hertz, NDSEG, Ford, NPSC, GEM, etc.)?
☑ 0-10% ☐ 11-20% ☐ 31-40% ☐ 41-50% ☐ 51-60%
☐ 61-70% ☐ 71-80% ☐ 81-90% ☐ 91-100%
This figure is an estimate ☑

12. In the past five years, approximately how many oral presentations at conferences have resulted from your program per year? 8

13. In the past five years, approximately how many poster presentations at conferences have resulted from your program per year? 20

14. Approximately, how many publications in refereed journals have resulted from your program per year? 2

15. What is the total annual budget for your program?
$220,000

16. Typically, how much direct financial support does your institution provide? $40,000

17. Typically, what kind of non-financial support does your institution provide to your program?
☑ waiver of overhead/reduction of normal overhead
☐ faculty salaries
☐ release time
☐ summer support for director
☑ clerical/administrative support (full or part-time)
☐ graduate tutoring
☑ school facilities (including office space)
☑ telephone/photocopying/postage
☑ computer use/computer support
☐ free food/entertainment and/or transportation for students
☐ free housing for students
☑ other: Ground transportation

18. Which agency is the principal source of your support (NSF, NSA, others)?
NSA and NSF

19. What measures do you use to assess the success of the program?
Long-term tracking of students through yearly questionnaires; end of summer questionnaires;

20. Does your program target a particular audience? ☑ Yes ☐ No
Please describe:
Hispanics/Latinos and Native Americans, the populations served by the Society for Advancement of Chicanos and Native Americans in Science (SACNAS)

21. Please describe any unusual features of your program.
Two institutions are involved.

22. Briefly describe the goals of your program.
To increase the number of U.S. Hispanics and Native Americans earning graduate degrees and pursuing careers in the mathematical sciences

2006 Survey of Undergraduate Programs in Mathematics

Please note: This survey is being conducted in conjunction with the Promoting Undergraduate Research in Mathematics (PURM) conference that was held on September 28-30, 2006. If this form is inappropriate for your program and if you are interested in submitting an article for publication in the conference proceedings, then please contact the proceedings editor, Joe Gallian (jgallian@d.umn.edu).

Name of respondent: Pam Ryan

Name of program: 2 programs: STEP and Mathematical Biology

Program held at what institution? Truman State University

Check here if this program is no longer active: ☐
If so, last year of operation:

1. Briefly describe the kind of program you direct:

☐ REU

☐ REU program for certain group:

☑ Other: multiple programs involving math majors

Describe: both programs are interdisciplinary; data provided will only be for math majors

2. How long has your program existed (including other directors)?

3 years - but only had students the past 2 yr

3. What is your adviser/student ratio? about 1:2

4. In all the years of program operation, what is the total number of students who have participated in the program? 13

5. Of the students who have attended your program in the past five years, what percent were women? 23%

If this figure is an estimate, check here: ☐

6. Of the students who have attended your program in the past five years, what percent were Hispanic/Latino, Native American or African American? 0%

If this figure is an estimate, check here: ☐

7. Is your program:

☐ instructional in nature
☑ research oriented
☐ both

Comments:

8. How many weeks does your program run?

☐ 1 ☐ 2 ☐ 3 ☐ 4 ☐ 5 ☐ 6 ☐ 7 ☐ 8 ☐ 9 ☑ 10+

9. If your program has an instructional component, are students asked to work (check both if necessary):

☑ in groups ☑ individually

If your program has a research component, do students work on their research projects (check both if necessary):

☑ in groups ☑ individually

Comments:

For questions 10-16, please give an estimate if exact figures are not available (Indicate when the figures are estimates).

10. Of the students whose academic career you are knowledgeable since they participated in your program, give the number who fit into the following categories:

11 are still undergraduates

2 are currently in a graduate program in the mathematical sciences

☐ have received a Masters in the mathematical sciences (and are no longer in graduate school)

☐ have received a Ph.D. in the mathematical sciences

☐ are currently in a graduate program in the sciences (excluding the mathematical sciences)

☐ have received a Masters in the sciences (excluding the mathematical sciences)

☐ have received a Ph.D. in the sciences (excluding the mathematical sciences)

☐ other (e.g., did not obtain an advanced degree, or are pursuing graduate studies outside of the sciences)

18. Which agency is the principal source of your support (NSF, NSA, others)?

NSF

19. What measures do you use to assess the success of the program?

number of student presentations, posters, and papers; graduation rates in math and science; post-graduate plans; student and faculty e-journals

20. Does your program target a particular audience? [✓] Yes [ ] No
Please describe:

traditionally underrepresented groups, especially minorities and 1st generation students; students who have never had a research experience

21. Please describe any unusual features of your program.

We have a lot of group interaction. All students and mentors meet in one big group on Monday mornings during the summer. We also have group "lunch and learn" meetings on Wednesdays and group lunches on Fridays. The goal is to create a "community of scholars".

22. Briefly describe the goals of your program.

goal of the STEP Program: to increase the number of students getting bachelor degrees in and going to graduate school in the mathematical sciences (and science in general)

goal of the Mathematical Biology program: to improve interdisciplinary education of mathematicians and biologists so that they are better able to work on questions at the intersection of mathematics and biology.

Comments:

11. In the last five years, what percentage of your program participants have received national graduate fellowships (NSF, Hertz, NDSEG, Ford, NPSC, GEM, etc.)?
[✓]0-10% [ ]11-20% [ ]31-40% [ ]41-50% [ ]51-60%
[ ]61-70% [ ]71-80% [ ]81-90% [ ]91-100%
This figure is an estimate [ ]

12. In the past five years, approximately how many oral presentations at conferences have resulted from your program per year? 20

13. In the past five years, approximately how many poster presentations at conferences have resulted from your program per year? 5

14. Approximately, how many publications in refereed journals have resulted from your program per year? 1

15. What is the total annual budget for your program?
$600,000 per year

16. Typically, how much direct financial support does your institution provide? about $10,000 in tuition wavers

17. Typically, what kind of non-financial support does your institution provide to your program?
[ ]waiver of overhead/reduction of normal overhead
[ ]faculty salaries
[ ]release time
[ ]summer support for director
[ ]clerical/administrative support (full or part-time)
[ ]graduate tutoring
[✓]school facilities (including office space)
[✓]telephone/photocopying/postage
[✓]computer use/computer support
[ ]free food/entertainment and/or transportation for students
[ ]free housing for students
[ ]other:

2006 Survey of Undergraduate Programs in Mathematics

Please note: This survey is being conducted in conjunction with the Promoting Undergraduate Research in Mathematics (PURM) conference that was held on September 28-30, 2006. If this form is inappropriate for your program and if you are interested in submitting an article for publication in the conference proceedings, then please contact the proceedings editor, Joe Gallian (jgallian@d.umn.edu).

Name of respondent: Cynthia J. Wyels

Name of program: CSUCI Mathematics REU

Program held at what institution? CSU Channel Islands (formerly at CA Lutheran)

Check here if this program is no longer active: ☐
If so, last year of operation:

1. Briefly describe the kind of program you direct:
☐ REU
☑ REU program for certain group: underrepresented minorities
☐ Other:

Describe:

2. How long has your program existed (including other directors)?
3 years

3. What is your adviser/student ratio? 1 to 4 (2 years); 1 to 3 (1 year)

4. In all the years of program operation, what is the total number of students who have participated in the program? 14

5. Of the students who have attended your program in the past five years, what percent were women? 29 %

If this figure is an estimate, check here: ☐

6. Of the students who have attended your program in the past five years, what percent were Hispanic/Latino, Native American or African American? 93 %

If this figure is an estimate, check here: ☐

---

7. Is your program:
☐ instructional in nature
☑ research oriented
☐ both
Comments:

8. How many weeks does your program run?
☐ 1 ☐ 2 ☐ 3 ☐ 4 ☐ 5 ☑ 6 ☐ 7 ☐ 8 ☐ 9 ☐ 10+

9. If your program has an instructional component, are students asked to work (check both if necessary):
☐ in groups   ☐ individually

If your program has a research component, do students work on their research projects (check both if necessary):
☐ in groups   ☑ individually

Comments:
The first two years students worked individually on research projects; last year they worked in pairs. Results were much better when they worked in pairs.

For questions 10-16, please give an estimate if exact figures are not available (indicate when the figures are estimates).

10. Of the students whose academic career you are knowledgeable since they participated in your program, give the number who fit into the following categories:

7 are still undergraduates

3 are currently in a graduate program in the mathematical sciences

☐ have received a Masters in the mathematical sciences (and are no longer in graduate school)

☐ have received a Ph.D. in the mathematical sciences

1 are currently in a graduate program in the sciences (excluding the mathematical sciences)

☐ have received a Masters in the sciences (excluding the mathematical sciences)

☐ have received a Ph.D. in the sciences (excluding the mathematical sciences)

2 other (e.g., did not obtain an advanced degree, or are pursuing graduate studies outside of the sciences)

Comments:

11. In the last five years, what percentage of your program participants have received national graduate fellowships (NSF, Hertz, NDSEG, Ford, NPSC, GEM, etc.)?
☑ 0-10% ☐ 11-20% ☐ 31-40% ☐ 41-50% ☐ 51-60%
☐ 61-70% ☐ 71-80% ☐ 81-90% ☐ 91-100%
This figure is an estimate ☐

12. In the past five years, approximately how many oral presentations at conferences have resulted from your program per year? | 8 |

13. In the past five years, approximately how many poster presentations at conferences have resulted from your program per year? | 4 |

14. Approximately, how many publications in refereed journals have resulted from your program per year? | 0 |

15. What is the total annual budget for your program?
$25,000

16. Typically, how much direct financial support does your institution provide? | $0 |

17. Typically, what kind of non-financial support does your institution provide to your program?
☑ waiver of overhead/reduction of normal overhead
☐ faculty salaries
☐ release time
☐ summer support for director
☐ clerical/administrative support (full or part-time)
☐ graduate tutoring
☑ school facilities (including office space)
☐ telephone/photocopying/postage
☑ computer use/computer support
☐ free food/entertainment and/or transportation for students
☐ free housing for students
☐ other:

18. Which agency is the principal source of your support (NSF, NSA, others)?
MAA's SUMMA program (channeling NSA, NSF and Moody's funds)

19. What measures do you use to assess the success of the program?
1) Assessment done for all MAA NREUP's by outside evaluator
2) interviews with students participants

20. Does your program target a particular audience? ☑ Yes ☐ No
Please describe:
regional minority mathematics students

21. Please describe any unusual features of your program.
The students who've participated in this program were generally at a point in their undergraduate studies -- or had GPAs -- that would not lead to their being accepted to other REUs.

22. Briefly describe the goals of your program.
• Raise the mathematical maturity level of the program's participants.
• Get participants excited about doing mathematical research.
• Create a learning community.
• Help participants develop the confidence to succeed in ongoing mathematical studies.
• Increase participants' skills in communicating mathematics.
• Extend the participants' abilities to read, understand, construct, and write proofs.
• Acquaint participants with the culture and activities of research mathematics.
• Develop participants' skills in reading professional-level mathematics.
• Give the participants technical tools for future mathematical learning and research. (LaTeX, MathSciNet, presentation software)

Part VI

Conference
Program

Promoting Undergraduate Research in Mathematics Program

September 28-30, 2006

The Westin O'Hare Hotel, Rosemont, Illinois

Program

Thursday, September 28, 2006		
4:00 - 6:00 pm	Conference check-in	
5:00 - 6:00 pm	Reception	
6:00 - 6:30 pm	**Welcome:**	John Ewing, AMS
		Jim Schatz, NSA
6:30 - 7:30 pm	Dinner	
7:30 - 8:30 pm	**Opening address:**	Joe Gallian, University of Minnesota-Duluth
		Aparna Higgins, University of Dayton

Friday, September 29, 2006	
7:30 - 8:30 am	Breakfast
8:30 - 9:45 am	**Panel: *The Spectrum of Undergraduate Research Programs*** Moderator: Robert Megginson, University of Michigan Peter May, University of Chicago Zsuzsanna Szaniszlo, Valparaiso University Erika Camacho, Loyola Marymount University Suzanne Weekes, Worcester Polytechnic Institute *Panelists will describe current forms of various undergraduate research programs: NSF-funded REUs, other summer undergraduate research programs, minority-serving programs, and academic year undergraduate research programs.*
9:45 - 10:30 a.m	Break

10:30 - 11:15 am	**Small groups: Overcoming Challenges**
	Facilitators:
	      Kurt Bryan, Rose-Hulman Institute of Technology
	      Doug Faires, Youngstown State University
	      Jim Morrow, University of Washington
	      Tim Pennings, Hope College
	      Judy Walker, University of Nebraska, Lincoln
	      Stephen Wirkus, California State Polytechnic University
	
	Participants will break up into 6 small groups, each with a facilitator. In each group, participants will discuss challenges they have faced in their undergraduate research activities, and how they have overcome these challenges. For example, discussions might center on finding sources of support, how to deal with burnout or how to get recognition from departments or institutions for undergraduate research involvement. Facilitators will report back to the full group.

11:15 - 12 noon	**Reports from small groups** Moderator: Suzanne Lenhart, University of Tennessee
12:00 - 1:30 pm	Lunch

1:30 - 3:00 pm	Panel: Diversity issues in undergraduate research
	      Moderator: Ivelisse Rubio, University of Puerto Rico-Humacao
	
	      Ricardo Cortez, Tulane University
	      Dennis Davenport, Miami University
	      Herbert Medina, Loyola Marymount University
	      Darren Narayan, Rochester Institute of Technology
	
	The panelists of this session will discuss the importance of involving minority communities in undergraduate research in mathematics in order to increase their representation in the area. They will also discuss effective recruiting techniques and successful strategies.

3:00 - 3:45 pm	Break
3:45 - 5:15 pm	**Open Forum** Moderator: Aparna Higgins, University of Dayton
6:30 - 8:00 pm	Dinner

Saturday, September 30, 2006	
7:30 - 8:30 am	Breakfast

8:30 - 9:45 am	**Panel: *Assessment of Programs***
	      Moderator: Anant Godbole, East Tennessee State University
	
	      Michelle Wagner, NSA
	      Frank Connolly, University of Notre Dame
	      Erika Camacho, Loyola Marymount University
	      Dan Isaksen, Wayne State University
	
	Speakers will summarize information gleaned from assessments now available to determine the impact that research programs have had on students. The assessment techniques that some REUs now use will be presented.

9:45 - 10:30 am	Break
10:30 - 12 noon.	**Panel:** *Perspectives form Students* Moderator: Carl Cowen, Indiana University-Purdue University Indianapolis Paul Gibson, Northwestern University Rana Mikkelson, Iowa State University David Uminsky, Boston University Melanie Wood, Princeton University *Current graduate students will describe their undergraduate research experiences and the impact these have had on their mathematical development. They will also discuss the way their time in an undergraduate research program affected their graduate experiences.*
12:00 - 1:30 pm	Lunch
1:30 - 2:20 pm	**Parallel sessions** *Each participant will choose one of the following sessions when registering for the conference. Facilitators for each session will report back to the full group in the final plenary session.* • Uniform acceptance date • Generating research problems • Interdisciplinary research • Working with local students • Using graduate students as advisors • Helping undergraduate students into graduate school • Undergraduate biology and mathematics programs
2:30 - 3:15 pm	**Reports from parallel sessions** Moderator: Charles Johnson, College of William and Mary
3:15 - 4:00 pm	**Closing panel:** Moderator: Barbara Deuink, NSA Frank Connolly, University of Notre Dame Joseph Gallian, University of Minnesota-Duluth Aparna Higgins, University of Dayton Ivelisse Rubio, University of Puerto Rico-Humacao

Program Committee:

Frank Connolly, University of Notre Dame
Joseph Gallian, University of Minnesota–Duluth
Aparna Higgins, University of Dayton
Ivelisse Rubio, University of Puerto Rico–Humacao